U0545511

深智數位
股份有限公司

深智數位
股份有限公司

自序

自 2022 年 11 月 30 日 OpenAI 發佈 ChatGPT 以來，大模型技術掀起了新一輪人工智慧浪潮。ChatGPT 在各個領域（如人機對話、文字摘要、內容生成、問題解答、識圖、數學計算、程式撰寫等）獲得了比之前演算法好得多的成績，在很多方面超越了人類專家的水準，特別是人機對話具備了一定的共情能力，這讓 AI 領域的從業者和普通大眾相信通用人工智慧（Artificial General Intelligence，AGI）時代馬上就要來臨了。

大模型除了對話能力達到了真正可以互動的水準，更厲害的是當模型參數達到一定規模（100B[①]以上）時，會湧現新的能力，即大模型具備舉一反三、任務分解、邏輯推理、解決未知任務的能力，這在之前的機器學習範式中是從未出現過的。之前的機器學習模型都是為某個具體任務建構的，只能完成特定的任務，對於新任務，必須訓練新的機器學習模型。

最近 7~8 年，沒有哪一項科技進步如 ChatGPT 這般吸引全球的目光[②]。除了媒體的大肆報導，國內外各類科技公司、科學研究機構、高等院校也在跟進大模型技術，基於大模型的創業公司如雨後春筍般層出不窮。

---

① 這裡的 B 代表 Billion（十億）。

② 上一次引發全球關注的 AI 大事件是 2016 年的 AlphaGo 戰勝人類頂尖圍棋手。

不到一年時間，國外就湧現了上百家大模型應用的創業公司，做得優秀的如 Midjourney、Runway、Inflection AI、Anthropic 等，都獲得了上億甚至幾十億美金的融資，估值達數十億、上百億美金。此外，一系列優秀的大模型相繼被發佈，如 Anthropic 的 Claude、Google 的 Bard、Meta 的 LLaMA 等。

中國大陸也不甘落後，各個大廠、創業公司、科學研究院校相繼發佈了大模型產品，如智譜 AI 的 ChatGLM、百度的文心一言、華為雲的盤古大模型、阿里雲的通義千問、科大訊飛的星火大模型等，也有不少「大佬」親自下場做大模型。

以 ChatGPT 為核心的大模型相關技術，可以應用於搜尋、對話、內容創作等許多領域。在推薦系統領域的應用也不例外，在這方面已經有廣泛的學術研究，學術界發表了大量的相關論文。我相信，在不久的將來（2024 年—2025 年），大模型相關技術會在工業界被大量用於推薦系統，並成為推薦系統的核心技術，就像 2016 年開始的深度學習技術對推薦系統的革新一樣。

ChatGPT 和大模型相關技術不能被任何人、任何行業忽視，它在各行各業一定會呈現「爆發式」的發展。從 2023 年年初開始，我一直關注大模型相關技術的進展及其在行業上的應用，特別是在推薦系統領域的應用。在幾乎每天都有大模型相關重磅突破發佈的當下，我們必須跟上技術發展的步伐。為此，我花了近一年的時間寫了這本書，希望能拋磚引玉，為普及、推廣大模型在推薦系統上的應用發揮作用。

在開啟我們的學習之旅前，我先用非常直白、淺顯的方式來簡單說明為什麼大模型能應用於推薦系統，有了這個基礎認識，我相信你可以更進一步地學習本書中的知識。

本書需要讀者具備一定的背景知識。舉例來說，熟悉推薦系統、有一定的程式設計能力，如果了解 NLP 就更好了。讀者如果在學習過程中發現對某個基礎知識不熟悉，可以自行補充學習。

大模型在底層建構 Transformer 架構，透過巨量的網際網路文字資訊預測下一個 token[③]出現的機率來預訓練模型[④]。由於有巨量的網際網路文字資料，模型的預訓練過程不需要人工標注（但需要對資料進行前置處理），一旦模型完成預訓練，就可以用於語言理解和語言生成任務。簡單來說，大模型基於巨量文字中的 token 序列中下一個 token 出現的機率進行統計建模，透過學習在替定語言部分後出現下一個 token 的機率來完成下游任務（例如文字摘要、翻譯、生成文字等）。

對於推薦系統，使用者過往的操作就是一個有序的序列，每個使用者的操作序列就像一篇文字，所有使用者的操作序列就像大模型的預訓練語料庫，預測使用者下一個操作就像預測詞序列的下一個 token[⑤]。透過這個簡單的類比，我們知道推薦系統可以被嵌入大模型的理論框架中。因此，直觀地看，大模型一定可以用於解決推薦系統問題。

上面的想法比較簡單，只用到了使用者與物品的互動資訊。實際上，推薦系統的資料來源更複雜，除了使用者歷史互動序列，還有人物誌、物品畫像資訊等。部分人物誌、物品畫像資訊，例如使用者的年齡、性別、偏好等，物品的標題、標籤、描述文字等，可以利用自然語言來呈現，使用者歷史互動序列、人物誌、物品畫像等資訊都可以被輸入大模型，為大模型提供更多的背景知識，最終的推薦結果會更加精準。

推薦系統涉及很多多模態資料，例如物品有描述文字、圖片，甚至視訊介紹等，這類異質的資訊對於推薦系統的效果相當重要。多模態資料可以被轉化為文字資訊供大模型使用，目前的多模態大模型可以直接處理多模態資料，因此也可以直接用於推薦系統。

---

③ token 既可能是一個單字，也可能是一個單字的一部分。

④ BERT 等模型基於左右兩邊的 token 預測中間的 token，這屬於模型架構上的不同。

⑤ 這裡推薦系統的物品類似語言模型中的一個 token。

目前，即使不使用圖片、視訊等多模態資料，利用好文字資料也能讓大模型很強大。大模型的強大之處是具備 zero-shot、few-shot 的能力[⑥]，很多推薦機制都利用了大模型的這個能力，只不過需要在使用大模型的過程中設計一些提示詞（prompt）和範本（template）來啟動大模型的推薦能力。

說一下我個人對啟動的理解。大模型有上百億、上千億，甚至上兆個參數，是一個非常龐大的神經網路。當用一些提示詞告訴大模型作為推薦系統進行個性化推薦時，就啟動了深度神經網路中的某些連接，這些連接是神經網路的某個子網路，而這個子網路具備進行個性化推薦的能力，這個過程非常類似人類大腦神經元的工作機制。例如當你看到美食時，就會啟動大腦中負責進食的區域，這個區域是大腦整個複雜神經元網路的子網路，導致可能產生流口水、吞咽等行為，這裡「看到美食」就類似大模型的提示詞。另外，我們在進行腦力激盪時，突然被別人啟發想到某個絕妙的創意也是一種啟動過程。few-shot 更複雜一些，需要在提示詞中告訴大模型一些過往的推薦案例，例如使用者看了 A、B、C 三個視訊後，會看另一個視訊 D，讓它臨時學習如何推薦。

提示詞學習沒有改變大模型的參數，即沒有進行梯度下降的反向傳播訓練，但為什麼具備 zero-shot、few-shot 的能力呢？提示詞作為一個整體，啟動了大模型神經網路的某個功能區域。大模型具備多輪對話能力的道理也是類似的，我們可以將多輪對話作為一個整體，這個整體啟動了大模型在某個對話主題下的功能區域，導致大模型能「記住」多輪對話之前的資訊[⑦]。由於目前的大模型不具備增量學習[⑧]的能力，對話完成後，對話中的新資訊並沒有被大模型學習到。

---

⑥ zero-shot 指預訓練後可以直接完成未知的下游任務。few-shot 指舉出幾個範例，大模型可以解決類似的問題，即所謂的上下文學習能力，也就是舉一反三的能力。

⑦ 這個對話是作為整體輸入大模型的，或者可以視為整個對話過程就是一次連貫的語言生成過程，只不過部分話語是人類舉出的，模型接著人類的話語繼續生成。

⑧ 增量學習指遇到一個新資訊馬上學習到模型的參數中，人是具備增量學習能力的，增量學習肯定是大模型未來最重要的一個研究方向。

除了直接利用大模型的 zero-shot、few-shot 能力進行推薦，我們還可以按照大模型的輸入、輸出範式準備推薦系統的相關資料，然後透過監督學習微調大模型，讓大模型更進一步地調配具體的推薦場景，這也是將大模型應用於推薦系統的非常有價值的方向。

另外，大模型壓縮的世界知識、大模型的湧現能力、大模型的自然對話能力都可以極佳地被用於推薦系統，解決深度學習推薦系統很難解決的問題，下面舉兩個例子說明。

首先，大模型有助緩解資料稀疏問題，特別是冷開機問題[9]，這是當前深度學習推薦系統的主要瓶頸。透過從在不同模型架構中學習的預訓練模型中提取和遷移知識，可以提高推薦系統的通用性、稀疏性、效率和有效性等性能。

其次，大模型一個很大的優勢是可以利用對話的方式跟使用者互動，就像 ChatGPT 所呈現的那樣。如果能將推薦系統設計成一個跟使用者互動的對話式推薦引擎，那麼大模型可以利用自然語言回應使用者的個性化需求，從而提升使用者的整體體驗和參與度。

透過前面的介紹，相信你已經大致知道為什麼大模型可以被應用於推薦系統了，也知道了將大模型應用於推薦系統的獨特優勢，那麼如何將大模型應用於推薦系統呢？這就是本書的核心內容，你將從這本書中找到答案。

⑨ 大模型學習的是巨量的網際網路知識，對於新物品、新使用者都可以極佳地進行知識遷移。

前言

為什麼寫作本書

我從 2010 年開始研究、實踐推薦系統，屬於最早從事推薦系統工作的一批人。過去 15 年，我有過至少 4 次從零開始成功建構推薦系統的經歷，曾經負責建構的推薦系統最高有超過千萬日活躍使用者數量（Daily Active User，DAU）。我出版過兩本推薦系統相關圖書，分別是《建構企業級推薦系統：演算法、工程實現與案例分析》（2021.09）和《推薦系統：演算法、案例與大模型》（2024.04）。過去十幾年的經歷讓我見證了推薦系統技術的完整發展過程，我也一直跟隨推薦系統的發展大勢進行學習、實踐。

2022 年 11 月底 ChatGPT 發佈後，全世界掀起了新一輪的人工智慧浪潮，以 ChatGPT 為代表的大模型技術逐步滲透到各個應用場景和領域，當然也包括推薦系統。將大模型應用在推薦系統上的相關學術研究非常多，截至 2024 年 5 月，已有上百篇相關的研究論文發表。大模型在產業上的應用也逐步開始：阿里巴巴在淘寶上內測了淘寶問問——一個對話式推薦引擎；Meta 在嘗試利用大模型技術實現兆級參數的新一代推薦系統；百度正在利用大模型重構底層的核心搜尋、推薦模組……

ChatGPT 和大模型帶來的影響是空前的，過去沒有哪一項 AI 技術對全球的衝擊像 ChatGPT 和大模型這麼大。我預見到 ChatGPT 和大模型會革新推薦行業，因此在過去的一年多時間裡，我閱讀了上百篇大模型、大模型推薦系統相關的論文，並且跟行業專家進行了密切的交流。

另外，我從 2023 年 4 月開始創業，公司業務為 B 端的數智化轉型，主要方向是精細化營運、大模型搜尋推薦、大模型智慧知識顧問等。過去一年，我針對將大模型應用於推薦、搜尋等話題與相關的企業進行交流並付諸實踐。

結合過往的學習、交流、實踐，我準備將自己掌握的大模型推薦系統的知識框架進行系統整理，整理成一本系統介紹大模型推薦系統的圖書，希望為推薦行業提供一套完整的、基於大模型的方法論和實踐指南。

希望我的經驗和經歷可以幫助想學習大模型在推薦系統上應用的讀者系統化了解並快速掌握大模型推薦系統。

目標讀者

本書主要講解大模型在推薦系統中的應用，既有演算法原理，又有程式實現，聚焦於如何利用最新的大模型技術賦能、革新、重構現有的推薦系統。本書需要讀者有一定的推薦系統基礎知識，了解大模型的一些基本原理，本書適合以下讀者。

1. 推薦系統開發人員及推薦演算法研究人員。

2. 期望從事推薦系統相關工作的學生。

3. 在大專院校從事推薦演算法研究，希望對大模型在推薦系統中的應用有更加了解的科學研究人員。

4. 對大模型在推薦、搜尋中的應用感興趣的產品和營運人員。

5. 期望將大模型引入推薦、搜尋產品的公司管理層人員。

如何閱讀本書

本書分為 9 章，包含大模型基本原理介紹、將大模型應用於推薦系統的想法和方法、在電子商務推薦場景中使用大模型等。下面分別對各個章節進行簡單介紹。

第 1~3 章是準備部分。第 1 章介紹大模型的基礎知識，包括大模型的發展歷史、資料資源、資料前置處理、大模型預訓練、大模型微調、大模型推理、大模型部署、相關軟體和框架等，這一章是後續章節的理論基礎，是為沒有大模型基礎的讀者準備的。如果你已經非常熟悉大模型，那麼可以略過這一章。第 2 章對後續章節用到的資料和開發環境介紹，本書的程式實現基於微軟的 MIND 資料集和 Amazon 電子商務資料集，開發環境包括 Python 沙箱、CUDA 和 MacBook。第 3 章將大模型在推薦系統中的應用抽象為 4 種範式——生成範式、預訓練範式、微調範式、直接推薦範式。

第 4~7 章詳細說明第 3 章介紹的 4 種範式，每章既包含演算法原理，又包含相關的案例說明，同時會基於 MIND 資料集舉出對應的程式實現。

第 8 章是一個完整的實戰案例，基於 Amazon 商品評論資料，利用大模型解決電子商務推薦問題。本章講解如何利用大模型來解決生成使用者興趣畫像、生成個性化商品描述資訊、猜你喜歡推薦、連結推薦、冷啟動、推薦解釋、對話式推薦 7 類問題，針對每類問題都提供完整的步驟及對應的程式實現。其中，前 6 類問題是大模型對傳統推薦演算法解決方案的有效補充和突破，有了大模型的支援，傳統推薦演算法有更新穎、更好的解決方案；而對話式推薦是基於傳統推薦系統的一種新的推薦形態，借助大模型的自然語言對話能力，我們可以採用對話的方式為使用者進行個性化推薦，在提升使用者體驗的同時，帶來更高的商業價值。

第 9 章為將大模型推薦系統更進一步地應用於工業界提供了一些想法和方法，包含大模型的高效預訓練、高效推理，以及真實業務場景中的一些問題和建議。

本書是專門為方便讀者快速上手大模型推薦系統而準備的。本書主要有三個特點：一是將大模型應用於推薦系統抽象為 4 種範式，在這個統一的框架下，你可以系統化地學習大模型推薦系統；二是包含電子商務場景的最佳實踐，針對每種問題，提供完整的利用大模型解決推薦問題的想法、步驟和方法；三是包含程式實戰內容，第 2~8 章都有完整的程式實現。**本書程式碼可至深智官網下載，或是你也可以到 GitHub 的程式倉庫 liuq4360/llm4rec_abc 下載，作者可能會有更新的程式碼**

希望本書能夠成為有志於利用大模型技術更進一步地賦能傳統推薦、搜尋系統的同好和從業者的方法論和落地實戰指南！

勘誤和支持

由於作者水準和寫作時間有限，書中難免有所紕漏，懇請讀者批評指正。你可以將書中描述不準確的地方或錯誤告訴我，以便本書重印或再版時更正。你可以透過微信 liuq4360 與我取得聯繫，或發送郵件至電子郵件 891391257@qq.com，我很期待看到你真摯的回饋。

致謝

首先，感謝 AI 行業的技術信仰者，正是他們的不懈努力為大模型技術帶來快速的發展和廣闊的前景。同時，感謝過去一年來給予我信任的個人和企業，透過與他們合作，我更進一步地理解和實踐了大模型相關技術在產業上的應用。

其次，感謝電子工業出版社張爽老師，在她的耐心指導和建議下，我一步步最佳化了本書的結構和內容，讓這本書的品質獲得了保證。

最後，感謝我的父母和家人，是他們的無私付出讓我有足夠的時間學習、實踐、創業！

謹以此書，獻給所有懂我、關心我、支持我的家人和朋友！

劉強

目錄

第 1 章 基礎知識

第 2 章 資料準備與開發環境準備

第 3 章 大模型推薦系統的資料來源、一般想法和 4 種範式

第 4 章 生成範式：大模型生成特徵、訓練資料與物品

第 5 章 預訓練範式：透過大模型預訓練進行推薦

第 6 章 微調範式：微調大模型進行個性化推薦

第 7 章 直接推薦範式：利用大模型的上下文學習進行推薦

第 8 章 實戰案例：大模型在電子商務推薦中的應用

第 9 章　專案實踐：大模型落地真實業務場景

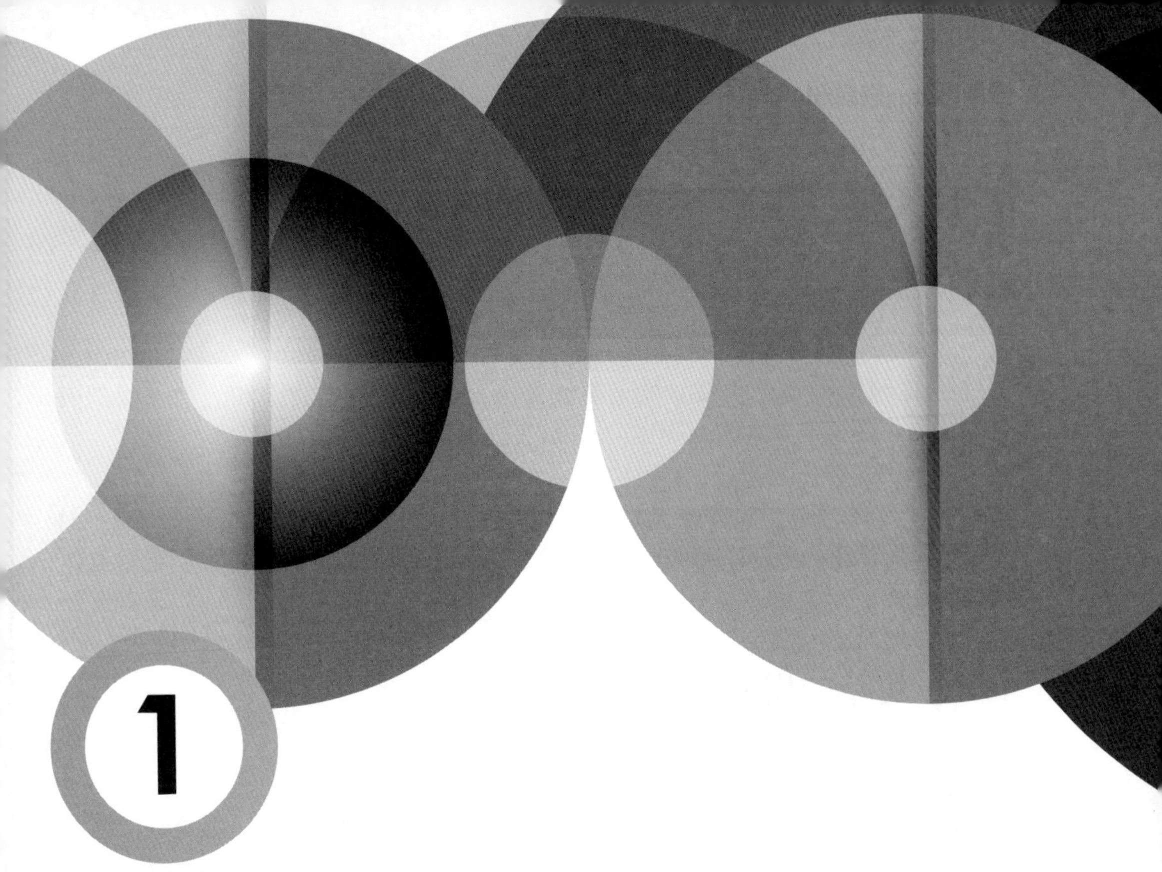

1 基礎知識

大模型的全稱是大語言模型（Large Language Model，LLM），它能夠應用於推薦系統，並且極有可能像深度學習那樣革新推薦系統。大模型是一個有相當高的門檻並且快速發展的領域，在我們正式學習大模型如何應用於推薦系統之前，有必要先了解大模型的一些基礎知識，以便更進一步地學習後續知識。

本章涵蓋大模型相關資源、大模型預訓練、指令微調、對齊微調、上下文學習和高效推理等核心基礎知識。

1.1 大模型相關資源

預訓練參數量在千億等級的大模型是一個複雜的系統工程，需要巨量的資料及龐大計算的資源。只有 Google、微軟、百度等大型企業，或 OpenAI、Anthropic 等獲得了足夠資金的明星創業公司才有足夠的資金和人力去探索與嘗試。

好在隨著 ChatGPT 的爆紅，競爭進入白熱化，很多公司都開放原始碼了中小模型（例如 Baichuan、Qwen、Yi 等），個人和小型創業公司也可以在垂直領域進行大模型的嘗試和探索。

1.1.1 可用的模型及 API

目前國內外開放原始碼的參數量在 1B 到 100B 的大模型非常多，有智譜 AI 的 ChatGLM、百川智慧的 Baichuan 系列、阿里雲的通義千問等，但是最紅的、生態最好的還是 Meta 開放原始碼的 LLaMA，圖 1-1 就是基於 LLaMA 進行延伸開發的大模型生態，從這張圖中可以感受到 LLaMA 生態之大。

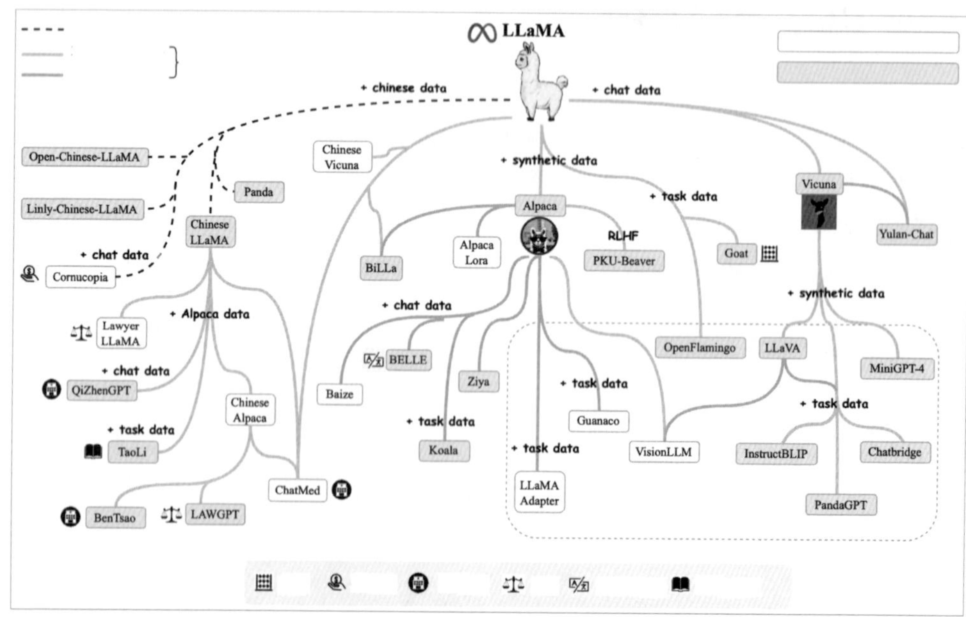

▲ 圖 1-1 基於 LLaMA 的生態

LLaMA 2 開放原始碼了 3 個大小分別為 7B、13B、65B 的大模型。2024 年 4 月，Meta 公司開放原始碼了 LLaMA 3。

這些開放原始碼大模型經過了預訓練或微調，我們可以從 Hugging Face 社區 [1] 中獲取。Hugging Face 是大模型時代的 GitHub，是全球最大、最成熟的大模型社區，裡面的資源非常豐富，值得探索、學習。

圖 1-2 是筆者在 Hugging Face 上搜尋 LLaMA 的結果，基於其開發的大模型應有盡有。

由阿里達摩院開放原始碼的 ModelScope（ModelScope）[2] 也是一個不錯的獲取大模型及相關資料的來源，如圖 1-3 所示。

上面介紹的是預訓練好的、可以使用的大模型，需要下載並部署（或微調）才能使用。如果你想直接使用現成的大模型，就需要呼叫商業的大模型 API，或用 Ollama、FastGPT 架設一個類似 OpenAI 的大模型 API。

如果你想親自嘗試，體驗一下從零開始預訓練大模型帶來的挑戰和樂趣，就需要獲取預訓練大模型的資料。

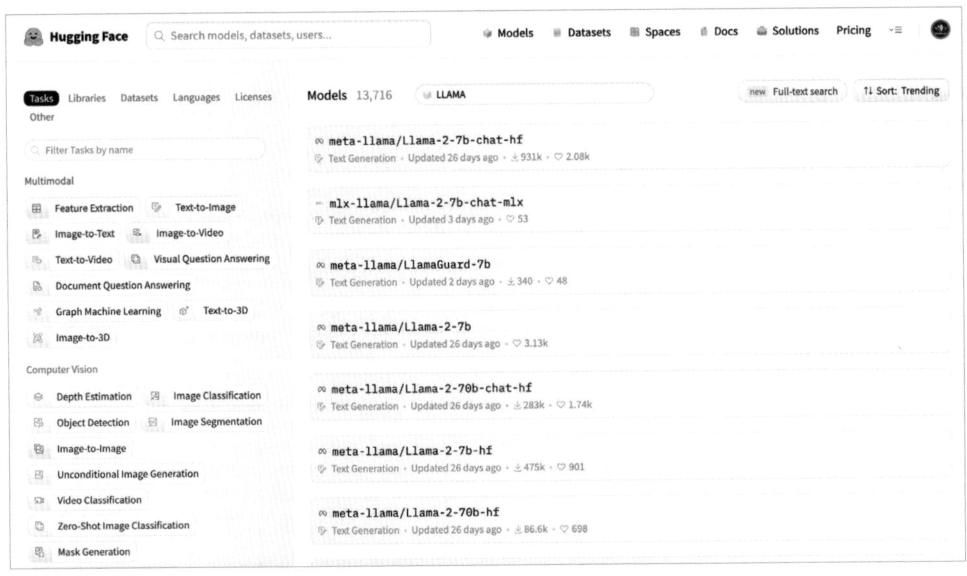

▲ 圖 1-2 基於 LLaMA 開發的大模型

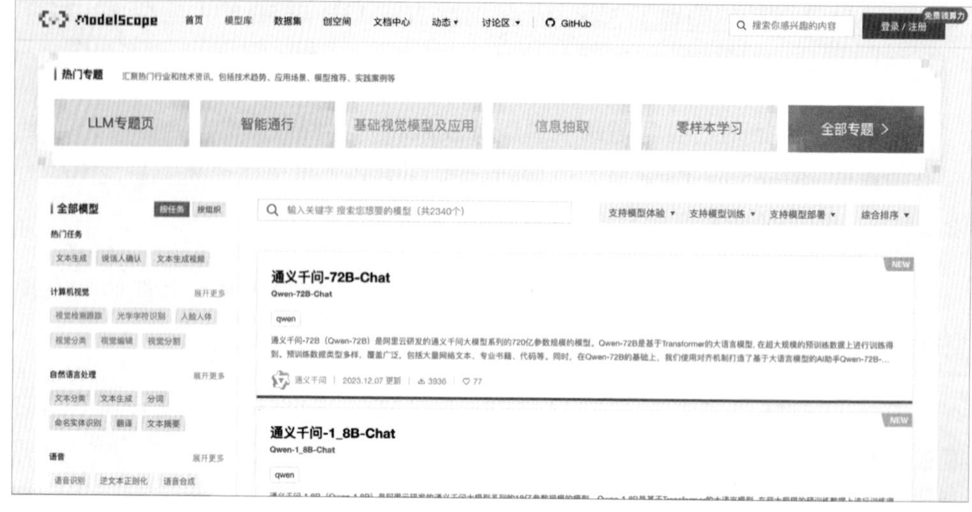

▲ 圖 1-3 ModelScope 上的大模型資源 (編按：本圖例為簡體中文介面)

1.1.2 資料資源

Hugging Face 和 ModelScope 上有非常多的資料資源，截至本書寫作時，Hugging Face 社區上有 85158 個資料集，覆蓋多種語言，其中英文資料集居多，如圖 1-4 所示。

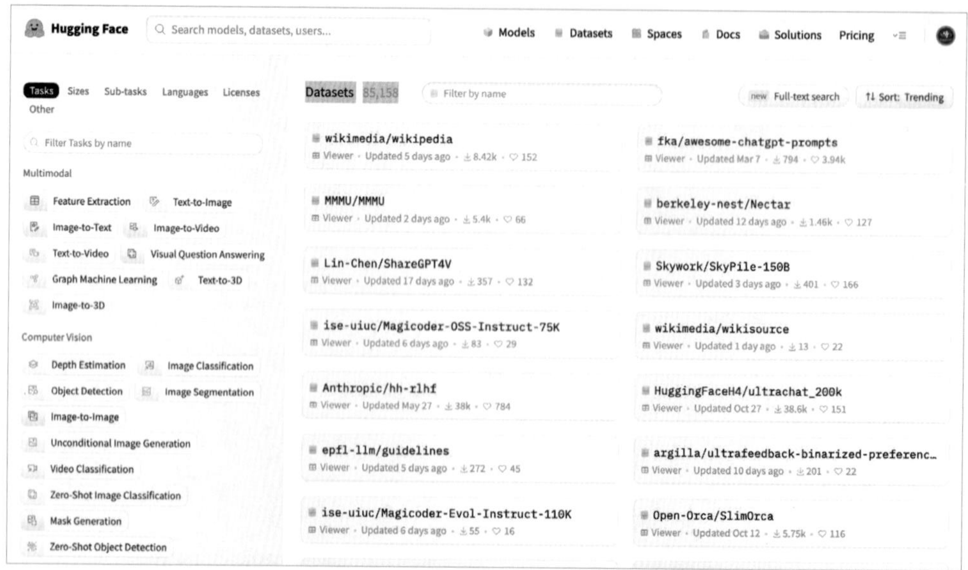

▲ 圖 1-4 Hugging Face 上的資料資源

ModelScope 上的資源覆蓋非常多的中文語料庫，如圖 1-5 所示。我們可以透過關鍵字搜尋相關的資料。

▲ 圖 1-5 ModelScope 上的資料資源 (編按：本圖例為簡體中文介面)

前面提到的都是通用資料，而真正有價值的資料是垂直行業的私有資料，有了這些資料，就可以預訓練垂直行業的大模型，本書講解的大模型推薦系統就是垂直領域的應用。

1.1.3 軟體資源

大模型是基於 Transformer 架構的深度學習模型，需要有成熟的軟體生態輔助大模型預訓練、微調、執行。這裡為讀者提供兩個比較好的資源。

基礎深度學習函式庫

目前最流行的深度學習函式庫是 PyTorch、TensorFlow、JAX，大模型是基於這些深度學習函式庫預訓練的，你只要熟悉一個即可。這裡建議學習 PyTorch，它更容易上手，生態也非常豐富。

大模型庫

Hugging Face 社區開放原始碼的 Transformers 函式庫對 Transformer 架構做了非常好的封裝，透過簡單的 API 就可以使用大模型。同時，該函式庫與 Hugging Face 社區深度綁定，可以直接使用資料名稱、模型名稱，非常方便。本書中的程式實戰部分會經常使用該函式庫。

1.1.4 硬體資源

預訓練、微調大模型需要較多的硬體資源，表 1-1 是不同參數量的 LLaMA 模型在 CPU 上執行時期需要的資源。即使是 7B 的模型，也至少需要 3.9GB 記憶體才能執行。

▼ 表 1-1 不同參數量的 LLaMA 模型在 CPU 上執行時期需要的資源

模型參數量	佔用空間	4 bit 量化佔用空間
7B	13GB	3.9GB
13B	24GB	7.8GB
30B	60GB	19.5GB
65B	120GB	38.5GB

1.2 大模型預訓練

大模型預訓練主要包括 4 個步驟：資料收集與前置處理、確定模型架構、確定目標函式和解碼。

1.2.1 資料收集與前置處理

大模型相關的預訓練資料（簡單起見，後面將預訓練資料統一簡稱為資料）來源主要包括 5 大類：網頁、對話、圖書和新聞、科學知識、程式。前面提到的 Hugging Face 和 ModelScope 中都有相關的資料。

預訓練一個效果好的大模型需要將上面的 5 類資料按照不同的比例搭配，有點類似中醫的藥材配比，非常講究。比例不好，可能就沒有那麼好的「療效」，甚至產生副作用，需要進行非常多的嘗試。

舉例來說，LLaMA 65B 開放原始碼模型的資料有 87% 來自網頁、2% 來自對話、3% 來自圖書和新聞、3% 來自科學知識、5% 來自程式；而 GPT-3 的預訓練資料有 84% 來自網頁、16% 來自圖書和新聞。

資料只是基礎，還需要對資料進行前置處理。資料前置處理的主要步驟如圖 1-6 所示。

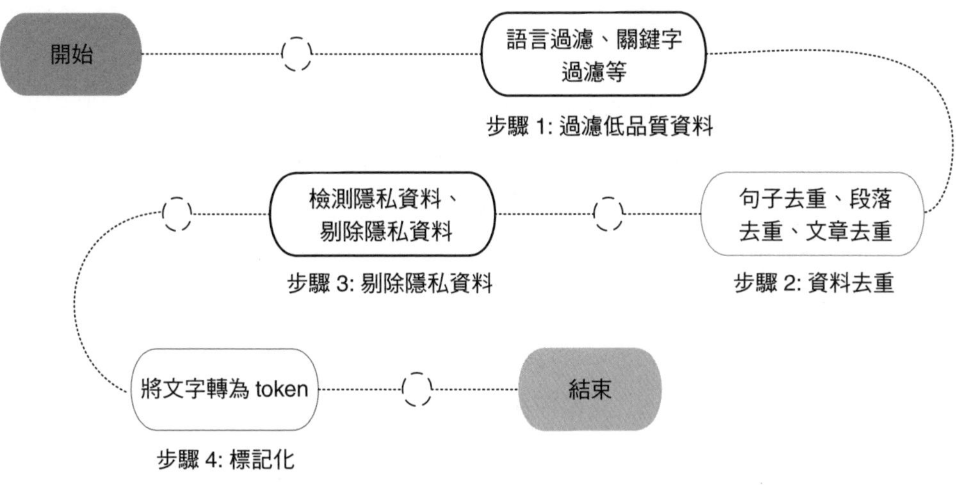

▲ 圖 1-6 資料前置處理的主要步驟

1. 過濾低品質數據

需要過濾掉從網頁上爬取的資料、HTML 標記、品質低或錯別字多的圖書、命名混亂或模組不清晰的程式等。

2. 資料去重

因為收集的資料來自多個資料來源，不可避免地會出現重複，而重復資料會極大影響大模型的性能，所以需要採用一些方法進行去重。例如可以採用詞頻、嵌入向量相似等策略去重。

3. 剔除隱私資料

需要剔除電話號碼、住址、身份證字號、帳號、密碼等隱私資料，否則一旦大模型被攻擊，就容易導致資訊洩露。OpenAI 就出現過此類事件。

4. 標記化（Tokenization）

透過上面 3 個步驟，文字已經準備好了。大模型採用 Transformer 架構，需要將文字標記化，即把文字分割為 token。針對英文，token 可以是一個單字或單字的部分（例如詞根）；針對中文，token 可以是單一中文字或片語。

1.2.2 確定模型架構

圖 1-7 是 GPT-1 的 Transformer 架構，其中包括 12 個 Transformer 層，GPT-2、GPT-3、GPT-4 與其類似，只不過層數更多。GPT 採用的是單向遮罩的注意力架構，即只能利用文字前面的詞來預測後面的詞。而更早的 BERT 模型可以利用兩邊的詞來預測中間的詞。

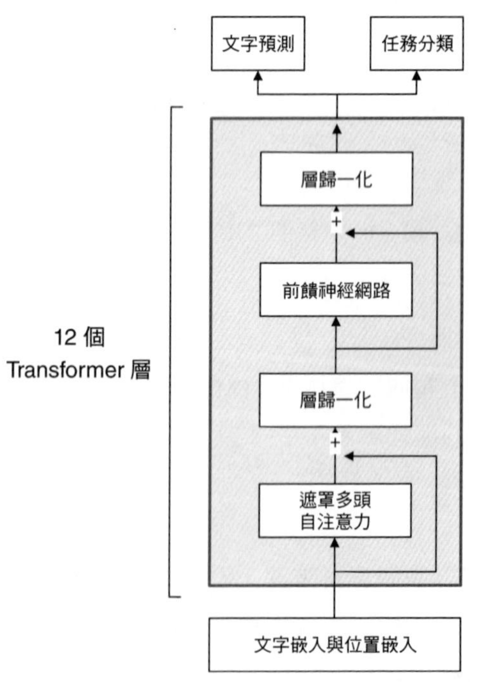

▲ 圖 1-7 GPT-1 的 Transformer 架構

圖 1-7 中的文字嵌入與位置嵌入是指語料庫中 token 的嵌入及對應的位置嵌入。token 嵌入容易理解，是為了對語言進行建模。而位置嵌入的主要目的是學習詞之前的位置依賴關係。一般用下面的三角函式表示 token 的位置。

$$\mathrm{PE}(\mathrm{pos}, 2i) = \sin\left(\frac{\mathrm{pos}}{10000^{2i/d}}\right)$$

$$\mathrm{PE}(\mathrm{pos}, 2i+1) = cos\left(\frac{\mathrm{pos}}{10000^{2i/d}}\right)$$

其中，PE 是位置嵌入，pos 表示單字所在的位置，$2i$ 和 $2i+1$ 表示位置編碼向量中的對應維度，d 則對應位置編碼的總維度。透過上面這種方式計算位置編碼有兩個好處：首先，正餘弦函式的範圍為 [−1,1]，算出的位置編碼與原詞嵌入相加不會使得結果偏離過遠而破壞原有單字的語義資訊。其次，依據三角函式的基本性質，第 $\mathrm{pos}+k$ 個位置的編碼是第 pos 個位置的編碼的線性組合，這就表示位置編碼中蘊含著單字之間的距離資訊，也就是包含單字之前的語義關係資訊。

圖 1-7 中的層歸一化是指進行分層歸一化。具體而言，計算每層所有啟動函式的平均值和方差，然後對啟動函式進行正態分佈歸一化。進行層歸一化的目的是更穩定地預訓練大模型。大模型有很多層，所以每一層都需要進行歸一化。

圖 1-7 中的前饋神經網路是一個以 Relu 函式為啟動函式的 2 層全連接神經網路，具體公式如下。

$$\mathrm{FFN}(x) = \mathrm{Relu}(xW_1 + b_1)W_2 + b_2$$

圖 1-8 是圖 1-7 中的遮罩多頭自注意力（Masked Multi Self Attention）部分，它是 Transformer 架構的核心，類似於人的眼睛可以關注視線中間及週邊的環境，從而對周圍環境產生認知。

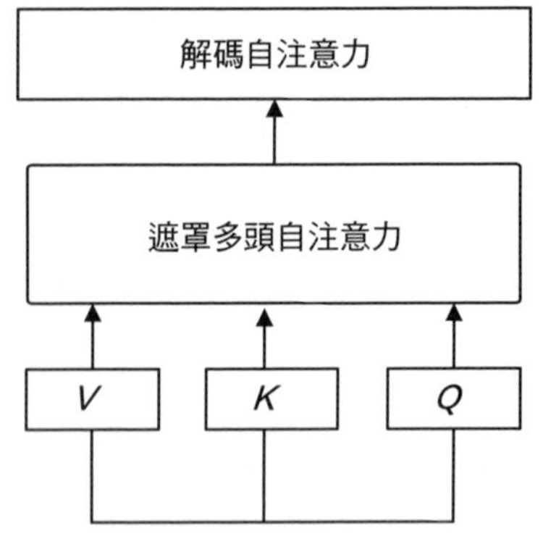

▲ 圖 1-8 遮罩多頭自注意力

注意力是查詢（$\boldsymbol{Q}$）和一組鍵（$\boldsymbol{K}$）、值（$\boldsymbol{V}$）對的函式，可以用下面的公式表示。

$$\text{Attention}\left(\boldsymbol{Q},\boldsymbol{K},\boldsymbol{V}\right)=\text{softmax}\left(\frac{\boldsymbol{Q}\boldsymbol{K}^{\mathrm{T}}}{\sqrt{d_k}}\right)\boldsymbol{V}$$

可以透過 $\boldsymbol{W}^{Q}\in\mathbf{R}^{d\times d_q}$ 、$\boldsymbol{W}^{K}\in\mathbf{R}^{d\times d_k}$ 和 $\boldsymbol{W}^{V}\in\mathbf{R}^{d\times d_v}$ 3 個線性變換將輸入序列中的每個單字表示 x_i 都轉為其對應的 $q_i\in\mathbf{R}^{d_q}$ 、$k_i\in\mathbf{R}^{d_k}$ 和 $v_i\in\mathbf{R}^{d_v}$ 向量。圖 1-9 是針對 Machine、Learning 兩個單字做對應的線性變換得到的向量。關於 Transformer 的詳細介紹，可以閱讀參考文獻 [2]。

輸入 Machine Learning

嵌入表示 x_1 x_2

查詢 q_1 q_2 W^Q

鍵 k_1 k_2 W^K

值向量 v_1 v_2 W^V

▲ 圖 1-9 多頭自注意力與 $\boldsymbol{Q}$、$\boldsymbol{K}$、$\boldsymbol{V}$ 的關係

1.2.3 確定目標函式及預訓練

給定一個無監督的 token 語料庫，$\mathcal{U}=\{u_1,u_2,\cdots,u_n\}$使用標準的語言建模目標函式最大化以下似然函式。

$$L_1(\mathcal{U})=\sum_i \ln P\left(u_i \mid u_{i-k},u_{i-(k-1)},\cdots,u_{i-1};\Theta\right) \tag{1-1}$$

其中，k 是上下文視窗的大小[①]，條件機率 P 使用參數為 θ 的神經網路建模。這些參數使用隨機梯度下降法進行預訓練。一般將 Transformer 作為語言模型（P）的建構組件，式（1-1）展開如下。

$$\boldsymbol{h}_0=\boldsymbol{U}\boldsymbol{W}_{\mathrm{e}}+\boldsymbol{W}_{\mathrm{p}}$$

$$\boldsymbol{h}_l=\text{transformer\_block}(\boldsymbol{h}_{l-1})\ \forall l\in[1,n]$$

$$P(u)=\text{softmax}(h_{\mathrm{n}}W_{\mathrm{e}}^{\mathrm{T}})$$

這裡$\boldsymbol{U}=\left(u_{-k},u_{-(k-1)},\cdots,u_{-1}\right)$是上下文的 token 向量，$n$ 是 Transformer 的層，transformer_block 就是圖 1-7 中的 Transformer 架構，softmax 是最上層預測部分的啟動函式。$\boldsymbol{W}_{\mathrm{e}}$ 是 token 的嵌入矩陣，而 $\boldsymbol{W}_{\mathrm{p}}$ 是位置嵌入矩陣，$\boldsymbol{W}_{\mathrm{e}}$ 和 $\boldsymbol{W}_{\mathrm{p}}$ 分別對應圖 1-7 中的文字嵌入和位置嵌入。

語料庫 $\mathcal{U}$ 通常是文字文件，因此預訓練最最佳化模型不需要標注資料，可以直接使用巨量的文字作為預訓練樣本，即利用文字中前面的 token 來預測下一個 token，這個過程就是預訓練。

當然，真正的大模型語料庫都非常大，Transformer 層也非常多，有非常多的超參數需要處理，下面列舉 3 個最重要的。

- 批大小。

批大小（Batch Size）指大模型一次處理的 token 規模，批越大，模型的吞吐量越好，預訓練一次的時間越短，對系統網路、讀寫、記憶體頻寬等要求也越高。

---

① 基於前面 k 個 token 預測下一個 token。

- 學習率。

學習率（Learning Rate）是模型在利用反向傳播演算法進行迭代最佳化過程中更新梯度的幅度。針對大模型，一般開始時用一個比較大的學習率，然後隨著模型逐步收斂，調整學習率，讓學習率逐步下降到最初學習率的固定比例（例如 10%）。

- 最佳化器。

最佳化器（Optimizer）即大模型預訓練的最佳化方法，不同最佳化器有各自的優缺點和適用情形，預訓練大模型一般用 Adam 或 AdamW 最佳化器。

圖 1-10 列出了業界主流大模型對應的最佳化參數。

模型	批次大小 (#tokens)	學習率	預熱	衰減方法	最佳化器	精度類型	權重衰減	梯度截斷	Dropout
GPT3 (175B)	32K→3.2M	6×10^{-5}	yes	cosine decay to 10%	Adam	FP16	0.1	1.0	-
PanGu-α (200B)	-	2×10^{-5}	-	-	Adam	-	0.1	-	-
OPT (175B)	2M	1.2×10^{-4}	yes	manual decay	AdamW	FP16	0.1	-	0.1
PaLM (540B)	1M→4M	1×10^{-2}	no	inverse square root	Adafactor	BF16	lr^2	1.0	0.1
BLOOM (176B)	4M	6×10^{-5}	yes	cosine decay to 10%	Adam	BF16	0.1	1.0	0.0
MT-NLG (530B)	64 K→3.75M	5×10^{-5}	yes	cosine decay to 10%	Adam	BF16	0.1	1.0	-
Gopher (280B)	3M→6M	4×10^{-5}	yes	cosine decay to 10%	Adam	BF16	-	1.0	-
Chinchilla (70B)	1.5M→3M	1×10^{-4}	yes	cosine decay to 10%	AdamW	BF16	-	-	-
Galactica (120B)	2M	7×10^{-6}	yes	linear decay to 10%	AdamW	-	0.1	1.0	0.1
LaMDA (137B)	256K	-	-	-	-	BF16	-	-	-
Jurassic-1 (178B)	32 K→3.2M	6×10^{-5}	yes	-	-	-	-	-	-
LLaMA (65B)	4M	1.5×10^{-4}	yes	cosine decay to 10%	AdamW	-	0.1	1.0	-
LLaMA 2 (70B)	4M	1.5×10^{-4}	yes	cosine decay to 10%	AdamW	-	0.1	1.0	-
Falcon (40B)	2M	1.85×10^{-4}	yes	cosine decay to 10%	AdamW	BF16	0.1	-	-
GLM (130B)	0.4M→8.25M	8×10^{-5}	yes	cosine decay to 10%	AdamW	FP16	0.1	1.0	0.1
T5 (11B)	64K	1×10^{-2}	no	inverse square root	AdaFactor	-	-	-	0.1
ERNIE 3.0 Titan (260B)	-	1×10^{-4}	-	-	Adam	FP16	0.1	1.0	-
PanGu-Σ (1.085T)	0.5M	2×10^{-5}	yes	-	Adam	FP16	-	-	-

▲ 圖 1-10 業界主流大模型對應的最佳化參數

除了上面提到的參數，資料如何並行、如何進行分散式預訓練，都是需要關注的重點問題，這些是更深入的課題，讀者可以自行學習。

1.2.4 解碼策略

大模型預訓練好之後，如何利用呢？這就不得不說解碼策略了。所謂解碼策略，就是在預訓練的大模型基礎上，如何基於文字生成對應的後續文字。用數學語言來表述，就是機率預估問題。

$$x_i = \arg\max{}_x P(x \mid x < i)$$

這裡的 P 是大模型預測的條件機率，$x<i$ 是第 i 個單字前面的單字，即輸入大模型的語料，x_i 即讓 P 的值最大的那個詞。圖 1-11 是一個例子。

I am sleepy. I start a pot of _____

coffee	0.661	strong	0.008	soup	0.005
water	0.119	black	0.008	...	...
tea	0.057	hot	0.007	happy	4.3e-6
rice	0.017	oat	0.006	Boh	4.3e-6
chai	0.012	beans	0.006	...	...

▲ 圖 1-11 基於前面的詞預測下一個詞

針對這個例子，$x<i$ =「I am sleepy. I start a pot of 」，x_i =coffee，因為 coffee 出現的機率是 0.661，最大。

上面是貪婪搜尋（Greedy Search）方法，也就是窮盡大模型詞庫中的所有單字，每一步都找到機率最大的那個詞，直到遇到結束符號。根據貪婪搜尋方法，圖 1-12 最終獲得的預測序列是 ABC。

時間步驟	1	2	3	4
A	0.5	0.1	0.2	0.0
B	0.2	0.4	0.2	0.2
C	0.2	0.3	0.4	0.2
<eos>	0.1	0.2	0.2	0.6

▲ 圖 1-12 不同時間步驟的不同 token 出現的機率

很明顯，這樣做將原來指數等級的求解空間直接壓縮到了與長度線性相關的大小。在每一步中選擇具有最高機率的 token，可能導致忽略具有較高整體機率但較低局部機率的句子。這種關注當下的策略丟棄了絕大多數可能解，無法保證最終得到的序列機率是最佳的。

解決這個問題的一種方法是集束搜尋（Beam Search），它對貪婪策略進行了改進。集束搜尋的想法很簡單，就是稍微放寬考察的範圍。在每個時間步，不再只保留當前機率最高的 1 個輸出，而是保留 n 個（n beam 的大小）。當 n=1 時，集束搜尋就退化為貪婪搜尋。

圖 1-13 是一個集束搜尋範例，每個時間步有 A、B、C、D、E 共 5 種可能的輸出，n=2，也就是說每個時間步都會保留到當前步為止條件機率最佳的兩個序列。

在時間步 1，A 和 C 是最佳的，因此獲得了 A 和 C 兩個結果，其他 3 個被拋棄。時間步 2 會基於這兩個結果繼續生成，在 A 這個分支可以得到 5 個候選集，分別是 AA、AB、AC、AD、AE。同理，分支 C 也得到 5 個候選集。此時對這 10 個候選集進行統一排名，保留最佳的兩個，即 AB 和 CE。時間步 3 從新的 10 個候選集裡挑選最好的兩個，獲得了 ABD、CED 兩個結果。可以發現，集束搜尋在每個時間步需要考察的候選集數量是貪婪搜尋的 n 倍，因此是一種以時間換性能的方法。

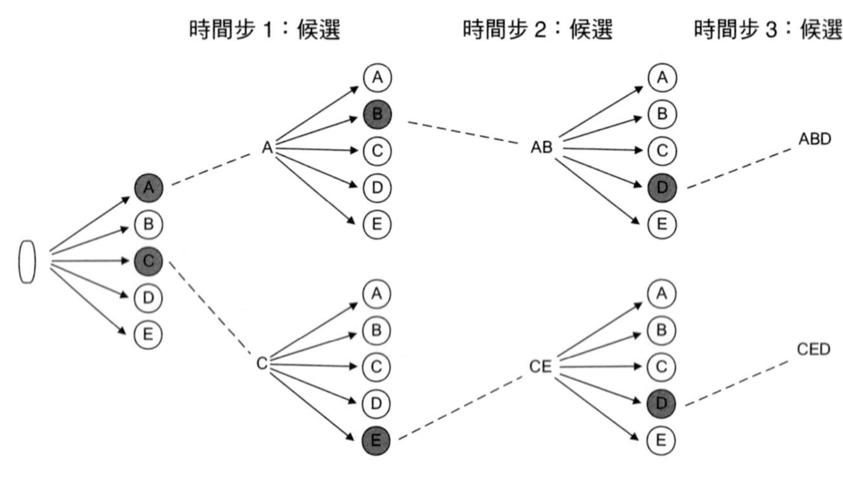

▲ 圖 1-13 集束搜尋範例

為了提升大模型輸出的隨機性（創新性），可以採用機率抽樣的方式生成 x_i，即在樣本中出現的機率越大，被抽到的機率就越大。在這種情況下，不是每次預測的結果都是 x_i，只不過 x_i 出現的機率最大。常見的抽樣方法有以下 3 種。

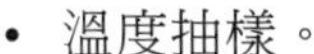

- 溫度抽樣。

為了控制採樣的隨機性，溫度抽樣（Temperature Sampling）透過調整 softmax 函式的溫度係數，計算詞彙表上第 j 個 token 出現的機率。下式中 l_j 是某個 token 的 logit，t 是溫度參數。

$$P(x_j \mid x_{<i}) = \frac{\exp(l_j/t)}{\sum_{j'} \exp(l_{j'}/t)}$$

t 的範圍在 0 到 1 之間，降低溫度 t 增加了選擇具有高機率的單字的機會，同時減少了選擇具有低機率單字的機會。當 t 被設置為 1 時，它將成為預設的隨機採樣；當 t 接近 0 時，相當於貪婪搜尋。此外，當 t 變為無限大時，它退化為均勻採樣。

- top-*k* 抽樣。

與溫度抽樣不同，top-*k* 抽樣直接截斷機率較低的 token，並且僅從具有 top-*k* 機率的 token 中採樣。舉例來說，在前面的例子中，top-5 將從單字 coffee、water、tea、rice 和 chai 的歸一化機率中進行採樣。

- top-*p* 抽樣。

由於 top-*k* 抽樣不考慮整體可能性分佈，因此恒定的 k 值可能不適合不同的上下文。基於此，從累積機率大於或等於 p 的最小集合中進行採樣的 top-*p* 抽樣（也稱核心採樣）被提出。在實踐中，可以透過向按機率降冪排列的詞彙表中逐漸增加 token 來建構最小集合，直到它們的累積機率值超過 p。

1.3 大模型微調

大模型預訓練一般是基於通用文字進行的，大模型學習到的是通用能力（包括語言生成、常識知識、邏輯推理等）。雖然超大規模的大模型（例如 ChatGPT、GPT-4 等）也可以用於解決一般的下游任務，效果也不錯，但一些非常專業的場景，例如金融、法律、醫療，對知識、技能的要求更高，往往需要大模型在專業知識上重新進行「歷練」。如果是更小的模型（例如 1B ～ 100B），就更需要在垂直領域進行二次學習，才能更進一步地解決下游任務。

指令微調和對齊微呼叫於在特定領域知識上對大模型進行二次訓練。具體來說，指令微調是一種監督學習過程，模型的輸入、輸出都是自然語言（例如輸入是中文的「我愛你」，輸出是英文的「I love you」，這是從中文到英文的翻譯問題）；而對齊微調將人類整合到大模型的學習過程中，一般採用人類專家標注的樣本學習一個價值函式，然後用強化學習的想法讓價值函式監督大模型的預訓練，讓大模型的輸出與人類的價值觀對齊（例如輸出的是準確的、無害的、有用的資訊）。

1.3.1 微調原理

經過預訓練，大模型可以獲得解決各種任務的一般能力，但為了在特定問題上或領域中有更好的表現，需要對預訓練模型進行微調。在用式（1-1）中的目標函式預訓練模型後，透過監督學習任務對參數進行微調。

假設有一個標記的資料集 $\mathcal{C}$，其中每個實例由輸入 token 序列 $x^1,x^2,\cdots,x^m$，以及標籤 y 組成。將資料登錄預訓練模型，以獲得最後一個隱藏層的啟動函式 h_l^m，然後將其輸入具有參數 W_y 的、增加了一層線性輸出層的模型預測 y [①]。

$$P(y \mid x^1,x^2,\cdots,x^m) = \mathrm{softmax}\left(h_l^m W_y\right)$$

這樣就獲得了以下的求最大值的目標函式。

$$L_2(\mathcal{C}) = \sum_{(x,y)} \ln P(y \mid x^1,x^2,\cdots,x^m)$$

實踐經驗表明，將語言建模過程作為微調的輔助目標有助學習，加入輔助目標既可以提高監督模型的泛化能力，又可以加快模型的收斂速度[②]。具體而言，就是最佳化以下目標。

$$L_3(\mathcal{C}) = L_2(\mathcal{C}) + \lambda L_1(\mathcal{C})$$

① 這裡用多分類任務來說明，所以輸出層用了 softmax 函式，其實監督過程也可以是序列建模任務等其他類型的監督學習任務。

其中，λ為權重。

1.3.2 指令微調

與預訓練不同，指令微調通常更有效，只需要中等數量的樣本用於訓練。指令微調是有監督的訓練過程，其最佳化過程與預訓練在某些方面有所不同，例如目標函式（如序列到序列的 loss）和最佳化的超參數（如較小的批大小和學習率）。

本質上，指令微調是在以自然語言表示的格式化樣本集合上微調經過預訓練的大模型的方法。為了執行指令微調，我們首先需要收集或構造滿足指令格式的樣本，然後使用這些格式化的樣本以監督學習的方式對大模型進行微調。在指令微調後，大模型在下游任務上會表現出卓越的性能，在多語言環境中也是如此。

指令微調一般分為兩步：第一步是以一定的格式構造監督資料集，資料集中的輸入和輸出標籤都是文字形式的，可以是已有的標注文字，也可以利用大模型生成標注[3]。第二步是對大模型進行監督微調，更進一步地解決下游任務，圖 1-14 是指令微調的一般流程。

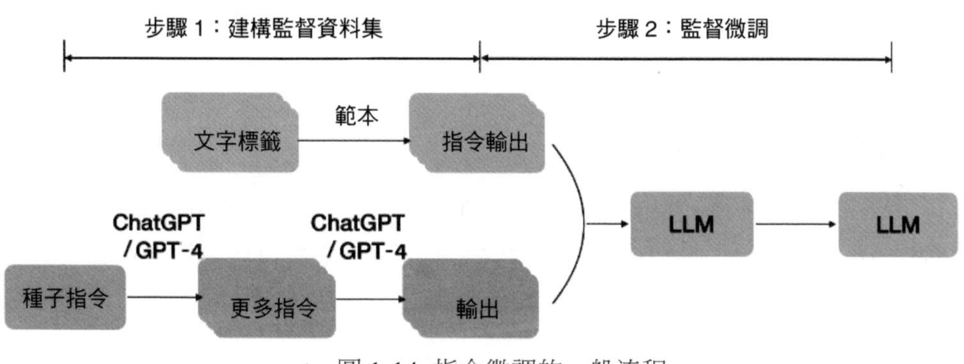

▲ 圖 1-14 指令微調的一般流程

② GPT-1 就是這麼做的。

③ 目前很多大模型就是利用 ChatGPT/GPT-4 來輔助標注的，即利用 ChatGPT/GPT-4 進行知識蒸餾。

通常透過以下 3 種方法建構監督資料集：基於開放原始碼的 NLP 資料集、基於人機對話及利用大模型合成資料，如圖 1-15 所示，下面詳細說明。

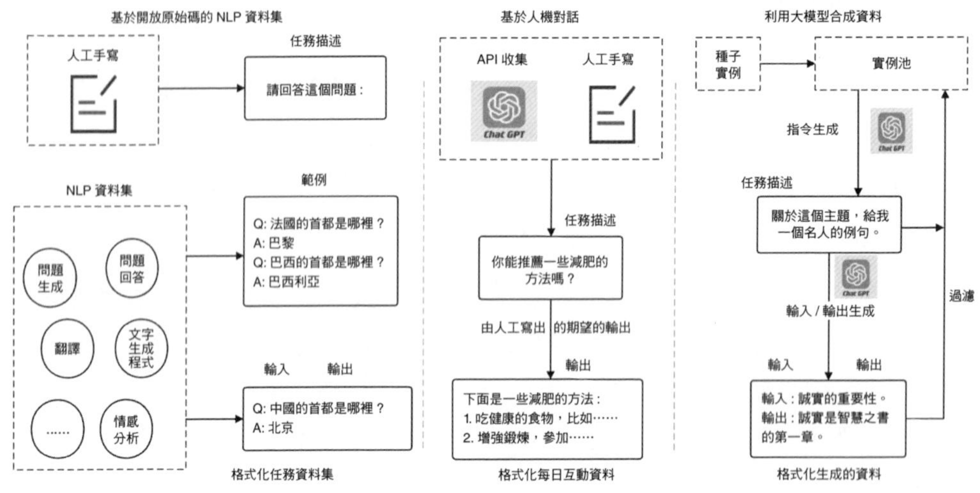

▲ 圖 1-15 建構監督資料集的 3 種方法

方法 1：基於開放原始碼的 NLP 資料集

機器學習領域有過很多 NLP 標注資料集，例如文字摘要、文字分類、翻譯等，利用這些 NLP 標注資料集，只需要人工增加一些任務描述，就可以建構出適合大模型處理的監督資料集，這為指令微調的監督樣本提供了便捷的、多樣的監督資料來源。

方法 2：基於人機對話

ChatGPT 發佈後，很多人向 ChatGPT 提了很多問題，這些問題就是監督樣本的輸入，這些輸入是真實的人類需求，可以請專家對這些問題進行專業答覆（例如專業減肥方法），那麼這些問題描述和對應的專家答覆就是一個訓練樣本對。

方法 3：利用大模型合成資料

這是一種半自動化的方法。需要先有一個實例池（這裡的實例就是監督樣本對，一般需要 100 個左右），然後隨機從實例池中取出一個實例，讓大模型生成任務描述，進一步讓大模型生成輸入、輸出，形成新的實例，並加入實例池。這個過程可以自動化執行，以拓展監督樣本，降低人工標注的成本。

基於大模型的指令微調這類監督學習任務與傳統的監督學習有所區別，如圖 1-16 所示。

傳統監督學習使用大量的標記範例表示任務語義，建構監督資料集的過程成本很高，因此很難推廣到新的任務中。圖 1-16 中基於情感分析建構的監督任務無法推廣到實體辨識領域。而利用指令微調建構的大模型的監督學習樣本更加通用，可以覆蓋更多的下游任務，泛化能力更好，也可以輕鬆應對新的下游任務。

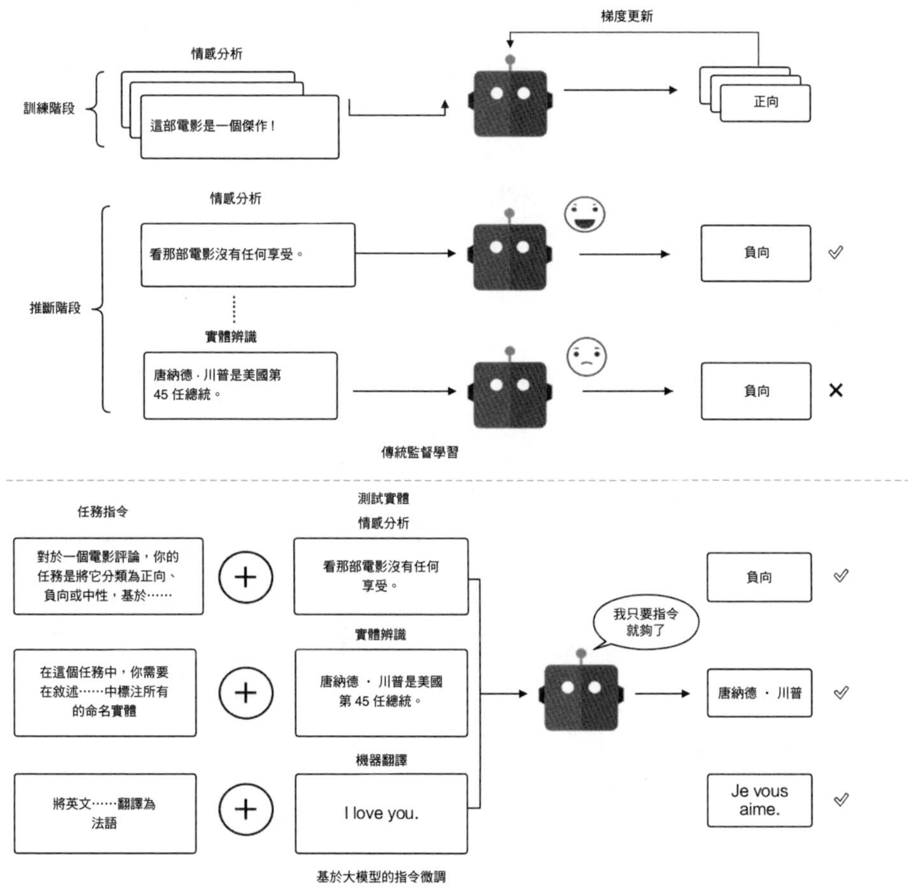

▲ 圖 1-16 傳統監督學習與基於大模型的指令微調的區別

有了監督微調的樣本，就可以進行微調了。在指令微調的具體實現上，至少有兩種方法，下面分別介紹。

第一種方法是微調整個模型，也就是說，採用梯度下降的反向傳播演算法更新整個模型的參數。由於大模型的參數通常較多，這種微調方法耗費時間和運算資源較多，成本太高。另外，微調可能影響模型已有的能力，使模型在某些任務上的表現變差，所以推薦使用第二種方法。

第二種方法是參數有效的微調，具體實現方法有多種，這裡主要講解 LoRA，這也是最流行的微調方法。LoRA 將大模型的每一層（到下一層）的矩陣（大模型的參數）表示為兩個低階矩陣的乘積，在微調時，大模型之前的矩陣保持不變，只微調 LoRA 在每層增加的兩個低階矩陣，參數量較第一種方法少了很多（可以減少 90% 以上），訓練成本低。同時，在微調時，LoRA 保持大模型之前的矩陣不變，微調好後，在微調的下游任務上使用低階矩陣，而在其他任務上使用大模型之前的矩陣，以保持大模型本來的能力。

圖 1-17 是 LoRA 的演算法架構，其中左側是大模型的網路，右側是低階矩陣 $\boldsymbol{A}$ 和 $\boldsymbol{B}$。後續章節中的微調程式實現大多採用 LoRA 方法。

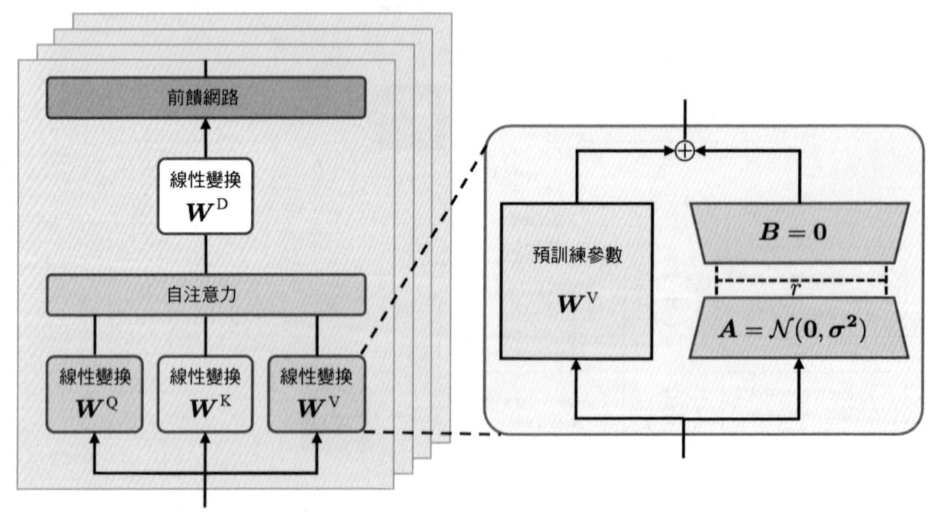

▲ 圖 1-17 LoRA 的演算法架構

關於指令微調的詳細介紹，參見參考文獻 [3] 和 [4]。後續章節會在推薦系統的框架下講解指令微調的方法，並舉出實戰案例。

1.3.3 對齊微調

大模型在廣泛的 NLP 任務中顯示出了非凡的能力。然而，這些模型有時可能出現意想不到的行為，例如編造虛假資訊、實現不正確的目標，以及產生有害、有誤導性和有偏見的言論。大模型的目標是透過 token 前置處理來預訓練模型參數，缺乏對人類價值觀或偏好的考慮。為了避免這些意想不到的行為，學界提出了人類對齊（Human Alignment）的方法，以使大模型的行為符合人類的期望。然而，與最初的預訓練和微調（例如指令微調）不同，這種調整需要考慮的標準各異（例如有用、誠實和無害）。研究證明，人類對齊在一定程度上損害了大模型的一般能力①，這種現象被稱為**對齊稅**（Alignment Tax）。

對齊的目的是讓大模型可以更進一步地服務於人類社會，產生正向的價值。人們越來越傾向於透過制定各種標準來規範大模型的行為。這裡以 3 個具有代表性的對齊標準（有用、誠實和無害）為例進行討論，這些標準已被廣泛採用。此外，大模型還有其他對齊標準，包括行為、意圖、激勵和內部特性，這些標準與上述 3 個標準相似②。

為了使大模型與人類價值觀保持一致，學術界提出了人類回饋強化學習（Reinforcement Learning from Human Feedback，RLHF）。在強化學習（Reinforcement Learning，RL）中，智慧體透過與環境互動（action）獲得環境的回饋（feedback），基於回饋形成新的對話模式與策略，最終透過多輪互動，可以更進一步地從環境中學習，獲得更多的綜合回報，如圖 1-18 所示。RLHF 利用人類的回饋資料對大模型進行微調，這有助進一步改進模型。RLHF 採用強化學習演算法（目前 PPO 演算法使用廣泛），透過學習獎勵模型使大模型適應人類回饋。

① 為了實現人類對齊，大模型在其他任務上的表現變差了。

② 或至少從技術處理上相似。

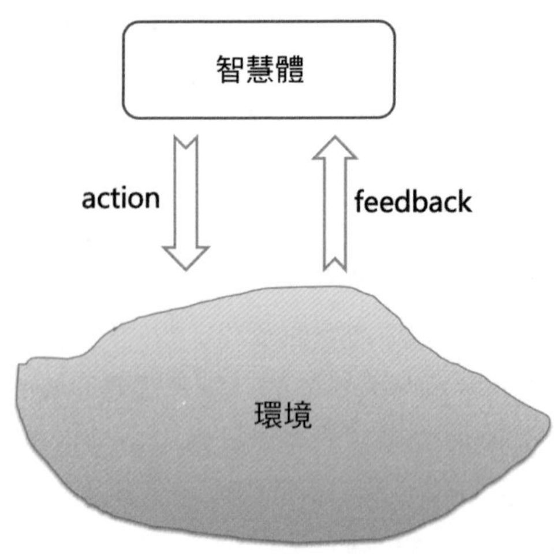

▲ 圖 1-18 強化學習範式

在 RLHF 語境下，強化學習的環境就是人類，透過人類的價值觀來訓練獎勵函式，利用獎勵函式來「調教」大模型，讓大模型的輸出滿足人類的價值偏好。這種方法將人類納入大模型的訓練迴圈，以開發良好對齊的大模型①。

RLHF 系統主要包括 3 個關鍵元件：預先訓練的待對齊的大模型、從人類回饋中學習的獎勵模型和訓練大模型的強化學習演算法。具體來說，預先訓練的待對齊的大模型通常是一個生成模型，它是用現有的預訓練的大模型參數初始化的。舉例來說，OpenAI 將 175B GPT-3 作為其 RLHF 模型（InstructGPT）待對齊的大模型，DeepMind 將擁有 2800 億個參數的模型 Gopher 作為其 GopherCite 模型的待對齊大模型。

此外，獎勵模型（RM）通常以純量值的形式（例如 0 ～ 1，值越大代表跟人類價值觀越匹配）提供反映人類對大模型生成的文字的偏好的監督訊號②。獎勵模型可以採用微調的大模型或使用人類偏好的資料從頭訓練的大模型。通常將與待對齊的大模型參數規模不同的大模型作為獎勵模型。舉例來說，OpenAI 使用 6B GPT-3、DeepMind 使用 7B Gopher 作為獎勵模型。

① InstructGPT 模型就將 RLHF 作為核心方法。

② 透過人類回饋來說明生成文字是否有用、是否誠實、是否無害。

為了使用來自獎勵模型的訊號來最佳化預訓練的待對齊的大模型，需要設計一種用於調整大模型的特定強化學習演算法。圖 1-19 是 RLHF 演算法的工作流，主要有以下 3 個步驟。

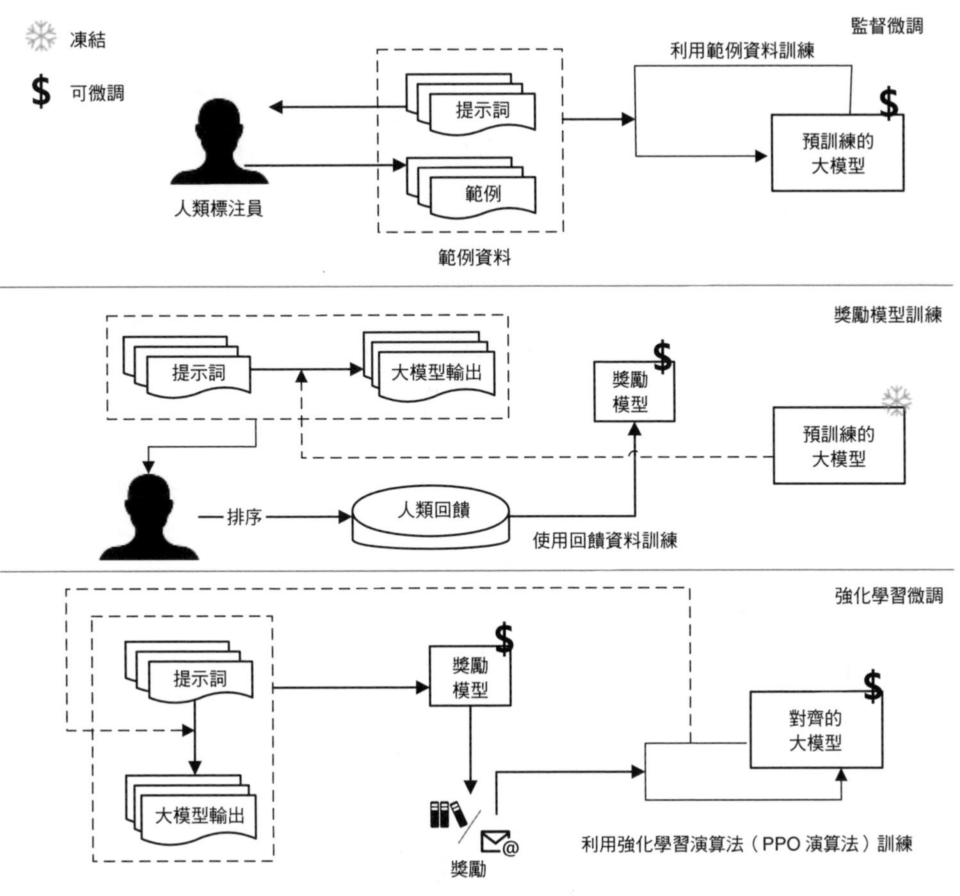

▲ 圖 1-19 RLHF 演算法的工作流

（1）監督微調。為了獲得大模型最初執行所需的行為，需要收集監督資料集，該資料集包含用於微調大模型的提示詞（指令）和輸出。這些提示詞和輸出可以是由人類標注員為某些特定任務撰寫的，同時需要確保任務的多樣性（主要目的是提升模型的泛化能力）。舉例來說，InstructGPT 要求人類標注員為一些生成性任務（如開放式 QA、腦力激盪、聊天和改寫）撰寫提示詞（如列出讓我對工作重新充滿熱情的 5 個想法）和輸出。這與指令微調的方法類似，在某些情況下可以省略。

（2）獎勵模型訓練。使用人類回饋資料來訓練獎勵模型。具體而言，我們首先將大模型用來抽樣的提示詞（來自監督資料集或人工生成的提示詞）作為輸入來生成一定數量的輸出文字。然後邀請人類標注員來評估這些輸入 - 輸出對的品質。標注過程可以以多種形式進行，一種常見的方法是對生成的多個候選文字分別評分排序，以減少標注者之間的不一致性。最後訓練獎勵模型，以預測滿足人類偏好的輸出。在 InstructGPT 中，標注者將模型生成的輸出從最好到最差進行排序，並訓練獎勵模型（6B 的 GPT-3）來預測排序。

（3）強化學習微調。將大模型的對齊微調形式化為強化學習問題。用預先訓練的大模型充當策略，該策略將提示詞作為輸入並傳回輸出文字，其動作空間是整個詞彙表，狀態是當前生成的 token 序列，並且由獎勵模型提供獎勵。為了避免與初始（調整前）大模型顯著偏離，懲罰項通常被加入獎勵函式。舉例來說，InstructGPT 使用 PPO 演算法針對獎勵模型最佳化大模型。對於每個輸入提示，InstructGPT 計算當前大模型和初始大模型的生成結果之間的 KL 偏差作為懲罰項。第 2 步和第 3 步可以多次迭代，以便更進一步地對齊大模型，使其符合人類的期望。

目前，OpenAI 內部正在進行一個特殊的專案——超級對齊。使用小模型來監督大模型（例如利用 GPT-2 監督已完成預訓練但沒有經過微調和 RLHF 的 GPT-4），並輔以置信損失等方法，激發大模型的能力，可以讓沒有被微調過的 GPT-4 至少達到微調後的 GPT-3.5 的水準，如圖 1-20 所示。

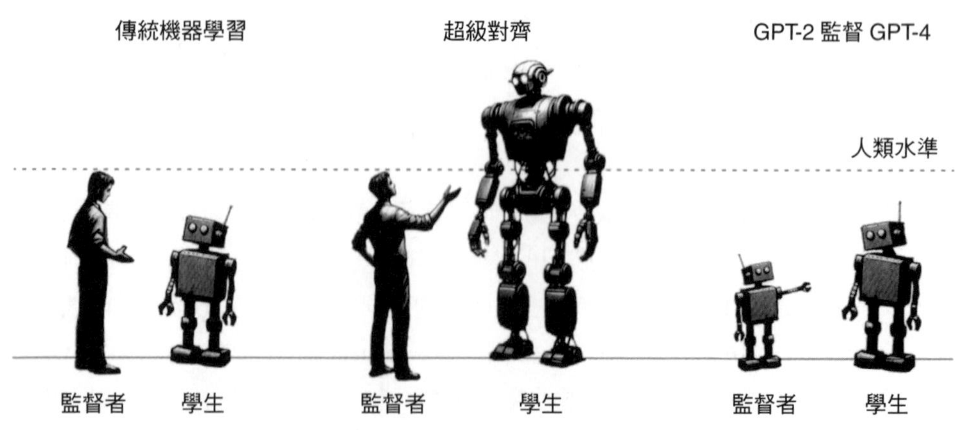

▲ 圖 1-20 3 種不同的對齊微調

關於 RLHF 的詳細介紹，可以從參考文獻 [5~7] 中獲得更多的資訊，第 6 章會基於推薦系統講解 RLHF 的具體案例。

1.4 大模型線上學習

一般來說，當大模型需要解決特定領域的問題或小模型想要產生更好的效果時，需要對模型進行微調。如果模型足夠大（例如 GPT-4 這種體量的模型），那麼它本身就能完成非常多的下游任務，並且效果相當驚豔，這也是大模型最吸引人的地方。最理想的情況是，透過一次預訓練，大模型就能完成所有的下游任務。本節就介紹這方面的嘗試：當大模型預訓練好後，直接利用大模型壓縮的世界知識進行推理，解決現實問題。

在預訓練或微調之後，可以透過合適的提示詞策略使用大模型。一種典型的提示方法是上下文學習（In-Context Learning，ICL），它以自然語言文字的形式描述任務或提供範例（Demonstration），然後輸入大模型獲得答案。還可以在提示中插入一系列中間推理步驟，即思維鏈（CoT），來增強大模型的上下文學習能力。此外，還有一種執行複雜任務的方法——規劃（Planning），該方法首先將任務分解為較小的子任務，然後生成一個行動計畫來一個一個完成這些子任務，最終獲得原始問題的答案。

1.4.1 提示詞

提示詞是使用大模型完成各種任務的主要方法，也是非專業人士使用大模型的唯一方法。這種方法可以讓我們使用自然語言與大模型互動，具備可操作性。由於提示詞的品質可以在很大程度上影響大模型在特定任務中的效果，因此可以使用某些原則和技巧手動建立合適的提示詞，讓大模型輸出更好的結果。

手動建立提示詞的過程也稱為提示詞工程（Prompt Engineering）。精心設計的提示詞有助引導大模型完成特定任務。為了幫助你寫出更合適的提示詞，下面從提示詞的組成部分和設計提示詞的原則兩個維度提供一些建議。

1. 提示詞的組成部分

一個功能完整的提示詞可以讓大模型更進一步地服務於我們的任務，產出更符合我們預期的結果。一般來說，提示詞包括以下 4 部分。

- 任務描述。

任務描述（Task Description）指需要大模型遵循的特定指令。一般來說，應該用自然語言清楚地描述任務目標。對於具有特殊輸入或輸出格式的任務，通常需要詳細地說明，並且可以進一步利用關鍵字來突出顯示特殊設置，以便更進一步地指導大模型完成任務。

- 輸入資料。

在常見情況下，用自然語言描述輸入資料（大模型要回答的問題）很簡單。對於特殊的輸入資料（Input Data），如知識圖譜和表格，有必要採用適當而方便的方式提升其可讀性。對於結構化資料，通常按順序將原始記錄（例如知識三元組）轉為序列。此外，程式語言（例如可執行程式）也被看作結構化資料，可以讓大模型利用外部工具（例如程式執行器）來產生精確的結果。

- 上下文資訊。

除了任務描述和輸入資料，上下文資訊（Contextual Information）或背景資訊對特定任務也是必不可少的，舉例來說，檢索到的文件對開放式問答非常有用。檢索到的文件的品質及其與問題的相關性會對生成的答案產生影響，因此需要在適當的提示詞範本或表達格式中包含上下文資訊。此外，上下文任務範例有助引導大模型完成複雜任務，它可以更進一步地描述任務目標、特殊的輸出格式，以及輸入和輸出之間的映射關係。

- 提示詞風格。

對不同的大模型，設計一種合適的提示詞風格（Prompt Style）來激發它們解決特定任務的能力也是很重要的。總的來說，提示詞應該是一個清晰的問題或詳細的說明，讓大模型能夠極佳地理解和回答。在某些情況下，可以透過增加首碼或尾碼更進一步地指導大模型。舉例來說，使用首碼「讓我們一步一步地思考」有助引導大模型逐步推理，使用首碼「你是該領域的專家」可以提升大模型在某些特定任務中的性能。此外，對於基於聊天的大模型（例如 ChatGPT），建議將提示詞分解為子任務的多個提示詞，然後透過多回合對話將其輸入大模型，而非直接輸入長的或複雜的任務提示詞。

2. 設計提示詞的原則

好的提示詞可以讓大模型輸出的結果更符合要求，但這需要經驗。不過，下面 4 個基本原則可以為你提供一些指導。

- 清晰地描述任務目標。

任務描述不應模棱兩可或不清楚，這可能導致不準確或不恰當的回答。清晰而詳細的描述應包含解釋任務的各種元素，包括任務目標、輸入/輸出資料（舉例來說，給定一個長文件，我希望你生成一個簡明的摘要）和回應約束（舉例來說，摘要的長度不能超過 50 字）。透過提供清晰的任務描述，大模型可以更有效地理解目標任務並生成所需的輸出。

- 分解為簡單的、詳細的子任務。

對複雜任務，重要的是將其分解為幾個更容易、更詳細的子任務，以幫助大模型逐步實現目標。舉例來說，我們可以以編號的形式明確地列出子任務。透過將目標任務分解為子任務，大模型可以專注於解決更容易的子任務，並最終對複雜任務舉出準確性更高的結果。

- 提供幾個範例。

大模型可以從解決複雜任務的上下文學習中獲得幫助，提示詞可以包含輸入 - 輸出對形式的少量任務範例，即 few-shot 範例[①]。few-shot 範例可以幫助大模型學習輸入和輸出之間的語義映射。在實踐中，建議為目標任務提供一些高品質的範例，這能有效促成最終任務的執行。

① 類似於教小學生兩位數乘法時，先給他們演示幾個案例。

- 使用對模型友善的格式。

由於大模型是在專門建構的資料集上預訓練的，因此有一些提示格式可以使大模型更進一步地理解指令。OpenAI 官方文件建議將 ### 或 """ 作為停止符號來分隔指令和上下文，以便大模型更進一步地理解。大多數現有的大模型使用英文執行任務的效果更好，因此使用英文指令來完成困難任務是更友善的。

以上基礎知識和技巧可以幫助你寫出不算太差的提示詞。寫出有效的提示詞需要多實踐，多累積相關經驗。希望你多用大模型，體會其中的奧妙。

1.4.2 上下文學習

上下文學習也叫情境學習 [8]，透過使用格式化的自然語言提示詞（包括任務描述或任務範例），讓大模型解決特定的問題。圖 1-21 左側是上下文學習的示意圖。首先提供任務描述，同時從任務資料集中選擇幾個樣本（few-shot 學習）或不選擇任何樣本（zero-shot 學習）作為範例；然後將它們以特定的順序組合在一起，形成範本化的自然語言提示詞；最後將待測試的查詢實例（一般是符合某種範本的自然語言）附加到範例中，作為大模型的輸入。基於任務範例，大模型可以在沒有顯式梯度更新的情況下辨識並執行新任務。

設 $D_k=\{f(x_1,y_1),f(x_2,y_2),\cdots,f(x_k,y_k)\}$ 表示一組具有 k 個範例的樣本集，其中 $f(x_k,y_k)$ 是將第 k 個任務範例轉為自然語言提示的提示函式。給定任務描述 I 、範例 D_k 和新的輸入查詢 x_{k+1} ，大模型生成的輸出 $\hat{y}_{k+1}$ 的預測可以表示為以下形式。

$$\mathrm{LLM}\left(I,\underbrace{f(x_1,y_1),f(x_2,y_2),\cdots,f(x_k,y_k)}_{\text{demonstrations}},f\left(\underbrace{x_{k+1}}_{\text{input}},\underbrace{\_}_{\text{answer}}\right)\right)\rightarrow\hat{y}_{k+1}$$

其中實際預測結果 $\hat{y}_{k+1}$ 被留白，待大模型預測。由於上下文學習的性能在很大程度上依賴範例，因此在提示詞中正確設計範例非常重要。

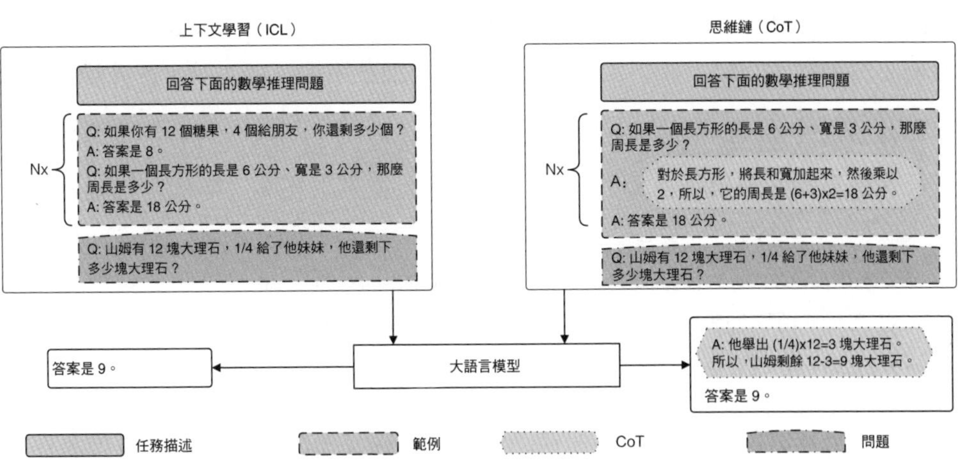

▲ 圖 1-21 上下文學習和思維鏈提示詞的比較（上下文學習用任務描述、幾個範例和一個查詢提示大模型，而思維鏈提示詞涉及一系列的中間推理步驟）

1.4.3 思維鏈提示詞

思維鏈 [9] 提示詞是一種改進的提示策略，用於提升大模型在複雜推理任務上的表現，如算術推理、常識推理和符號推理。思維鏈沒有像上下文學習那樣簡單地用輸入 - 輸出對建構提示詞，而是增加了中間推理步驟，用於引導最終輸出，如圖 1-21 右側所示。

下面詳細說明如何將上下文學習和思維鏈結合，以及思維鏈提示詞何時有效、為何有效。一般來說思維鏈可以以兩種方式與上下文學習一起使用，即少樣本思維鏈（few-shot CoT）和零樣本思維鏈（zero-shot CoT）。

少樣本思維鏈是上下文學習的一種特殊情況，它結合了思維鏈推理步驟，將每個範例的 < 輸入 , 輸出 > 擴充為 < 輸入 , 思維鏈 , 輸出 >。設計恰當的思維鏈提示詞對於激發大模型的複雜推理能力至關重要。身為直接的方法，使用不同的思維鏈（每個問題有多個推理路徑）可以有效提高學習能力（類似於透過題海戰術提高解題能力）。具有複雜推理路徑的提示詞更有可能激發大模型的推理能力（類似於學會了做大學數學題，再做高中數學題就不在話下），這可以提高生成答案的準確性。然而，所有這些方法都依賴帶標注的思維鏈資料集，

這限制了它們在實踐中的使用。為了克服這一局限性，Auto-CoT 利用零樣本思維鏈，透過特定的提示詞讓大模型生成思維鏈推理路徑，從而減少人工標注。

與少樣本思維鏈不同，零樣本思維鏈提示詞中不包括人工標注的任務範例。相反，它直接生成推理步驟思維鏈，然後據此推導出答案。零樣本思維鏈首先讓大模型在「讓我們一步一步思考」的提示詞下生成推理步驟，然後在「因此，答案是」的提示詞下得出最終答案。當模型大小超過一定規模時，這種策略會大大提高模型的效果，但對小模型無效，這表明大模型的湧現能力與模型規模高度相關。

儘管思維鏈能讓複雜推理任務的效果有所提高，但思維鏈提示詞仍然存在推理錯誤和不穩定等問題。其實，透過設計更好的思維鏈提示詞和增強的思維鏈生成策略，可以提升思維鏈的推理能力。圖 1-22（從左到右）是典型的思維鏈提示詞策略的演變過程，下面簡單介紹。

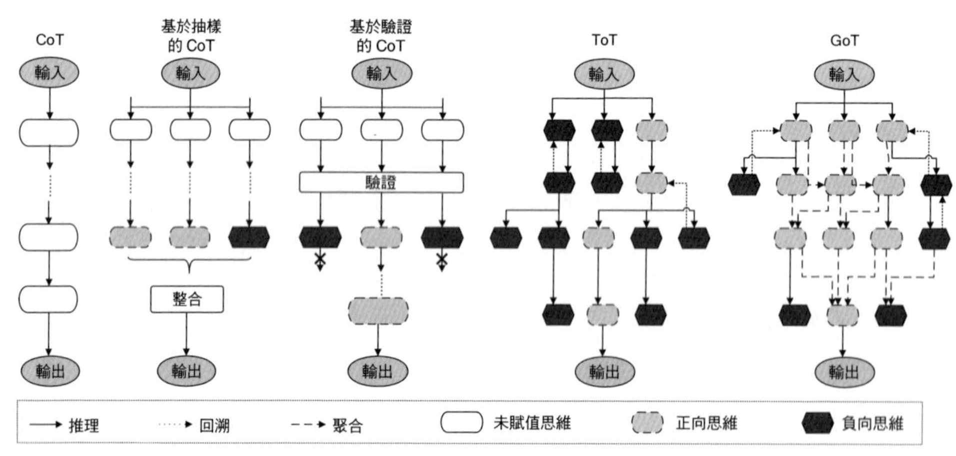

▲ 圖 1-22 典型的思維鏈提示詞策略的演變過程

1. 設計更好的提示詞

一般來說，使用多個推理步驟（類似於透過多種方法解答一道題），或用更長的推理步驟（類似於利用更高階的知識舉出更複雜的解題方法）可以激發

大模型的推理能力，讓思維鏈獲得更好的效果。但這依賴標注好的思維鏈資料集，限制了該方法使用的廣度。

2. 生成增強的思維鏈

大模型可能會產生不正確的推理步驟或不穩定的生成過程。一般可以用基於抽樣或驗證的方法來提升大模型的思維鏈能力。

- 基於抽樣的方法。

該方法對多個推理步驟分別生成結果，然後對多個結果進行投票，使用類似於整合演算法的想法獲得最終結果。為了讓結果更準確，還可以利用 K 個最複雜的推理步驟進行投票[①]。

- 基於驗證的方法。

推理步驟是一種管線的結構，前面步驟的錯誤會累積到後續步驟中，因此自然可以想到對前面的步驟進行驗證。可以基於新訓練的驗證器（一種機器學習演算法）進行驗證，或利用大模型本身進行驗證。

3. 推理結構延展

基礎的思維鏈的鏈式推理結構限制了它解決複雜問題的能力，因為解決複雜問題往往需要向前探索或回溯。為了解決此問題，可以最佳化傳統的鏈式思維鏈結構，建構思維樹（Tree of Thought，ToT）或思維圖（Graph of Thought，GoT）推理結構。

- 思維樹推理結構。

這種推理方法類似於層次的樹狀結構，在分層樹結構中確定推理過程，中間思維是節點。透過這種方式，大模型能夠並行探索多個推理路徑，並進一步支援向前探索和回溯，以形成更全面的決策。

① 這是建立在「複雜的推理步驟會獲得更好的結果」這一假設基礎上的。

• 思維圖推理結構。

儘管思維樹推理結構有助並行推理，但它也對推理過程施加了限制。更複雜的拓撲結構使思維圖在推理方面具有更大的靈活性，從而能夠表徵更複雜的關係和互動。GoT 將推理過程概念化為任意圖，其中頂點表示中間思維，邊表示這些思維之間的依存關係。與 ToT 相比，GOT 可以在生成新思維時進一步利用其他推理路徑的思維。然而，這種方法需要與大模型進行大量的互動，這使得思維探索過程效率非常低。

1.4.4 規劃

上下文學習和思維鏈提示詞的概念簡單，比較通用，但難以處理複雜的任務，如數學推理和多跳（multi-hop）問答。一種基於提示詞的規劃方法（prompt-based planning）由此被提出，這個方法將複雜的任務分解為更小的子任務，並生成完成任務的行動計畫。具體步驟可以參考圖 1-23，下面簡單說明具體細節。

在這個技術範式中通常有 3 個組成部分：任務規劃器（Task Planner）、計畫執行器（Plan Executor）和環境（Environment）。具體來說，由大模型扮演的任務規劃器旨在生成完成目標任務的整個計畫。計畫可以以不同的形式存在，舉例來說，自然語言形式的動作序列或用程式語言撰寫的可執行程式。計畫執行器則負責執行計畫中的行動，它既可以透過文字任務導向的大模型實現，也可以透過具體任務導向的機器人等物件實現。環境指計畫執行器執行任務的地方，可以根據具體任務進行設置，例如大模型本身或《我的世界》（Minecraft）等外部虛擬世界。它以自然語言或其他多模態訊號的形式向任務規劃器提供有關動作執行結果的回饋。

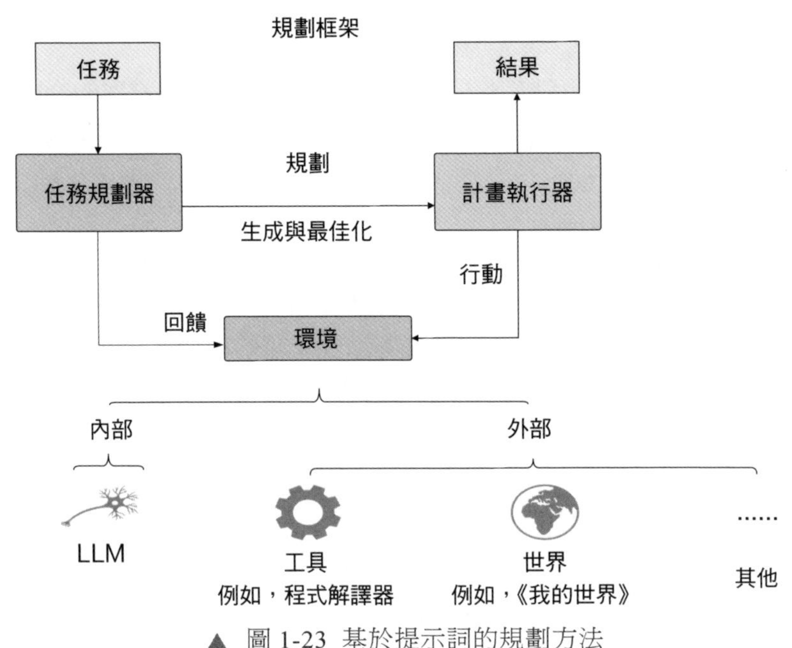

▲ 圖 1-23 基於提示詞的規劃方法

面對複雜任務，任務規劃器首先需要清楚地理解任務目標，並根據大模型的推理生成合理的計畫。然後，計畫執行器根據計畫行事，環境會為任務規劃器生成回饋。任務規劃器可以結合從環境中獲得的回饋完善其初始計畫，形成新的方案，並迭代執行上述步驟，以獲得更好的結果。

1.5 大模型推理

大模型具有強大的語言生成、知識記憶及推理能力，獲得了越來越多人的青睞，並逐步在商業上得到大規模應用。然而大模型的推理應用成本過高，大大阻礙了技術落地。因此，大模型的推理性能最佳化成為業界研究的熱點。

大模型推理面對運算資源和計算效率的挑戰，最佳化推理性能不僅可以減少硬體成本，還可以提高模型的即時回應速度。它使模型能夠更快速地執行自然語言理解、翻譯、文字生成等任務，從而改善使用者體驗，加速科學研究，推動各行各業應用的發展。

大模型推理服務特別注意兩個指標：輸送量和延遲。

輸送量:主要從系統的角度來看,即系統在單位時間內能處理的 token 數目。計算方法為系統處理完成的 token 數目除以對應耗時，其中，token 數目一般指輸入序列和輸出序列長度之和。輸送量越高，代表大模型服務系統的資源使用率越高，對應的系統成本越低。

延遲:主要從使用者的角度來看,即使用者平均收到每個 token 所需的時間。計算方法為使用者從發出請求到收到完整回應所需的時間除以生成序列長度。一般來講，當延遲不大於 50 ms 時，使用者使用體驗會比較流暢。

輸送量影響系統成本，輸送量高代表系統單位時間處理的請求多，系統使用率高。延遲影響使用者使用體驗，即傳回結果要快。這兩個指標一般情況下會相互影響，因此需要權衡。

高效推理是大模型應用中最關鍵的一環，對大模型在商業上的成功應用至關重要。目前有很多創業公司專注於高效推理，例如被收購的一流科技的原班人馬新成立了一家專門做大模型推理的公司——矽基流動。

1.5.1 高效推理技術

高效推理方面的學術探索非常多，這裡介紹幾個比較核心的技術，包括這些技術的基本原理、能解決的問題，以及為大模型推理帶來的價值。

1. FlashAttention 機制

Transformer 的計算複雜度和空間複雜度隨序列長度 N 的增長呈指數級增長，所以處理長 token 非常消耗顯示記憶體。FlashAttention 機制就是為解決該問題而提出的，它可以減少記憶體頻寬，以最小化 I/O 操作的數量，從而加快在 GPU 上處理的速度。身為最佳化的 Attention 實現，FlashAttention 利用運算元融合來減少記憶體頻寬瓶頸。

2. 低位元量化技術

在神經網路壓縮中，量化（Quantization）通常指從浮點數到整數的映射過程，尤其是 8 位元整數量化（INT8 量化）。對於神經網路模型，通常有兩種資料需要量化——權重（模型參數）和（隱藏層）啟動值，它們最初以浮點數表示。

為了說明模型量化的基本思想，引入一個簡單但流行的量化函式：$x_q = R(x/S) - Z$，它將浮點數 x 轉為量化值 x_q。在該函式中，S 和 Z 分別表示縮放因數（涉及決定削波範圍的參數 α 和 β）和零點因數（決定對稱或非對稱量化），$R(.)$ 表示將縮放後的浮點值映射為近似整數的捨入運算。

相反的過程為從量化值中恢復原始值：$\tilde{x} = S \cdot (x_q + Z)$。量化誤差被計算為原始值 x 和恢復值 $\tilde{x}$ 之間的數值差。範圍參數 α 和 β 對量化性能有很大影響，通常需要根據實際資料分佈情況以靜態（離線）或動態（執行時期）方式對量化性能進行校準。

3. 剪枝

剪枝（Pruning）是量化的一種補充後訓練技術（類似於決策樹的剪枝操作），目的是去除給定模型的部分權重，而不降低其性能。修剪有結構化和非結構化兩種模式：結構化稀疏模型用更小但密集的部分代替模型的密集部分；非結構化稀疏模型包含值為零的權重，修剪後不會影響網路的行為。

4. 混合專家

混合專家（MoE）結構通常由一組專家（模組）和一個門控網路組成，每個專家具有唯一的權重，門控網路確定哪個專家處理輸入。混合專家模型不會同時使用所有專家，而只啟動其中的子集，從而減少了推理時間。此外，在模型分散式設置中，可以透過將每個專家放在單獨的加速器上減少裝置之間的通訊：只有託管門控網路和相關專家的加速器需要通訊。

GPT-4 採用了混合專家結構，該結構可以在推理過程中節省大量的運算資源和成本。法國大模型初創公司 Mistral AI（創始人來自 Meta 和 DeepMind）開放原始碼的 Mixtral-8x7B MoE 也採用了混合專家結構。

5. 解碼策略

解碼策略可以極大地影響推理的計算成本。舉例來說，集束搜尋是對計算量和更好推理效果的一種權衡策略。計算成本較高的解碼方案的另一個例子是採樣並排序（Sample-and-Ranking）：先透過隨機採樣構造 N 個無關的 token 序列 $y_1, y_2, \cdots, y_N$ ，然後將最高機率序列作為最終輸出。

以延遲為導向的策略，如推測採樣（Speculative Sampling），首先使用較小的 draft 模型自回歸生成長度為 K 的草稿；然後使用較大的目標模型對 draft 模型進行評分，最後採用類似拒絕採樣的機制，從左到右接收 token 的子集作為最終模型的輸出。

1.5.2 高效推理軟體工具

上面提到的是一些重要的提升大模型推理性能的方法，這些方法已經在一些軟體工具中實現了，下面對兩個主流的高效推理軟體工具介紹。

1. DeepSpeed-MII

DeepSpeed-MII 是微軟開發的深度學習最佳化函式庫（與 PyTorch 相容），特別注意高輸送量、低延遲和成本效益。它的功能包括區塊 KV 快取、連續批次處理、動態分割使用、張量並行和高性能 CUDA 核心，以支援大模型（如 LLaMA 2-70B、Mixtral8x7B（MoE）和 Phi-2）的快速高產出的文字生成。

2. vLLM

vLLM 是一個快速、記憶體高效、易於使用的大模型推理和服務函式庫。為了實現快速推理，它經過了特別最佳化，具有高服務輸送量，使用 PagedAttention 的有效注意力記憶體管理、連續批次處理和最佳化的 CUDA 核心。此外，vLLM 還支援各種解碼演算法、張量並行和流式輸出。為了方便與其他系統的整合，vLLM 支援使用 Hugging Face 模型，還提供了與 OpenAI 相容的 API。

1.6 總結

本章介紹了大模型相關的基礎知識，這些知識有助你更進一步地理解大模型的概念和核心思想。

在後續進行大模型實戰時，大模型相關資源可以提供一些方向性的指導。大模型預訓練的流程和步驟是必須掌握的核心基礎知識。

同時，本章簡單整理了對大模型進行微調的技術原理，涵蓋指令微調和對齊微調。這兩類微調方法是大模型特有的方法，也是大模型能夠有如此強大學習能力的關鍵。指令微調可以讓大模型更進一步地完成下游任務，更進一步地按照人類自然語言的對話模式解決問題。對齊微調可以讓大模型的輸出更加可控、安全、準確，只有透過對齊，大模型才能真正為社會創造正向價值、造福人類。

大模型線上學習和高效推理也是本章的核心基礎知識。由於大模型壓縮了巨量世界知識，預訓練好後的大模型就具備直接用於推理的能力。由於大模型使用自然語言進行互動，我們可以透過自然語言的方式來解決下游問題。

提示詞是我們與大模型進行高效互動的關鍵，在大模型時代，每個人都需要掌握提示詞使用技巧。為了讓大模型的能力得到更大程度的發揮，我們可以採用上下文學習、思維鏈、規劃等高階的提示詞技巧，讓大模型從人類學習和解決問題的方式中找到原型。

由於參數規模巨大和 Transformer 架構的特殊性，大模型在推理時需要消耗大量資源，因此，高效推理是大模型應用的關鍵，目前有非常多的方法和工具供我們使用。

大模型能夠成功的最大原因是科學家一直堅信**可以透過增加樣本來更進一步地預測下一個詞，從而突破模型的能力**，這也是 OpenAI 堅守的第一性原理。這一原理相當樸素簡單，與機器學習中的「奧卡姆剃刀」原理（**如無必要，勿增實體**）有異曲同工之妙，然而，簡單的真理往往容易被人忽視。在堅持第一性原理的基礎上，剩下的就是如何利用巨量的資料、龐大的運算資源產生「大力出奇蹟」的效果，OpenAI 做到了，這才有了現在的大模型革命。

MEMO

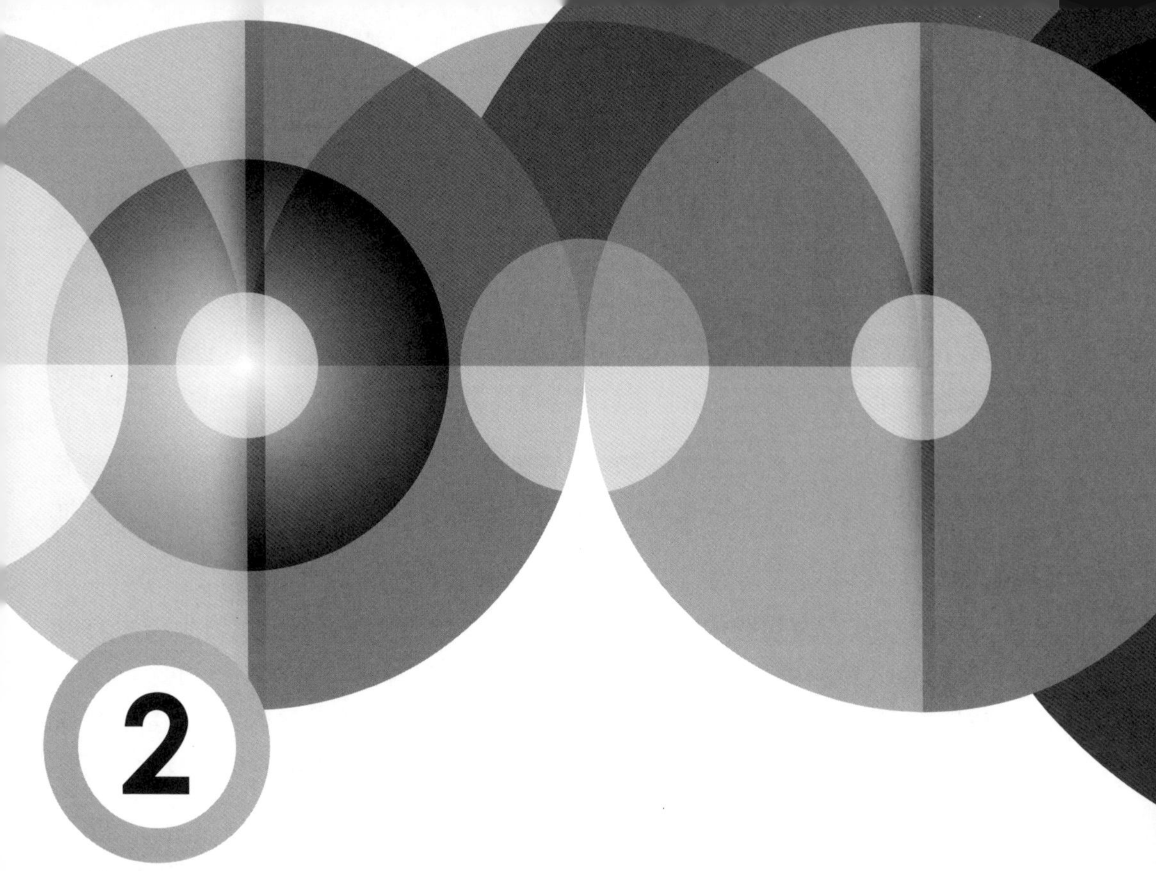

2 資料準備與開發環境準備

第 1 章講解了大模型的基礎知識。本章開始深入講解大模型推薦系統的技術原理、細節及程式實現，透過詳細的分析和程式案例，幫你掌握大模型推薦系統的核心知識。

在開啟大模型推薦系統探索之旅前，我們需要做一些準備工作。資料是大模型的原料，同時，需要提前架設好大模型推薦系統的開發平臺，方便後面的程式實戰。

2.1 MIND 資料集介紹

第 4 ～ 7 章將微軟開放原始碼的新聞資料集 MIND 作為資料原料，講解大模型推薦系統的 4 種範式。這個資料集的數量足夠大，應用場景也比較通用，希望你詳細了解並提前下載該資料集，最好事先對資料進行一些分析和探索。

MIND 是用於新聞推薦研究的大型態資料集 [1,2]，是從微軟新聞網站的匿名行為日誌中收集的。其任務是作為新聞推薦的基準資料集，促進新聞推薦和推薦系統領域的研究。

MIND 資料集包含約 16 萬篇英文新聞文章和 100 萬個使用者生成的 1500 多萬筆曝光日誌。每篇新聞文章都包含豐富的文字內容，包括標題、摘要、正文、類別和實體。每個曝光日誌都包含該使用者在此之前的點擊事件、未點擊事件和歷史新聞點擊行為。為了保護使用者隱私，每個使用者都會安全地散列到匿名 ID 下。

MIND 資料集包含 3 部分，分別是訓練集、驗證集、測試集，每個資料集的資料結構都是類似的，包含 4 個檔案，具體說明如下。

- news.tsv：新聞文章的資訊。
- behaviors.tsv：使用者的點擊歷史記錄和曝光日誌。
- entity_embedding.vec：從知識圖中提取新聞實體的嵌入表示。
- relation_embedding.vec：從知識圖中提取實體之間關係的嵌入表示。

下面分別對這 4 個檔案進行詳細說明。

1. news.tsv

news.tsv 包含 8 個欄位，是對新聞各個維度資訊的描述，如表 2-1 所示。

▼ 表 2-1 news.tsv 的詳細說明

欄位	說明	例子
News ID	新聞的唯一 ID	N23144
Category	新聞所屬的類目	health
Subcategory	子類目	weightloss
News Title	新聞標題	50 Worst Habits For Belly Fat
News Abstract	新聞摘要	These seemingly harmless habits are holding you back and keeping you from shedding that unwanted belly fat for good
News URL	新聞的 URL 地址	https://assets.XXX.com/labs/mind/AAB19MK.html
Entities in News Title	新聞標題中的實體	[{"Label": "Adipose tissue", "Type": "C", "WikidataId": "Q193583", "Confidence": 1.0, "OccurrenceOffsets": [20], "SurfaceForms": ["Belly Fat"]}]
Entities in News Abstract	新聞摘要中的實體	[{"Label": "Adipose tissue", "Type": "C", "WikidataId": "Q193583", "Confidence": 1.0, "OccurrenceOffsets": [97], "SurfaceForms": ["belly fat"]}]

下面舉出一筆真實的資料樣本，方便你更直觀地了解。

```
N23144  health  weightloss  50 Worst Habits For Belly Fat  These seemingly harmless
habits are holding you back and keeping you from shedding that unwanted belly fat for good.
  https://assets.msn.com/labs/mind/AAB19MK.html  [{"Label": "Adipose tissue",
"Type": "C", "WikidataId": "Q193583", "Confidence": 1.0, "OccurrenceOffsets": [20],
"SurfaceForms": ["Belly Fat"]}]  [{"Label": "Adipose tissue", "Type": "C",
"WikidataId": "Q193583", "Confidence": 1.0, "OccurrenceOffsets": [97], "SurfaceForms":
["belly fat"]}]
```

這裡的實體（表 2-1 中最後兩行的資料結構）是一個清單，代表多個實體，每個實體是一個字典，字典的鍵描述如下。

- Label：Wikidata knwoledge 圖中的實體名稱。
- Type：此實體在 Wikidata 中的類型。
- WikidataId：Wikidata 中的實體 ID。
- Confidence：實體連結的置信度。
- OccurrenceOffsets：標題或摘要文字中的字元級實體偏移量。
- SurfaceForms：原始文字中的原始實體名稱。

2. behaviors.tsv

behaviors.tsv 是新聞的曝光資料，每筆記錄都代表一次曝光，包含 5 個欄位，具體如表 2-2 所示。

▼ 表 2-2 behaviors.tsv 的詳細說明

欄位	說明	例子
Impression ID	曝光 ID	2232746
User ID	使用者唯一 ID	U151246
Impression Time	曝光時間	11/13/2019 12:42:51 PM
User Click History	使用者點擊歷史，是曝光時間之前使用者點擊過的新聞	N27587 N49668
Impression News	曝光的新聞，是曝光中顯示的新聞	N39887-1 N22811-0 N110709-1 N1923-0 N24001-1 N76677-0 N123968-0 N4504-0 N48188-0 N75618-0 N88472-0 N127572-1 N53474-0 N82233-0 N100261-0 N13761-0 N94594-0 N79044-0 N27862-0 N120573-0 N124677-0 N10285-1 N27028-0

這裡說明一下，曝光的新聞的格式如下。

```
[News ID 1] [label1] ... [News ID n] [labeln]
```

標籤表示使用者是否點擊了新聞，1 代表使用者點擊了，0 代表沒有點擊。使用者點擊歷史和曝光的新聞中的所有資訊都可以在新聞資料檔案中找到。

下面是一筆真實的曝光資料的樣本。

```
2232746  U151246  11/13/2019  12:42:51  PM  N27587  N49668  N39887-1  N22811-0
N110709-1  N1923-0  N24001-1  N76677-0  N123968-0  N4504-0  N48188-0  N75618-0  N88472-0
N127572-1  N53474-0  N82233-0  N100261-0  N13761-0  N94594-0  N79044-0  N27862-0
N120573-0  N124677-0  N10285-1  N27028-0
```

3. entity_embedding.vec 和 relation_embeddings.vec

entity_embedding.vec 和 relation_embeddings.vec 包含透過 TransE 方法從子圖（來自 WikiData 知識圖）中學習的實體和關係的 100 維嵌入表示。在這兩個檔案中，第一列是實體 / 關係的 ID，其他列是嵌入向量值。這些資料主要為知識感知類新聞推薦提供資料支援，本書不會用到。下面是一個真實的樣本。

實體 / 關係的 ID：Q42306013

嵌入值：0.014516 -0.106958 0.024590 ... -0.080382

2.2 Amazon 電子商務資料集介紹

本節詳細介紹 Amazon 電子商務資料集，第 8 章的實戰案例中會使用這個資料集。這個資料集覆蓋面廣，在學術論文中多次出現。

Amazon 電子商務資料集包含衣服、書、電子產品、遊戲等 29 個門類的資料，如圖 2-1 所示。

Amazon Fashion	reviews (883,636 reviews)	metadata (186,637 products)
All Beauty	reviews (371,345 reviews)	metadata (32,992 products)
Appliances	reviews (602,777 reviews)	metadata (30,459 products)
Arts Crafts and Sewing	reviews (2,875,917 reviews)	metadata (303,426 products)
Automotive	reviews (7,990,166 reviews)	metadata (932,019 products)
Books	reviews (51,311,621 reviews)	metadata (2,935,525 products)
CDs and Vinyl	reviews (4,543,369 reviews)	metadata (544,442 products)
Cell Phones and Accessories	reviews (10,063,255 reviews)	metadata (590,269 products)
Clothing Shoes and Jewelry	reviews (32,292,099 reviews)	metadata (2,685,059 products)
Digital Music	reviews (1,584,082 reviews)	metadata (465,392 products)
Electronics	reviews (20,994,353 reviews)	metadata (786,868 products)
Gift Cards	reviews (147,194 reviews)	metadata (1,548 products)
Grocery and Gourmet Food	reviews (5,074,160 reviews)	metadata (287,209 products)
Home and Kitchen	reviews (21,928,568 reviews)	metadata (1,301,225 products)
Industrial and Scientific	reviews (1,758,333 reviews)	metadata (167,524 products)
Kindle Store	reviews (5,722,988 reviews)	metadata (493,859 products)
Luxury Beauty	reviews (574,628 reviews)	metadata (12,308 products)
Magazine Subscriptions	reviews (89,689 reviews)	metadata (3,493 products)
Movies and TV	reviews (8,765,568 reviews)	metadata (203,970 products)
Musical Instruments	reviews (1,512,530 reviews)	metadata (120,400 products)
Office Products	reviews (5,581,313 reviews)	metadata (315,644 products)
Patio Lawn and Garden	reviews (5,236,058 reviews)	metadata (279,697 products)
Pet Supplies	reviews (6,542,483 reviews)	metadata (206,141 products)
Prime Pantry	reviews (471,614 reviews)	metadata (10,815 products)
Software	reviews (459,436 reviews)	metadata (26,815 products)
Sports and Outdoors	reviews (12,980,837 reviews)	metadata (962,876 products)
Tools and Home Improvement	reviews (9,015,203 reviews)	metadata (571,982 products)
Toys and Games	reviews (8,201,231 reviews)	metadata (634,414 products)
Video Games	reviews (2,565,349 reviews)	metadata (84,893 products)

▲ 圖 2-1 Amazon 電子商務資料集的 29 個門類

圖 2-1 的第 1 列是所有的門類，第 2 列是使用者評論資料（reviews），第 3 列是商品的中繼資料（metadata）。下面對使用者評論資料和中繼資料詳細說明。

使用者評論資料中包含 11 個欄位，具體如下。

- reviewerID：評論者（使用者）的 ID，如 A2SUAM1J3GNN3B。
- asin：商品 ID，如 0000013714。
- reviewerName：評論者的名字。
- vote：評論的贊成數（有多少人認同了使用者的評論）。

- style：產品中繼資料的字典，例如「格式」是「精裝本」。
- reviewText：評論的正文。
- overall：產品的評級。
- summary：評論的概要。
- unixReviewTime：評論的時間（UNIX 時間）。
- reviewTime：評論的時間。
- image：使用者收到產品後發佈到 Amazon 網上的圖片（評論的時候會附帶圖片）。

下面舉出一個 JSON 格式的使用者評論資料範例。


```
{
  "reviewerID": "A2SUAM1J3GNN3B",
  "asin": "0000013714",
  "reviewerName": "J. McDonald",
  "vote": 5,
  "style": {
    "Format:": "Hardcover"
  },
  "reviewText": "I bought this for my husband who plays the piano.  He is having a
wonderful time playing these old hymns.  The music  is at times hard to read because
we think the book was published for singing from more than playing from.  Great
purchase though!",
  "overall": 5.0,
  "summary": "Heavenly Highway Hymns",
  "unixReviewTime": 1252800000,
  "reviewTime": "09 13, 2009"
}
```


商品的中繼資料包含 14 個欄位，具體如下。

- asin：商品 ID，如 0000013714。
- title：商品名稱。
- feature：商品的特性說明。

- description：商品的描述資訊。
- price：商品價格（單位：美金）。
- imageURL：產品圖片的 URL。
- imageURLHighRes：產品高畫質圖片的 URL。
- related：相關產品（購買的其他商品、瀏覽的其他商品、一起購買的商品、瀏覽後再購買的商品）。
- salesRank：銷售排名資訊。
- brand：商品的品牌名稱。
- categories：商品所屬的類目。
- tech1：產品的第一個技術細節資料表。
- tech2：產品的第二個技術細節資料表。
- similar：相似的商品。

再來對照著看一下 JSON 格式的中繼資料範例，這裡請注意，不是所有欄位都有值。

```
{
  "asin": "0000031852",
  "title": "Girls Ballet Tutu Zebra Hot Pink",
  "feature": ["Botiquecutie Trademark exclusive Brand",
              "Hot Pink Layered Zebra Print Tutu",
              "Fits girls up to a size 4T",
              "Hand wash / Line Dry",
              "Includes a Botiquecutie TM Exclusive hair flower bow"],
  "description": "This tutu is great for dress up play for your little ballerina.
Botiquecute Trade Mark exclusive brand. Hot Pink Zebra print tutu.",
  "price": 3.17,
  "imageURL": "http://ecx.images-amazon.com/images/I/51fAmVkTbyL._SY300_.jpg",
  "imageURLHighRes": "http://ecx.images-amazon.com/images/I/51fAmVkTbyL.jpg",
  "also_buy": ["B00JHONN1S", "B002BZX8Z6", "B00D2K1M3O", "0000031909", "B00613WDTQ",
"B00D0WDS9A", "B00D0GCI8S", "0000031895", "B003AVKOP2", "B003AVEU6G", "B003IEDM9Q",
"B002R0FA24", "B00D23MC6W", "B00D2K0PA0", "B00538F5OK", "B00CEV86I6", "B002R0FABA",
"B00D10CLVW", "B003AVNY6I", "B002GZGI4E", "B001T9NUFS", "B002R0F7FE", "B00E1YRI4C",
```

```
"B008UBQZKU", "B00D103F8U", "B007R2RM8W"],
  "also_viewed": ["B002BZX8Z6", "B00JHONN1S", "B008F0SU0Y", "B00D23MC6W",
"B00AFDOPDA", "B00E1YRI4C", "B002GZGI4E", "B003AVKOP2", "B00D9C1WBM", "B00CEV8366",
"B00CEUX0D8", "B0079ME3KU", "B00CEUWY8K", "B004FOEEHC", "0000031895", "B00BC4GY9Y",
"B003XRKA7A", "B00K18LKX2", "B00EM7KAG6", "B00AMQ17JA", "B00D9C32NI", "B002C3Y6WG",
"B00JLL4L5Y", "B003AVNY6I", "B008UBQZKU", "B00D0WDS9A", "B00613WDTQ", "B00538F50K",
"B005C4Y4F6", "B004LHZ1NY", "B00CPHX76U", "B00CEUWUZC", "B00IJVASUE", "B00GOR07RE",
"B00J2GTM0W", "B00JHNSNSM", "B003IEDM9Q", "B00CYBU84G", "B008VV8NSQ", "B00CYBULSO",
"B00I2UHSZA", "B005F50FXC", "B007LCQI3S", "B00DP68AVW", "B009RXWNSI", "B003AVEU6G",
"B00HSOJB9M", "B00EHAGZNA", "B0046W9T8C", "B00E79VW6Q", "B00D10CLVW", "B00B0AVO54",
"B00E95LC8Q", "B00GOR92SO", "B007ZN5Y56", "B00AL2569W", "B00B608000", "B008F0SMUC",
"B00BFXLZ8M"],
  "salesRank": {"Toys & Games": 211836},
  "brand": "Coxlures",
  "categories": [["Sports & Outdoors", "Other Sports", "Dance"]]
}
```

你可以從參考文獻 [3] 中了解更多關於資料的細節。

2.3 開發環境準備

有了資料，預訓練、微調大模型還需要合適的硬體和軟體資源。本節的主要目的是幫助你架設一個自己的開發環境，以便對大模型進行預訓練、微調、呼叫。

在安裝大模型開發環境之前，我們需要建立 Python 沙箱①，從而得到一個穩定的環境。可以透過 miniconda 建立沙箱，透過連結 2-1 下載 conda 的安裝套件。

安裝好 conda 後，執行下面的命令，可以建立名稱為 llm 的 Python 沙箱環境，我們選擇的 Python 版本是 3.10。

```
conda create -n llm python=3.10
```

① 沙箱是一個隔離的環境，可以避免開發時出現軟體套件不相容的問題。可以將沙箱理解為一個獨立的容器環境。

在使用時需要啟動沙箱環境（下面的 llm 是環境的名稱，你可以選擇一個自己喜歡的名字）。

```
conda activate llm
```

如果要離開虛擬的沙箱環境，可以執行：

```
conda deactivate
```

虛擬環境建立完畢，可以透過 pip 或 conda 安裝相關的 Python 軟體套件。下面是執行大模型必需的軟體套件。

基礎軟體套件：

```
pip install numpy pandas scikit-learn
```

大模型相關的軟體套件：

```
pip install torch torchvision datasets accelerate peft bitsandbytes transformers trl
huggingface-hub
```

2.3.1 架設 CUDA 開發環境

這裡可以選擇具有輝達 GPU 的電腦， 或在雲端服務器上購買按小時付費的 GPU，例如選擇 A10 24GB 等型號的 GPU。

1. 安裝 CUDA

輝達 CUDA 工具套件為建立高性能 GPU 加速應用程式提供了一個開發環境。透過它可以在 GPU 加速的嵌入式系統、桌上型工作站、企業資料中心、基於雲端的平臺和超級電腦上開發、最佳化和部署應用程式。該工具套件包括 GPU 加速函式庫、偵錯和最佳化工具、C/C++ 編譯器和執行函式庫。可以從連結 2-2 中下載 12.x 版本的 CUDA。

對於 Windows 作業系統，可以透過 .exe 檔案進行安裝。對於 Linux 作業系統，可以透過下面的命令安裝（對應 CUDA 12.3）。

```
wget
https://developer.download.nvidia.com/compute/cuda/12.3.0/local_installers/cuda-repo-rhel7-12-3-local-12.3.0_545.23.06-1.x86_64.rpm
sudo rpm -i cuda-repo-rhel7-12-3-local-12.3.0_545.23.06-1.x86_64.rpm
sudo yum clean all
sudo yum -y install cuda-toolkit-12-3
```

可以透過下面的命令檢查是否安裝成功及對應的版本。

```
nvcc -V
```

還可以透過下面的命令查看顯示卡情況。

```
nvidia-smi
```

2. 安裝 cuDNN

輝達 CUDA 深度神經網路函式庫（cuDNN）是用於深度神經網路的 GPU 加速基礎函式庫。cuDNN 為標準常式提供了高度調優的實現，如前向和後向卷積、注意力、matmul、池化和歸一化等。可以從連結 2-3 中獲取對 cuDNN 的詳細介紹。同時，可以從連結 2-4 了解安裝 cuDNN 的細節，這裡不再贅述。

3. 下載 LLaMA 模型

我們可以從 Hugging Face 社區下載 LLaMA 模型，包括 chat 版本和普通版本。具體見連結 2-5 和連結 2-6。

這裡說明一下，git-lfs 用於處理大檔案，可以從連結 2-7 中獲得 git-lfs 的安裝指導。舉例來說，在 MacBook 中可以透過以下命令安裝。

```
brew install git-lfs
```

另外，也可以從 ModelScope 中下載 LLaMA 模型，詳見連結 2-8。

4. 直接利用基礎模型進行推斷

下載好 LLaMA 模型就可以利用它進行語言處理了，下面用 Transformers 函式庫的 pipeline 工具來實現大模型的知識問答。

```
# 需要登入 Hugging Face
# huggingface-cli login

# 匯入對應的套件
from transformers import AutoTokenizer
import transformers
import torch

# 模型名稱，如果之前沒有下載模型，那麼在建構 pipeline 的時候會自動下載，預設的儲存路徑與上面的一致。如果已經下載了模型，那麼可以填寫模型的儲存路徑。
model = "meta-llama/Llama-2-7b-chat-hf"

# tokenizer 方法
tokenizer = AutoTokenizer.from_pretrained(model)

# 構造 pipeline
pipeline = transformers.pipeline(
    "text-generation",
    model=model,
    torch_dtype=torch.float16,
    device_map="auto",
)

# 進行預測，生成應答
sequences = pipeline(
    'I liked "Breaking Bad" and "Band of Brothers". Do you have any recommendations of other shows I might like?\n',
    do_sample=True,
    top_k=10,
    num_return_sequences=1,
    eos_token_id=tokenizer.eos_token_id,
    max_length=200,
)
for seq in sequences:
    print(f"Result: {seq['generated_text']}")
```

執行上面的程式可以獲得以下結果。這裡需要注意，生成的結果具有一定的隨機性，你獲得的結果可能不完全一樣。

```
Result: I liked "Breaking Bad" and "Band of Brothers". Do you have any recommendations
of other shows I might like?
Answer:
Of course! If you enjoyed "Breaking Bad" and "Band of Brothers," here are some other
TV shows you might enjoy:
1. "The Sopranos" - This HBO series is a crime drama that explores the life of a New
Jersey mob boss, Tony Soprano, as he navigates the criminal underworld and deals with
personal and family issues.
2. "The Wire" - This HBO series is a gritty and realistic portrayal of the drug trade
in Baltimore, exploring the impact of drugs on individuals, communities, and the
criminal justice system.
3. "Mad Men" - Set in the 1960s, this AMC series follows the lives of advertising
executives on Madison Avenue, expl
```

從上面的過程中可以看到，透過 Transformers 函式庫使用大模型還是非常簡單的。

5. 微調基礎模型

如果希望基於開放原始碼的資料集或自己的資料集對 LLaMA 模型進行微調，那麼也可以透過指令稿來實現（基於程式層面更詳細的實現將在後續章節中講解）。

首先從連結 2-9 中下載 trl 函式庫，這是利用強化學習訓練大模型的工具。

```
# 進入下載目錄
cd trl

# 直接利用指令稿對模型進行微調
python trl/examples/scripts/sft_trainer.py \
    --model_name meta-llama/Llama-2-7b-hf \  # 模型目錄
    --dataset_name timdettmers/openassistant-guanaco \ # 資料集目錄
    --load_in_4bit \
    --use_peft \
    --batch_size 4 \
    --gradient_accumulation_steps 2
```

這裡使用的是之前下載的 LLaMA 模型，其中的資料集可以從 Hugging Face 上下載，詳見連結 2-10。

可以使用以下命令下載。

```
# Make sure you have git-lfs installed (https://git-lfs.com)
git lfs install
git clone
```

2.3.2 架設 MacBook 開發環境

雖然 MacBook 上搭載的不是輝達的 GPU，但是也可以建構大模型，這要感謝開放原始碼社區貢獻的 llama.cpp 和 llama2.c 等優秀開放原始碼軟體。這裡以 llama.cpp 為例來說明如何在 MacBook 上架設大模型開發環境，llama.cpp 對 MacBook 有很好的支援，除了 LLaMA，還支援 Alpaca、Falcon、Baichuan、MPT、Yi 和 Mistral AI 等國內外主流的開放原始碼大模型。

建議使用配備 M 系列晶片的 MacBook，共用記憶體[①] 24GB 左右。執行 7B 的 LLaMA 模型最少需要 13GB 記憶體，因此，如果執行量化版本或參數量更少的模型，也可以配置 16GB 記憶體。

具體步驟如下。

1. 拉取程式

在 MacBook 上安裝 git 工具，然後透過以下命令獲取 llama.cpp 的原始程式碼。

```
git clone https://github.com/ggerganov/llama.cpp.git
```

2. 編譯程式

拉取程式後，進入 llama.cpp 目錄，執行 make 進行編譯。

① 該系列 MacBook 的 CPU 和 GPU 共用記憶體。

```
cd llama.cpp
make
```

如果需要利用 MacBook 的 GPU，那麼可以使用以下命令。

```
# Build it with GPU
make clean
LLAMA_METAL=1 make
```

3. 安裝相依檔案

llama.cpp 目錄中的 requirements.txt 中包含相關的相依項，可以使用下面的命令安裝。

```
pip install -r requirements.txt
```

4. 下載模型

可以從 Hugging Face 社區中下載開放原始碼的 LLaMA 中文微調模型，具體如下。

```
# Make sure you have git-lfs installed (https://git-lfs.com)
git lfs install
git clone https://huggingface.co/hfl/chinese-alpaca-2-7b
```

上面的模型 chinese-alpaca-2-7b 是基於 LLaMA、利用中文進行微調的。

5. 進行模型量化處理

為了節省記憶體和提升執行速度，我們可以對模型進行量化處理，具體如下。

```
convert.py models/chinese-alpaca-2-7b-hf/
./quantize ./models/chinese-alpaca-2-7b-hf/ggml-model-f16.gguf ./models/chinese-
alpaca-2-7b-hf/ggml-model-q4_0.gguf q4_0
```

6. 執行

模型量化完成後就可以執行了，方便起見，我們建立一個 shell 指令稿 chat.sh。

```
#!/bin/bash
# temporary script to chat with Chinese Alpaca-2 model
# usage: ./chat.sh alpaca2-ggml-model-path your-first-instruction
SYSTEM='You are a helpful assistant. 你是一個樂於助人的幫手。'
FIRST_INSTRUCTION=$2
./main -m $1 \
--color -i -c 4096 -t 8 --temp 0.5 --top_k 40 --top_p 0.9 --repeat_penalty 1.1 \
--in-prefix-bos --in-prefix ' [INST] ' --in-suffix ' [/INST]' -p \
"[INST] <>
$SYSTEM

<>
$FIRST_INSTRUCTION [/INST]"
```

對上面的指令稿設置可執行許可權。

```
chmod +x chat.sh
```

然後就可以利用大模型進行對話了。

```
./chat.sh models/chinese-alpaca-2-7b-hf/ggml-model-q4_0.gguf '作為一個 AI 幫手，你可以幫助人類做哪些事情？'
```

你還可以直接安裝 llama.cpp 的 Python 綁定，這樣也可以直接在 Python 程式中使用 llama.cpp 提供的能力。

```
pip install llama-cpp-python
```

2.4 總結

2.1 節、2.2 節分別介紹了 MIND 資料集和 Amazon 電子商務資料集，後續章節會用到這些內容，讀者需要掌握各資料欄位及含義。

2.3 節介紹了開發環境準備，主要圍繞輝達的 CUDA 生態和蘋果的 MacBook 展開，讀者可以根據自己的實際情況選擇開發環境。在進入後面的章節之前，希望你準備好自己的開發環境，並嘗試動手實踐上面提到的簡單案例。

MEMO

大模型推薦系統的資料來源、一般想法和 4 種範式

第 1 章介紹了大模型的基礎理論知識，包括資料前置處理、預訓練、微調、RLHF、上下文學習、思維鏈、高效推理等。這些技術是大模型區別於傳統機器學習的部分，也是大模型賦能推薦系統的核心武器。

本章以新聞推薦場景為例，介紹大模型推薦系統的統一框架。無論是新聞、電子商務、音樂，還是短新聞等場景的推薦，都可以基於這個框架去實現或延伸。本章後面的實戰案例都是基於本章提供的方法論實現的。

本章先簡單介紹大模型推薦系統的資料來源，然後整理將大模型應用於推薦的一般想法，最後總結出將大模型應用於推薦系統的 4 種常用範式。

3.1 大模型推薦系統的資料來源

大模型的能力雖然很強，但「巧婦難為無米之炊」：沒有好的資料，大模型就無法生成好的輸出，無法保證滿足業務的預期。將大模型應用於推薦系統也一樣，需要有完整的資料支撐。

我們知道，建構大模型和推薦系統所需要的資料不太一樣，大模型需要的是文字類資料（包括程式），而傳統推薦系統更多依賴使用者行為資料和 ID 資料。那麼，大模型和推薦系統之間的資料要怎麼聯繫起來呢？

這就需要詳細整理一下大模型和推薦系統依賴的到底是什麼資料，以及將資料應用於大模型推薦系統中的具體方法。

3.1.1 大模型相關的資料

大模型依賴的兩類主要資料如圖 3-1 所示。

首先，文字類資料。這類資料包括從網頁爬取的資料、論文、電子書、程式等，資料量非常大，來源複雜多樣。如果希望將大模型應用在推薦系統中，就需要在預訓練大模型時先清洗資料，保證資料品質。這裡還要注意，不同種類資料的配比對大模型最終的能力有很大影響，選擇什麼類型的資料、資料之間的比例、剔除重復資料等都要慎重。

具體到新聞大模型推薦系統的文字類資料，主要指新聞的標題、標籤、來源、新聞正文、使用者對新聞的評論等。另外，使用者偏好、特性、畫像的文字資訊也屬於文字類資料。在大模型預訓練過程中，文字類資料已經被注入大模型，例如 ChatGPT 預訓練的資料來源中包含知識、常識類文字資訊。

其次，人工標注資料。大模型會出現幻覺，也可能生成不真實、有害（例如性別歧視）的內容，所以需要將大模型的輸出與人類的價值觀對齊。也就是說，我們需要在各種場景下進行資料標注，然後對大模型進行微調。這類資料一般是與專業領域（社會、倫理等）相關的，需要有一定的專業知識才能完成資料標注。

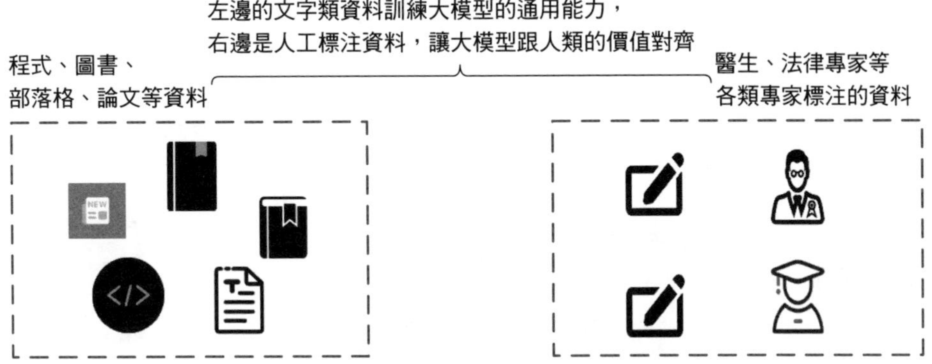
左邊的文字類資料訓練大模型的通用能力，
右邊是人工標注資料，讓大模型跟人類的價值對齊
程式、圖書、
部落格、論文等資料
醫生、法律專家等
各類專家標注的資料
NEW
</>

▲ 圖 3-1 大模型依賴的資料

對新聞推薦來說，軟色情、暴力、恐怖、種族歧視、錯誤價值觀引導、政治敏感性等維度的資訊，都需要透過人標注，使其與人類的價值對齊。

3.1.2 新聞推薦系統相關的資料

傳統的新聞推薦系統（內容推薦、協作過濾等）主要是基於使用者相關、新聞相關、場景相關、行為相關等資料建構演算法模型，進行個性化推薦的，如圖 3-2 所示。

使用者相關資料是必備的。這類資料主要包括使用者唯一 ID、使用者自身屬性資料（如年齡、性別、學歷等）、使用者偏好資料。其中，使用者偏好可以是自己展示的偏好，例如使用者透過關鍵字輸入的，或基於使用者歷史行為挖掘的興趣偏好。

新聞相關資料主要包括新聞唯一 ID、新聞自身屬性資料（如標題、標籤、內容等）、新聞的評論、評價資料。新聞的評論、評價資料是使用者賦予的，不同使用者的評論或評價也會不同。

場景相關資料是跟具體新聞、場景相關的，例如日期、時間、地理位置、使用者瀏覽時刻的產品路徑（例如瀏覽新聞）等。

行為相關資料是使用者跟新聞的互動資料，例如瀏覽、閱讀、分享等，這類資料是協作過濾、深度學習演算法依賴的最重要的資料，也是對推薦效果貢獻價值最大的一類資料。

▲ 圖 3-2 新聞推薦系統相關的資料

透過對大模型和新聞推薦系統依賴的相關資料進行整理，我們會發現，模型處理、生成的都是語言文字，而推薦系統的資料來源並不全是文字（例如使用者唯一 ID、新聞唯一 ID、使用者行為資料等），這就涉及一個關鍵問題：**如何將推薦系統依賴的資料編碼為可以供大模型使用的資料？**

3.1.3 將推薦資料編碼為大模型可用資料

我們能夠想到的最簡單、最直接的方法是不直接使用非文字資料。很多推薦系統相關的資料本身就是文字，利用好這類資料也可以進行推薦。你可能會有顧慮：這樣太簡單粗暴了，遺失了核心資料，肯定會影響推薦效果。

實際上並非如此。大模型具有壓縮的世界知識及強大的泛化能力，當將非文字資料以文字的形式輸入大模型時，也能造成比傳統新聞推薦演算法更好的效果。

如果不直接使用非文字資料，我們能想到的最簡單的方法就是不直接使用使用者唯一 ID、新聞唯一 ID，可以在推薦時在上下文中注入使用者資訊，而新聞唯一 ID 可以透過新聞的標題等資訊注入。圖片、新聞、音訊等多模態資料等也可以暫不考慮。現在很多大模型是多模態模型，這些模型可以自然將多模態資訊利用起來。

除此之外，還有一種方法——**將非文字資料轉為文字資料**。舉例來說，新聞唯一 ID 可以用標題替代，使用者也可以基於興趣表示，如廣義的新聞有訊息、通訊、新聞評論、新聞特寫、調查報告、專訪等，可以用這類標籤來表示使用者。還有一種方法是直接將使用者唯一 ID、新聞唯一 ID 作為語言模型的 token，這樣就可以把 ID 資訊利用起來了。

這裡還要注意，在模型輸出時也需要做一些處理，使用一些策略。舉例來說，大模型推薦的新聞可能不存在，或與提供的新聞文字表述有差異。對於這種情況，可以在將推薦新聞提供給大模型之前對其進行編號，讓大模型在推薦時輸出編號。

3.2 將大模型用於推薦的一般想法

基於第 1 章介紹的大模型的通用能力，我們可以深入思考一下怎樣將大模型應用於推薦系統。下面 4 個想法是可行的，這 4 個想法正是本章重點介紹的 4 個範式的來源。

首先，可以利用大模型的語言生成能力為使用者生成個性化新聞。例如基於已經發生的新聞事件，按照使用者更喜歡的表述方式來組織內容（是內容創造的過程），而非從已有新聞中選擇內容推薦給使用者。大模型也可以生成嵌入表示、token 和中間結果，供傳統推薦演算法使用。

其次，可以將使用者的興趣、候選新聞作為文字灌入模型，輸出使用者可能喜歡的新聞。注意，這裡的輸出指從候選集中選擇新聞輸出給使用者，與創造式生成不同。這裡主要利用了大模型的指令對齊和泛化能力，新聞推薦問題被抽象成文字生成問題。可以利用大模型的預訓練、微調等核心能力訓練大模型推薦系統，或利用大模型直接進行推薦。具體的實現過程在後面的案例部分會重點講解。

再次，從資料維度來看，大模型依賴文字資料，那麼推薦系統中使用者、新聞等相關的文字資訊都可以作為大模型的輸入資訊。文字資料可以造成在不同問題、不同場景下的橋樑作用，**幫助大模型更進一步地解決傳統推薦不擅長的問題**。傳統推薦模型更多使用使用者 ID、新聞 ID、使用者瀏覽、閱讀行為等資訊，透過在模型中增加人物誌、使用者評論、新聞內容相關的文字資訊，可以讓大模型利用到比傳統新聞推薦演算法更多的資訊，輔助大模型獲得更精準的推薦，甚至解決推薦冷啟動和推薦解釋問題。

最後，可以**將推薦、搜尋、知識獲取等使用者需求融合到統一的對話方塊架下**，形成一種新的推薦系統解決方案。這裡用到的是大模型的對話能力，將在第 8 章講解。

3.3 將大模型應用於推薦的 4 種範式

基於將大模型應用於推薦系統的一般想法，筆者整理出將大模型（包括 BERT 等預訓練語言模型）應用於新聞推薦系統的 4 種範式，下面分別介紹。

3.3.1 基於大模型的生成範式

在這種範式下，大模型直接生成新聞，並將生成的新聞推薦給使用者。這是一個重新創造的過程，生成的新聞不是之前已有的新聞，而是「無中生有」的。注意，新聞事件不是大模型捏造的，這裡的生成指借助大模型的生成能力將不同的真實事件進行重新組織，以更符合使用者的興趣偏好和閱讀習慣。

一般可以將使用者的興趣偏好作為提示詞／指令輸入大模型，然後由大模型自動生成使用者需要的新聞，如圖 3-3 所示。

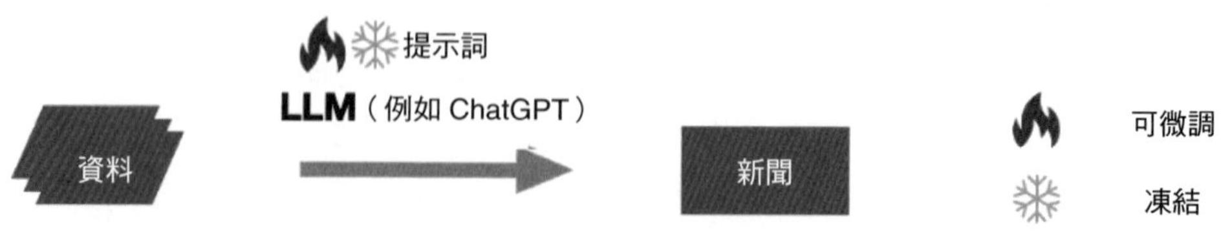

▲ 圖 3-3　基於大模型的生成範式

在短新聞、元宇宙、遊戲創作等領域，可以借助大模型的這種生成能力，臨時創作出使用者需要的新聞內容。這與傳統的基於已有新聞進行個性化推薦很不一樣。

這裡既可以直接使用參數量在千億級的超大規模大模型，也可以基於開放原始碼或自研的大模型進行領域適應性微調再使用。當然，如果模型參數太多，微調時間、費用成本都非常多，就需要評估之後再做決定。

目前，OpenAI 的 ChatGPT、GPT-4 是效果最好的大模型，但是其在中文場景下的效果不如在英文場景下好。對於中文場景，我們可以使用 ChatGLM、Qwen、Yi 等國產大模型。

3.3.2 基於 PLM 的預訓練範式

這種範式基於 PLM[①]、利用推薦系統特定的資料進行預訓練，預訓練好後直接用於新聞推薦任務，如圖 3-4 所示。這與傳統的新聞推薦演算法沒有什麼兩樣，只不過利用了 Transformer 語言模型的生成式架構來實現演算法，BERT4Rec、P5 就屬於這類演算法。

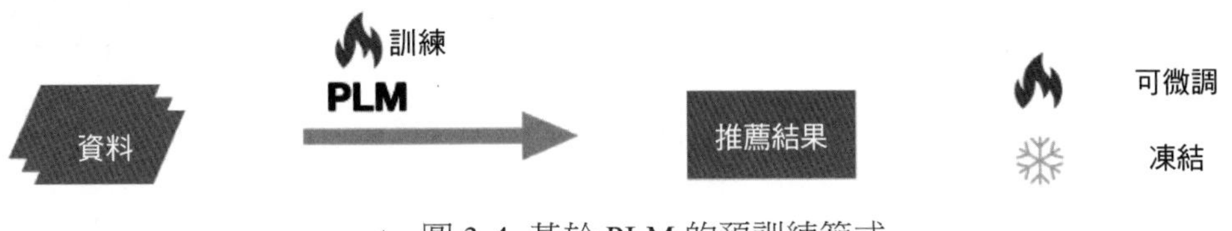

▲ 圖 3-4 基於 PLM 的預訓練範式

我們可以將單一使用者的資料，例如使用者資料、行為資料、新聞資料等資訊文字，按照時間順序及一定的範本編排成文字資訊。這樣一個使用者的資料就形成了一篇文件，所有使用者的此類資料就形成了知識庫，然後使用 Transformer 語言模型進行預訓練。預訓練之後，就可以利用大模型的生成能力進行個性化推薦。

① 不一定是大模型，即參數可能較小，例如 100M~1B，如 BERT。

這種範式的適用範圍非常廣，只要推薦系統相關的資料中包含文字資訊，就可以利用這一方式來實現。

注意，由於推薦系統相關的資料量可能很小，與預訓練 ChatGPT 這類大模型所需要的資料量無法相比，所以這種範式一般選擇參數較少的模型，否則容易過擬合。但這也不是絕對的，像 YouTube、抖音這樣有巨量使用者和物品的產品，使用者行為資料的體量也是非常大的。

3.3.3 基於大模型的微調範式

這種範式基於中等規模的大模型（如 LLaMA、T5 等）、使用新聞推薦系統相關的資料對大模型進行微調，讓微調後的大模型具備調配個性化新聞推薦的能力，如圖 3-5 所示。

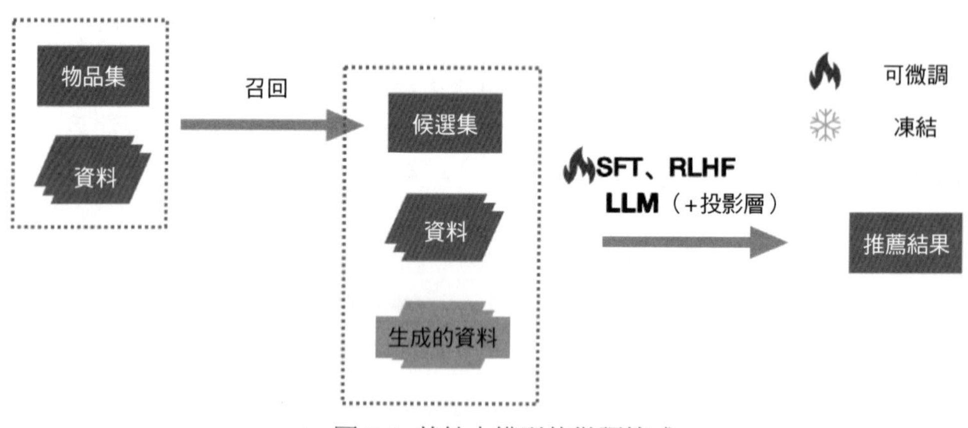

▲ 圖 3-5 基於大模型的微調範式

由於大模型推理的成本較高，輸入的 token 有限，通常透過傳統的召回演算法對新聞集進行篩選，獲得數量有限的候選集讓大模型進行二次篩選，二次篩選既可以是對候選集進行排序、也可以是進行評分預測，甚至是多分類，需要根據具體的實現方式決定，例如是否要增加投影層、投影層的結構是什麼等，具體實現會在實戰案例部分詳細說明。

這種範式最重要的是需要對大模型進行微調，既可以是指令微調，也可以是對齊微調。

指令微調（Instruction Finetune）指在下游任務（這裡是新聞推薦任務）中建構 <樣本, label> 對，然後利用監督學習對大模型進行微調。這裡的樣本、label 都是自然語言形式的指令，方便大模型進行處理。

對齊微調（RLHF）是建構人工 <樣本, label> 對，然後選擇 Reward 模型①，採用強化學習的想法微調大模型，讓大模型的輸出符合人類的需求。

這裡需要提到的是，我們既可以對整個模型微調，也可以透過一定的策略（例如 LoRA）只對模型的（額外）部分參數進行微調。由於大模型參數較多，基於 LoRA 等參數的微調方法是更高效的實現方式。

大模型具備一定的通用能力，通常使用比較少的樣本（例如 100 個）就可以獲得較好的微調效果。為了讓大模型更進一步地根據人類的指令進行回饋，通常可以選擇將經過指令微調的模型作為推薦大模型的基底模型，或在對推薦大模型進行微調之前進行一輪指令微調（可以借助一些開放原始碼的指令微調資料進行指令微調）。

圖 3-5 中生成的資料是基於前面提到的資料基礎生成的中間資料（例如利用大模型生成的使用者興趣描述文字等），這些資料可以作為輸入大模型的一部分指令。

最後提醒一下，基於 LLaMA、T5 等開放原始碼大模型進行微調，需要根據公司的硬體資源進行選擇，如果沒有足夠的 GPU，那麼可以選擇參數更少的同類模型進行微調②。

3.3.4 基於大模型的直接推薦範式

這種範式一般基於超大的大模型（如 ChatGPT、Claude 等），利用模型已經學習到的世界知識進行新聞推薦。因為大模型已經包含新聞相關的知識，所以不需要更新大模型的參數（一般認為大模型的參數中儲存了世界知識）。這

① 一般是基於人類的需要對大模型的輸出進行評估的函式。

② LLaMA 有多種選擇，T5 也有 3 種大小可供選擇。

時可以將推薦任務以格式化指令的方式輸入大模型，獲得最終需要的推薦結果，如圖 3-6 所示。

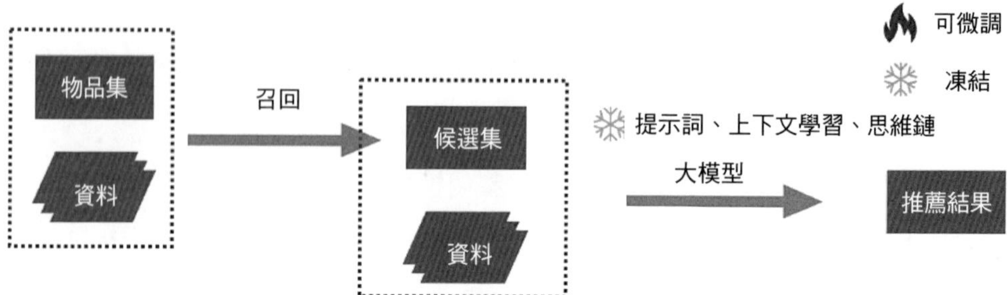

▲ 圖 3-6 基於大模型的直接推薦範式

指令可以以 3 種方式輸入大模型。

第一種形式是提示詞。這種方式將使用者的操作行為、使用者的興趣偏好，以及需要大模型解決的問題這 3 類重要資訊以格式化的方式輸入大模型，讓大模型按照某種規則輸出答案。

你是一個新聞推薦系統專家，現在需要按照以下條件為使用者進行個性化推薦。

使用者的新聞瀏覽歷史是：……

候選新聞集是：……

你需要基於使用者的瀏覽歷史，從候選集中選擇 3 個新聞作為輸出，選擇的新聞需要跟使用者的瀏覽歷史具備相關性。你只需要輸出結果就可以了，不需要進行多餘的解釋。

第二種形式是 few-shot ICL，也就是需要在輸入大模型的提示詞中舉出具體的案例，讓大模型基於舉出的案例現場學習、活學活用，一般可以獲得更好的推薦效果。例如：

你是一個新聞推薦系統專家，現在需要你使用案例說明的方法，基於使用者瀏覽歷史和候選集為使用者進行個性化推薦。

案例如下。

使用者的新聞瀏覽歷史是：……

候選新聞集是：……

使用者下一個會看的新聞是：……

現在使用者的瀏覽歷史是：……，那麼使用者下一個會看的新聞是什麼？你需要從候選新聞集……中選擇一個最合適的，你只需要輸出結果就可以了，不需要進行多餘的解釋。

最後一種形式是思維鏈。思維鏈在上下文學習的基礎上更進一步，透過告訴大模型將問題拆解後進行處理，讓大模型按照一定邏輯和步驟解決問題，從而獲得更好的新聞推薦效果。

你是一個新聞推薦系統專家，現在需要按照以下條件為使用者進行個性化推薦。

使用者的新聞瀏覽歷史是：……

候選新聞集是：……

你需要基於使用者的瀏覽歷史，從候選集中選擇 3 個新聞作為輸出。

你可以先基於使用者的瀏覽歷史，提取出使用者最喜歡的新聞標籤，這就是使用者的興趣畫像，然後基於使用者的興趣畫像從候選新聞集中選出 3 個跟使用者興趣最匹配的作為最終的答案。

你只需要輸出結果就可以了，不需要進行多餘的解釋。

3.4 總結

本章以新聞推薦場景為例，重點講解了大模型推薦系統的 3 個重要主題：大模型推薦系統的資料來源，包含大模型相關的資料和新聞推薦系統相關的資料；將大模型應用於推薦的一般想法，主要的方法和策略；將大模型應用於推薦的 4 種範式。

本章講解的 4 種大模型推薦範式非常重要，也是後續章節的指導框架，你需要熟悉每種範式的特點。

MEMO

4 生成範式：大模型生成特徵、訓練資料與物品

本章重點講解大模型的生成範式。

首先，大模型可以生成傳統推薦系統依賴的特徵、嵌入特徵和文字特徵，作為工具幫助推薦系統獲得更好、更全的特徵，讓推薦系統的效果更上一層樓。

其次，大模型還可以生成表格類[①]的訓練資料和待推薦的物品。現在就讓我們正式進入大模型推薦系統的探索之旅。

① 利用 Excel、MySQL 等表格儲存的資料，這也是推薦系統中最核心的一類資料，人物誌、物品畫像都可以以這種形式存放。

說明：從本章開始，絕大部分程式可以在本書同步的程式倉庫（連結 4-1）中找到。筆者在介紹相關程式時也會舉出對應的目錄。書中呈現的是核心程式，更完整的程式請參考程式倉庫。

4.1 大模型生成嵌入特徵

嵌入方法是過去 10 來年最有價值的演算法之一，已經在搜尋、廣告、推薦、NLP、CV 等領域獲得了大規模應用，效果非常好。我們自然想到，是否可以利用 PLM，甚至大模型進行嵌入呢？答案是肯定的。

4.1.1 嵌入的價值

利用大模型生成嵌入特徵，就是利用新聞的文字資訊和使用者瀏覽新聞的文字資訊（或人物誌等其他文字資訊）生成新聞或使用者的向量表示。新聞的向量表示可以用於計算新聞之間的相似性，以及新聞連結推薦，使用者瀏覽的新聞向量透過適當的「平均 / 池化」操作可以獲得使用者的向量表示。利用使用者向量和候選新聞向量可以估計使用者對候選新聞的點擊機率，這就是推薦系統的 CTR 預測問題，圖 4-1 是利用嵌入進行推薦的一般框架。

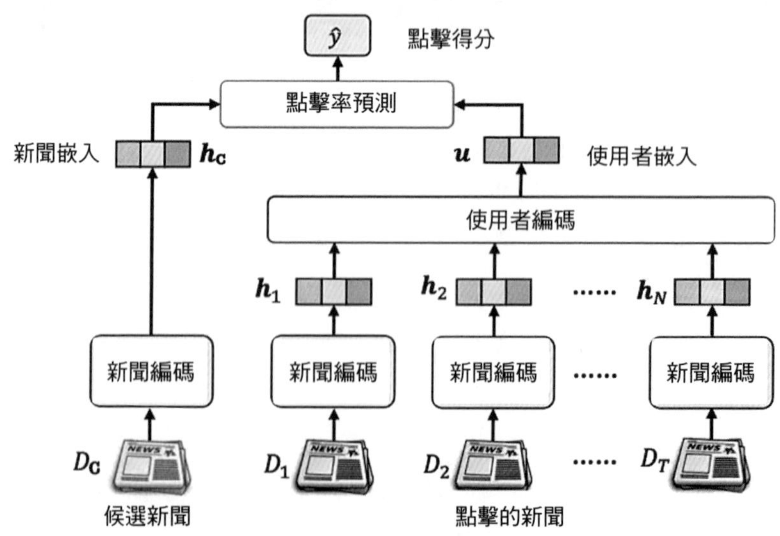

▲ 圖 4-1 利用嵌入進行推薦的一般框架

圖 4-1 是一般的方法，可以配合多種演算法實現。其中有 3 個模組是可以利用不同的方法實現的，下面詳細說明。參考文獻 [1,2] 裡有多種實現方案，你可以參考學習。

- 新聞嵌入。

新聞嵌入（News Encoder）有多種實現方式，可以只使用新聞標題嵌入，或考慮新聞的標籤、分類等，可以使用的技術有 CNN、BERT、基於 Transformer 架構的更複雜的演算法等。

- 使用者嵌入。

不同使用者點擊過的新聞數量不一致，所以使用者嵌入（User Encoder）需要將新聞嵌入向量進行平均 / 池化，以形成唯一的使用者嵌入表示。池化的方法有很多，最簡單的是加權平均，稍微複雜的有 GRU、注意力機制等，更複雜的包括多層 Transformer 模型等。

- 點擊率預測。

點擊率預測（Click Prediction）希望將使用者嵌入向量和物品嵌入向量透過一個函式轉為純量值，該值代表了使用者對候選新聞的興趣度或點擊率。可以採用內積、FM、神經網路等方法。

4.1.2 嵌入方法介紹

本節透過案例說明嵌入的具體的實現想法。

1. 利用 sentence-transformers 框架嵌入

這裡利用開放原始碼的 sentence-transformers 框架 [3,4] 實現新聞的嵌入表示，用 MIND 資料集的新聞標題代表新聞。

首先，執行下面的程式安裝 sentence-transformers 框架。

```
pip install -U sentence-transformers
```

安裝好後，選擇 3 筆新聞進行嵌入表示，具體程式如下。

```
from sentence_transformers import SentenceTransformer

# all-MiniLM-L6-v2 是一個文字嵌入模型，執行時期會下載這個模型
model = SentenceTransformer('all-MiniLM-L6-v2')

# 下面是 MIND 資料集中的 3 個新聞標題
sentences = ["Dispose of unwanted prescription drugs during the DEA's Take Back Day",
    "Chile: Three die in supermarket fire amid protests",
    "As Eagles take their bye, a look at how the defense has improved lately | Early
Birds"]

# 呼叫 model.encode() 對新聞標題進行嵌入表示
embeddings = model.encode(sentences)

# 列印嵌入結果 ,all-MiniLM-L6-v2 模型獲得的是 384 維的稠密向量表示
for sentence, embedding in zip(sentences, embeddings):
    print("Sentence:", sentence)
    print("Embedding:", embedding)
```

有了新聞的嵌入表示，就可以計算它們之間的相似性，並且按照相似性降冪排列。

```
from sentence_transformers import util

# 計算任意兩個嵌入的餘弦相似性
cos_sim = util.cos_sim(embeddings, embeddings)

# 將兩個新聞及其對應的餘弦相似性放入 list
all_news_combinations = []
for i in range(len(cos_sim)-1):
    for j in range(i+1, len(cos_sim)):
        all_news_combinations.append([cos_sim[i][j], i, j])

# 按照餘弦相似性從高到低排序
all_news_combinations = sorted(all_news_combinations, key=lambda x: x[0],
reverse=True)
```

```
# 輸出新聞及對應的相似性
for score, i, j in all_news_combinations[0:]:
    print("{} \t {} \t {:.4f}".format(sentences[i], sentences[j], cos_sim[i][j]))
```

sentence_transformers 是一個非常好用的框架。下面以 MIND 資料集為例，實現一個最簡單的個性化推薦：將使用者點擊過的新聞的嵌入向量的平均值作為使用者嵌入，將新聞嵌入與使用者嵌入的餘弦相似性作為使用者對新聞的偏好得分。具體的程式實現以下①。

```
"""
利用 sentence_transformers 框架實現一個最簡單的個性化推薦
1. 使用者嵌入：使用者瀏覽過的新聞嵌入的平均值
2. 預測：利用使用者嵌入與新聞嵌入的餘弦
"""

from sentence_transformers import SentenceTransformer, util
import pandas as pd
import numpy as np
col_spliter = "\t"
DIMS = 384  # all-MiniLM-L6-v2 模型的維數
TOP_N = 10  # 為每個使用者生成 10 個新聞推薦

df_news = pd.read_csv("./data/mind/MINDsmall_train/news.tsv", sep=col_spliter)
df_news.columns = ['news_id', 'category', 'subcategory', 'title', 'abstract', 'url',
'title_entity', 'abstract_entity']

df_behavior = pd.read_csv("./data/mind/MINDsmall_train/behaviors.tsv", sep=col_
spliter)
df_behavior.columns = ['impression_id', 'user_id', 'time', 'click_history', 'news']

model = SentenceTransformer('all-MiniLM-L6-v2')

# 獲取每個新聞及對應的嵌入向量
news_embeddings = {}
for _, row in df_news.iterrows():
    news_id = row['news_id']
```

① 程式：src/basic_skills/embedding/sentence_transformers_4_rec.py。

```
    title = row['title']
    embedding = model.encode(title)
    news_embeddings[news_id] = embedding

"""
為單一使用者生成 TOP_N 個推薦
"""
def rec_4_one_user(click_history):
    emb = np.zeros(DIMS, dtype=float)
    for news in click_history:
        emb = np.add(emb, news_embeddings[news])
    emb = emb/len(click_history)
    emb = emb.astype(np.float32)
    res = []
    for news_id, emb_ in news_embeddings.items():
        cos_sim = float(util.cos_sim(emb, emb_)[0][0])
        res.append((news_id, cos_sim))
    rec = sorted(res, key=lambda x: x[1], reverse=True)[:TOP_N]
    return rec

"""
為所有使用者生成推薦
"""
user_rec = {}
for _, row in df_behavior.iterrows():
    user_id = row['user_id']
    click_history = row['click_history'].split(' ')
    rec = rec_4_one_user(click_history)
    user_rec[user_id] = rec
```

上面的程式實現都是非常基礎、非常簡單的，你可以在自己的環境中嘗試一下。可以找一兩個典型使用者，透過看他的新聞瀏覽歷史和推薦的新聞，觀察一下推薦是否合理。

2. 其他嵌入方法

這裡再提供兩個比較有意思的實現想法，供你參考學習。

案例 1：UNBERT

參考文獻 [5] 中提供了一個將新聞和使用者瀏覽歷史一起生成嵌入表示並進行個性化推薦的框架，輸入嵌入是 token 嵌入、部分嵌入、位置嵌入和新聞部分嵌入的和，具體嵌入的架構如圖 4-2 所示。

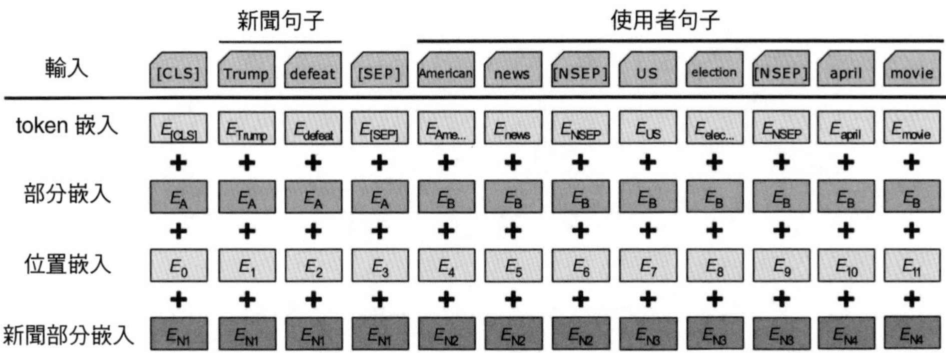

▲ 圖 4-2 輸入嵌入是 token 嵌入、部分嵌入、位置嵌入和新聞部分嵌入的和

這裡的部分（Segment）嵌入是新聞或使用者瀏覽歷史這兩個部分的嵌入，主要目的是區分新聞和使用者。新聞部分嵌入針對每個新聞（使用者可能看過多個新聞）進行嵌入，目的是區分不同新聞包含的語義資訊。

有了嵌入表示，可以採用與圖 4-1 中類似的框架來實現個性化推薦。UNBERT 模型採用了多層 Transformer 框架，如圖 4-3 所示，不僅利用具有豐富語言知識的預訓練模型來增強文字表示，還捕捉單字等級和新聞等級的多粒度使用者、新聞匹配訊號。

圖 4-3 中的 e_w 可以看作單字層面的嵌入表示，e_n 可以看作新聞層面的嵌入表示①。通過點擊率預測模組將它們拼接起來，可以整合單字層面和新聞層面的語義資訊，然後利用單層的網路進行預測，具體公式以下（W^c、b^c 是模型待學習的參數）。

$$y = \text{softmax}([e_w; e_n] \times W^c + b^c)$$

① CLS 是一個特殊的 token，業界普遍採用這種特殊 token 的嵌入來表示整個句子的嵌入表示。

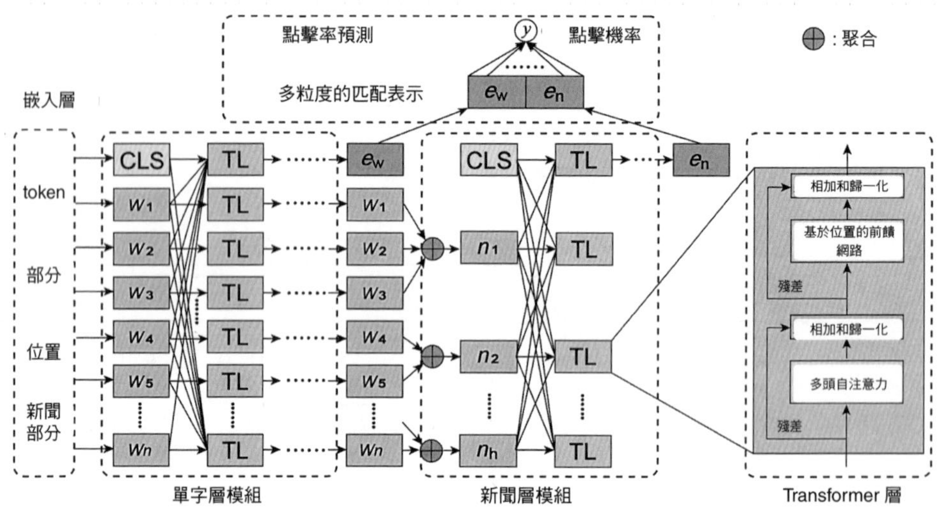

▲ 圖 4-3 UNBERT 模型（TL 代表 Transfomer Layer）

案例 2：PREC

PREC[6] 採用圖 4-4 所示的框架對新聞和使用者進行嵌入表示。嵌入包含 3 個層次：token 嵌入、位置嵌入和 view 嵌入。這裡的 view 是新聞或使用者包含的特徵維度：對於新聞，維度有標題、實體、摘要等；對於使用者，維度有使用者瀏覽的新聞、使用者的位置、使用者的畫像等，最終的嵌入是這 3 個層次的嵌入之和。

$$E = E_{\text{token}} + E_{\text{position}} + E_{\text{view}}$$

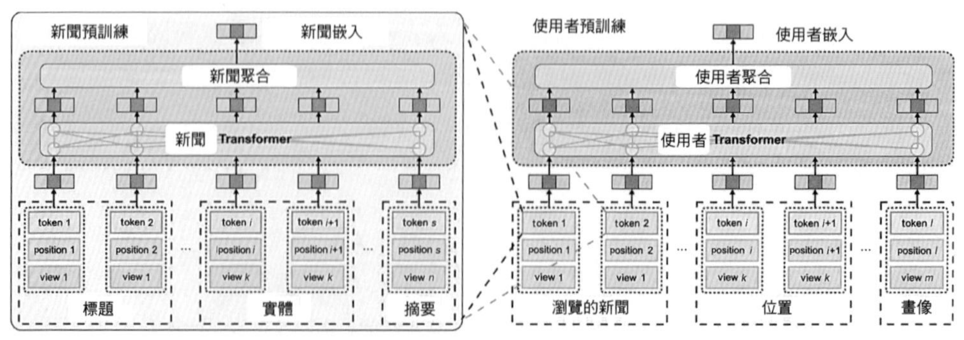

▲ 圖 4-4 PREC 嵌入框架

4.2 大模型生成文字特徵

大模型優秀的語言理解和生成能力可以用於生成描述新聞及使用者的興趣偏好的文字，進而用於新聞的個性化推薦中。

4.2.1 生成文字特徵

按照參考文獻 [7] 的想法，我們透過兩個程式案例來說明如何利用大模型為 MIND 資料集的新聞生成一個新的、更有表達力的標題，以及基於使用者瀏覽過的新聞生成使用者興趣畫像。這些生成的文字可以進一步被大模型「消化」，從而生成更精準的個性化推薦。

4.2.1.1 生成新聞標題

本節利用 Qwen1.5-72B-chat 大模型，基於新聞的標題、摘要、類目生成一個更有表現力的新標題，由於舉出了新聞的摘要和類目，有了更多的資訊注入，同時大模型本身包含足夠多的世界知識，因此是可以對標題進行最佳化並獲得更好的效果的。下面是具體的程式實現①。

```
import os
import time
import pandas as pd
from langchain.callbacks.manager import CallbackManager
from langchain.callbacks.streaming_stdout import StreamingStdOutCallbackHandler
from langchain.chains import LLMChain
from langchain.llms import LlamaCpp
from langchain.prompts import PromptTemplate
from tqdm import tqdm
MIN_INTERVAL = 1.5
# 新聞資料中包含的欄位說明
keys = dict(
    title='title',
    abstract='abs',
    category='cat',
```

① 程式：src/basic_skills/generative/generate_portrait/news_title.py。

```
    subcategory='subcat',
)
current_path = os.getcwd()
# 將新聞讀到 dataframe 中
news_df = pd.read_csv(
    filepath_or_buffer=os.path.join(current_path + '/data/news.tsv'),
    sep='\t',
    header=0,
)
# 建構新聞清單，元素是元組，元組前面是新聞 ID，後面是 dict，dict 是新聞相關資訊，下面是一個資料樣本
# ('N55528', {'title': 'The Brands Queen Elizabeth, Prince Charles, and Prince Philip Swear By',
#   'abstract': "Shop the notebooks, jackets, and more that the royals can't live without.",
# 'category': 'lifestyle', 'subcategory': 'lifestyleroyals', 'newtitle': ''})
news_list = []
for news in tqdm(news_df.iterrows()):
    dic = {}
    for key in keys:
        dic[key] = news[1][keys[key]]
    news_list.append((news[1]['nid'], dic))
# 提示詞範本
prompt_template = """You are asked to act as a news title enhancer. I will provide you a piece of news, with its original title, category, subcategory, and abstract (if exists). The news format is as below:
[title] {title}
[abstract] {abstract}
[category] {category}
[subcategory] {subcategory}
where title, abstract, category, and subcategory in the brace will be filled with content. You can only response a rephrased news title which should be clear, complete, objective and neutral. You can expand the title according to the above requirements. You are not allowed to response any other words for any explanation. Your response format should be:
[newtitle]
where [newtitle] should be filled with the enhanced title. Now, your role of news title enhancer formally begins. Any other information should not disturb your role."""
# 生成的新的新聞標題的儲存路徑
```

```
save_path = current_path + '/output/news_summarizer.log'
# 下面是呼叫 LLaMA 大模型的語法
callback_manager = CallbackManager([StreamingStdOutCallbackHandler()])
llm = LlamaCpp(
    model_path="/Users/liuqiang/Desktop/code/llm/models/gguf/qwen1.5-72b-chat-q5_k_
m.gguf",
    temperature=0.8,
    top_p=0.8,
    n_ctx=6000,
    callback_manager=callback_manager,
    verbose=True,
    stop=["<|im_end|>"]  # 生成的答案中遇到這些詞就停止生成
)

prompt = PromptTemplate(
    input_variables=["title", "abstract", "category", "subcategory"],
    template=prompt_template,
)
chain = LLMChain(llm=llm, prompt=prompt)
# 先統計出哪些已經計算了，避免後面重複計算
exist_set = set()
with open(save_path, 'r') as f:
    for line in f:
        if line and line.startswith('N'):
            exist_set.add(line.split('\t')[0])
# 呼叫大模型迭代計算新聞標題
for nid, content in tqdm(news_list):
    start_time = time.time()
    if nid in exist_set:
        continue
    try:
        title = content['title']
        abstract = content['abstract']
        category = content['category']
        subcategory = content['subcategory']
        enhanced = chain.run(title=title, abstract=abstract, category=category,
subcategory=subcategory)
        enhanced = enhanced.rstrip('\n')
        with open(save_path, 'a') as f:
```

```
            f.write(f'{nid}\t{enhanced}\n')
    except Exception as e:
        print(e)
    interval = time.time() - start_time
    if interval <= MIN_INTERVAL:
        time.sleep(MIN_INTERVAL - interval)
```

執行上述程式，可以獲得以下輸出，以 N 開頭的是新聞 ID， 新聞 ID 下面是生成的新的新聞標題。這些標題基本符合要求，但是在指令跟隨方面有所欠缺，有些標題有中括號，有些沒有。你可以嘗試調整提示詞或呼叫第三方 API（例如月之暗面）加以修正。

```
N55528
[The Royal Family's Favorite Brands: From Notebooks to Jackets
N19639
newtitle] Habits to Avoid for Successful Belly Fat Reduction
N61837
Aid Freeze Impact: In Ukraine's War Trenches, Lt. Molchanets Reveals the Human Cost
N53526
[newtitle] The Impact of Being an NBA Wife on Mental Health
N38324
newtitle] Expert Advice: How Dermatologists 推薦去除皮膚贅生物的方法
N2073
[newtitle] NFL Player Criticism Fines Spark Controversy
N49186
[It's an Unusually Warm October in Orlando, But Cooler Days Are Ahead]
```

4.2.1.2 生成使用者興趣畫像

本節基於使用者看過的所有新聞的標題，為使用者生成興趣 topic 和 region 資訊，其原理與生成新聞標題類似。topic 是新聞的類目，代表使用者的主題偏好。region 是新聞說明的事件發生 / 關注的地理位置資訊，代表使用者的地域偏好。下面是具體的程式實現①。

① 程式：src/basic_skills/generative/generate_portrait/user_portrait.py。

```
import json
import os
import time
import pandas as pd
from UniTok import UniDep
from langchain.chains import LLMChain
from langchain.callbacks.manager import CallbackManager
from langchain.callbacks.streaming_stdout import StreamingStdOutCallbackHandler
from langchain.llms import LlamaCpp
from langchain.prompts import ChatPromptTemplate
from langchain_core.prompts import PromptTemplate
from tqdm import tqdm
from prompter import MindPrompter, MindUser
MIN_INTERVAL = 0
current_path = os.getcwd()
mind_prompter = MindPrompter(current_path + '/data/news.tsv')
user_list = MindUser(current_path + '/data/user', mind_prompter).stringify()
# 將新聞讀到 dataframe 中
news_df = pd.read_csv(
    filepath_or_buffer=os.path.join(current_path + '/data/news.tsv'),
    sep='\t',
    header=0,
)
# 建構新聞字典，key 為新聞 ID，value 為新聞的標題
news_dict = {}
for news in tqdm(news_df.iterrows()):
    news_dict[news[1]['nid']] = news[1]['title']
depot = UniDep(current_path + '/data/user', silent=True)
nid = depot.vocabs('nid')
system = """You are asked to describe user interest based on his/her browsed news
title list, the format of which is as below:
{input}
You can only response the user interests with the following format to describe the
[topics] and [regions] of the user's interest
[topics]
- topic1
- topic2
...
[region] (optional)
- region1
```

```
- region2
...
where topic is limited to the following options:
(1) health
(2) education
(3) travel
(4) religion
(5) culture
(6) food
(7) fashion
(8) technology
(9) social media
(10) gender and sexuality
(11) race and ethnicity
(12) history
(13) economy
(14) finance
(15) real estate
(16) transportation
(17) weather
(18) disasters
(19) international news
and the region should be limited to each state of the US.
Only [topics] and [region] can be appeared in your response. If you think region are
hard to predict, leave it blank. Your response topic/region list should be ordered,
that the first several options should be most related to the user's interest. You are
not allowed to response any other words for any explanation or note. Now, the task
formally begins. Any other information should not disturb you."""

# 生成的使用者興趣畫像的儲存路徑
save_path = current_path + '/output/user_profiler.log'

# 下面是呼叫 LLaMA 大模型的語法
callback_manager = CallbackManager([StreamingStdOutCallbackHandler()])
llm = LlamaCpp(

model_path="/Users/liuqiang/Desktop/code/llm/models/gguf/qwen1.5-72b-chat-q5_k_m.gguf
",
```

```
    temperature=0.8,
    top_p=0.8,
    n_ctx=6000,
    callback_manager=callback_manager,
    verbose=True,
)
# 先統計出哪些已經計算了，避免後面重複計算
exist_set = set()
with open(save_path, 'r') as f:
    for line in f:
        data = json.loads(line)
        exist_set.add(data['uid'])
empty_count = 0
# 呼叫大模型迭代計算使用者興趣畫像
for uid, content in tqdm(user_list):
    start_time = time.time()
    if uid in exist_set:
        continue
    if not content:
        empty_count += 1
        continue
    try:
        prompt = PromptTemplate(
            input_variables=["input"],
            template=system,
        )
        chain = LLMChain(llm=llm, prompt=prompt)
        enhanced = chain.run(input=content)
        enhanced = enhanced.rstrip('\n')
        with open(save_path, 'a') as f:
            f.write(json.dumps({'uid': uid, 'interest': enhanced}) + '\n')
    except Exception as e:
        print(e)
    interval = time.time() - start_time
    if interval <= MIN_INTERVAL:
        time.sleep(MIN_INTERVAL - interval)
print('empty count: ', empty_count)
```

執行上述程式，可以獲得下面的結果（下面是兩個樣例）。可以看到，輸出的結果的格式是滿足要求的，但是部分 topic（例如 entertainment）不在要求的範圍內，說明大模型產生了幻覺，這可能是提示詞需要最佳化或該模型的指令跟隨能力不足以解決這個問題導致的。

```
{"uid": 0,
 "interest":
 [topics]
- entertainment
- sports
- politics
- celebrity gossip
- crime
- human interest
[region] (optional)
- USA}

{"uid": 1,
"interest":
[topics]
- health
- crime and law enforcement
- education
- food
- culture
- religion
- history
- international news

[region] (optional)
- Kentucky
- Minnesota
- California
- Brazil}
```

以上生成新的新聞標題和生成使用者興趣畫像的程式都比較基礎、簡單，你可以在自己的開發環境中執行一下（如果資源有限，可以選擇更小的模型）。

4.2.2 生成文字特徵的其他方法

下面透過兩個案例來詳細說明在特定場景下生成文字特徵的實現方法。

案例 1：RLMRec 生成物品與人物誌

RLMRec[8] 提供了一種利用大模型範本生成物品與人物誌的框架，如圖 4-5 所示。由於範本的指令非常完整，要求相對較高，需要一個效果很好的模型才能實現，所以本案例是使用 ChatGPT 實現的。

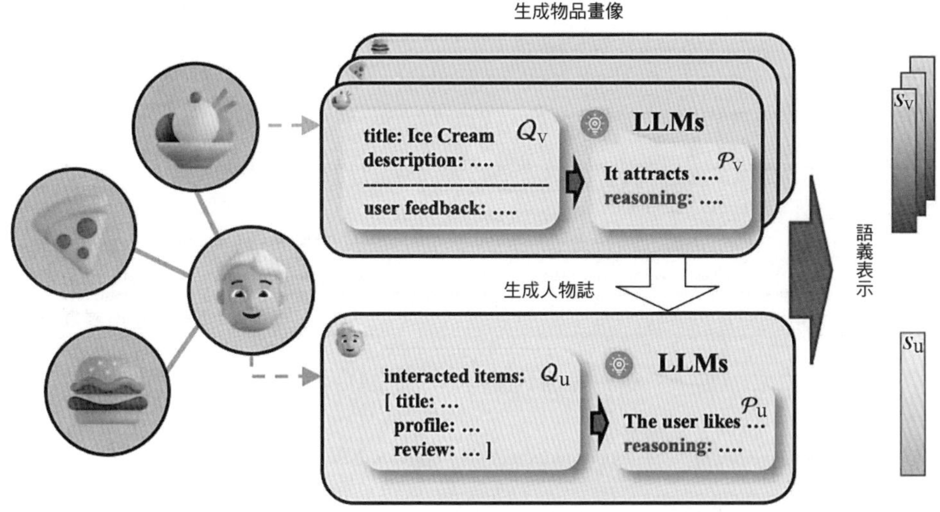

▲ 圖 4-5 RLMRec 中生成物品與人物誌的框架

人物誌應包含使用者喜歡的特定類型物品的描述，從而全面反映他們的品味和偏好。圖 4-6 是生成人物誌的範本，範本中對輸出的資料結構、欄位等提出了非常明確的要求。

物品畫像應該明確地表明它最容易吸引的使用者類型，並提供與這些使用者的偏好和興趣相一致的物品特徵和品質的清晰描述。圖 4-7 是生成物品畫像的範本，範本同樣對輸出的資料結構、欄位、字數等提出了明確的要求。

You will serve as an assistant to help me determine **which types of books a specific user is likely to enjoy.**
I will provide you with information about books that the user has purchased, as well as his or her reviews of those books.
Here are the instructions:
1. Each purchased book will be described in JSON format, with the following attributes:
{ "title": "the title of the book", (if there is no title, I will set this value to "None")
"description": "a description of **what types of users will like this book**",
"review": "the user's review on the book" (if there is no review, I will set this value to "None") }
2. The information I will give you:
PURCHASED ITEMS: a list of JSON strings describing the items that the user has purchased.

Requirements:
1. Please provide your decision in **JSON format**, following this structure:
{ "summarization": "A summarization of what types of books this user is likely to enjoy" (if you are unable to summarize it, please set this value to "None")
"reasoning": "briefly explain your reasoning for the summarization" }
2. Please ensure that the "summarization" is no longer than 100 words.
3. The "reasoning" has no word limits.
4. Do not provided any other text outside the JSON string.

指令

PURCHASED ITEMS: [
{ "title": "Croak",
"description": "**Young adult readers who enjoy paranormal and fantasy themes** would enjoy Croak.",
"review": "**Loved the writing style**, was different than most of what I have read. The narrative was like a storyteller, you could hear someone telling the story to you. …"}

{ "title": "Deadly Cool (Hartley Featherstone)",
"description": "Teenage girls who **enjoy a mix of humor, mystery, and high school drama** would enjoy Deadly Cool by Gemma Halliday. With plenty of red herrings and a quick pace, this book will keep …",
"review": "**I really enjoyed reading this**, was laughing out loud in the middle of the night. ..."}

{ "title": "Stitch (Stitch Trilogy, Book 1)",
"description": "**Fans of young adult paranormal romance novels** with a dash of mystery and suspense would enjoy Stitch (Stitch Trilogy, Book 1).",
"review": "…. **Book started out really well, had me totally hooked from the start**. Love me a good ghost story, …. "}
… (*Omitted due to page limit*)]

提示詞

{ "summarization": "This user **enjoys young adult fiction that** blends paranormal or supernatural elements with romance, mystery, humor, and coming-of-age themes. They also appreciate stories with complex world-building … .",
"reasoning": "Based on the reviews and descriptions of the purchased items, **the user seems to be drawn to young adult fiction that features paranormal or supernatural elements, such as ghosts and magical powers.** They also enjoy a mix of genres, including romance, mystery, humor, and coming-of-age themes. … ." }

生成的使用者畫像

▲ 圖 4-6 生成人物誌的範本

You will serve as an assistant to help me summarize **which types of users would enjoy a specific book.**
I will provide you with the title and a description of the book. Here are the instructions:
1. I will provide you with information in the form of a JSON string that describes the book:
{ "title": "the title of the book", (if there is no title, I will set this value to "None")
"description": "a description of the book", (if there is no description, I will set this value to "None") }

Requirements:
1. Please provide your answer in **JSON format**, following this structure:
{ "summarization": "A summarization of what types of users would enjoy this book" (if you are unable to summarize it, please set this value to "None")
"reasoning": "briefly explain your reasoning for the summarization" }
2. Please ensure that the "summarization" is no longer than 200 words.
3. Please ensure that the "reasoning" is no longer than 200 words.
4. Do not provide any other text outside the JSON string.

指令

{ "title": "The Bell Jar: A Novel (Perennial Classics)",
"description": "Plath was an excellent poet but is known to many for this largely autobiographical novel. The Bell Jartells the story of a gifted young woman's mental breakdown beginning during a summer internship as a junior editor at a magazine in New York City in the early 1950s. The real Plath committed suicide in 1963 and left behind this scathingly sad, honest and perfectly-written book, which remains one of the best-told tales of a woman's descent into insanity.--This text refers to the Hard coveredition. "}

提示詞

{ "summarization": "The Bell Jar would appeal to those interested in reading about **mental health and women's experiences**. Specifically, readers who **enjoy raw and honest depictions of mental illness and its effects on a young woman's life** would appreciate this book.",
"reasoning": "The Bell Jar delves into the mental breakdown of a young woman and her experiences navigating mental health and societal expectations as a woman in the 1950s. The book's autobiographical nature and raw, honest depiction of mental illness make it a compelling read for those interested in exploring these themes. Additionally, readers looking for works that examine the intersection of gender and mental illness would find The Bell Jar particularly thought-provoking." }

生成的物品畫像

▲ 圖 4-7 生成物品畫像的範本

案例 2：PALR 框架生成使用者興趣畫像

PALR 框架 [9] 透過大模型生成使用者興趣畫像，見圖 4-8 中間部分，然後基於生成的自然語言表示的人物誌、互動歷史序列及推薦候選集（召回），為使用者生成推薦（排序）。

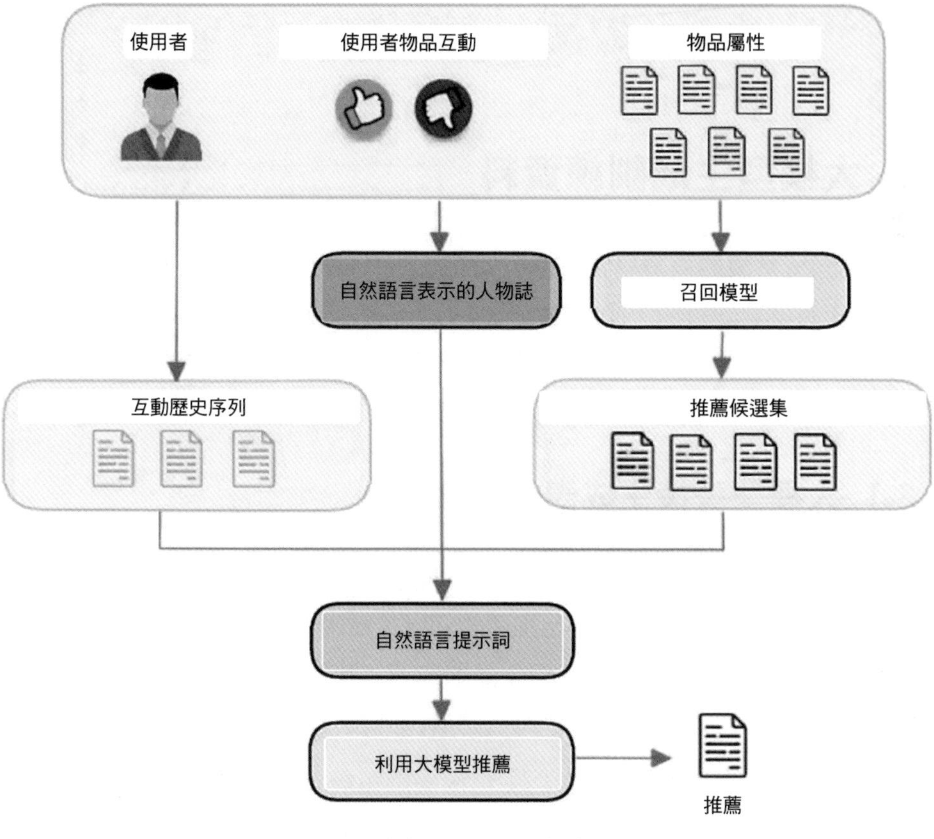

▲ 圖 4-8 PALR 框架

為了利用大模型生成使用者興趣畫像，需要使用生成使用者興趣畫像的語言範本對使用者行為進行編碼，如圖 4-9 所示。然後利用大模型對模型進行微調，微調後的模型就具備了生成使用者興趣畫像的能力。這裡將電影的關鍵字（標籤）作為使用者興趣畫像的表述。

Input: Your task is to use two keywords to summarize user's preference based on history interactions. The Output is an itemized list based on importance. The output template is: {1. KEY_WORD_1: "HISTORY_MOVIE_1", "HISTORY_MOVIE_2"; 2, KEY_WORD_2: "HISTORY_MOVIE_3"}
The history movies and their keywords are:
"Gigi": Romantic, Musical, Classic;
"Pokémon: The First Movie": Pokémon, Adventure, Action;
"Star Wars: Episode IV - A New Hope": Science Fiction, Epic, Space Opera;
"The Terminator": Terminator, Time Travel, Action;
"Back to the Future": Time-travel, Adventure, Friendship;
"Quest for Camelot": Adventure, Fantasy, Friendship;

Output: The user enjoys adventure and science-fiction movies.

▲ 圖 4-9 生成使用者興趣畫像的語言範本

4.3 大模型生成訓練資料

我們知道常規的機器學習模型（如 logistics 回歸、FM、深度學習等）以表格特徵①作為模型的特徵，而大模型不是利用無監督學習進行訓練，就是利用以文字形式呈現的 < 輸入，輸出 > 樣本對進行監督微調（SFT）。這兩類訓練資料都可以透過大模型生成，下面詳細說明。

4.3.1 大模型直接生成表格類資料

大模型具備生成文字的能力，這是大模型最直接、最容易被大眾感知的能力。文字生成可以直接用於個性化推薦的資料生成中。我們知道在很多場景下，推薦系統存在資料不足的問題（如產品剛發佈，沒有太多使用者和使用者行為資料），那麼利用大模型輔助生成資料就是一個非常樸素的想法。

參考文獻 [10] 提供了一種基於大模型微調生成表格資料的樣本的方法——GReaT。先將表格資料轉為文字輸入大模型進行微調，如圖 4-10 所示。圖中步驟 (a) 表示將表格資料轉為有意義的文字，步驟 (b) 表示特徵順序置換，步驟 (c) 表示將獲得的句子用於大模型微調。

① 使用表格儲存特徵，表中的一列對應一個特徵。

(a) →

"Age is 39, Education is Bachelors, Occupation is Adm-clerical, Gender is Male, Income is ≤ 50K.", "Age is 50, Education is HS-grad, Occupation is Exec-managerial, Gender is Female, Income is ≥ 50K.", "Age is 53, Education is 11th, Occupation is Handler-cleaners, Gender is Female, Income is ≥ 50K."

(b) ↓

"Education is Bachelors, Income is ≤ 50K, Age is 39, Occupation is Adm-clerical, Gender is Male.", "Income is ≥ 50K, Occupation is Exec-managerial, Age is 50, Education is HS-grad, Gender is Female.", "Occupation is Handler-cleaners, Education is 11th, Age is 53, Income is ≥ 50K, Gender is Female."

← (c)

▲ 圖 4-10 將表格資料轉為文字輸入大模型進行微調

微調後的模型可以基於一定的策略生成新樣本，如圖 4-11 所示。圖中步驟 (a) 表示將單一特徵名稱或特徵值對的任意組合轉為文字，從而使用預訓練的大模型生成新的資料，步驟 (b) 表示將資料登錄微調後的大模型完成採樣，步驟 (c) 表示轉為表格。

該方法可以保證生成的樣本與原始樣本分佈一致，這對於資料量不足的推薦場景是比較好的補充。與其他方法相比，GReaT 允許在沒有對模型進行重新訓練（只需要在大模型上進行微調）的情況下，將特徵子集任意組合[②]進行資料採樣。除了可以用來生成新的樣本資料，還可以對資料的遺漏值進行補充。

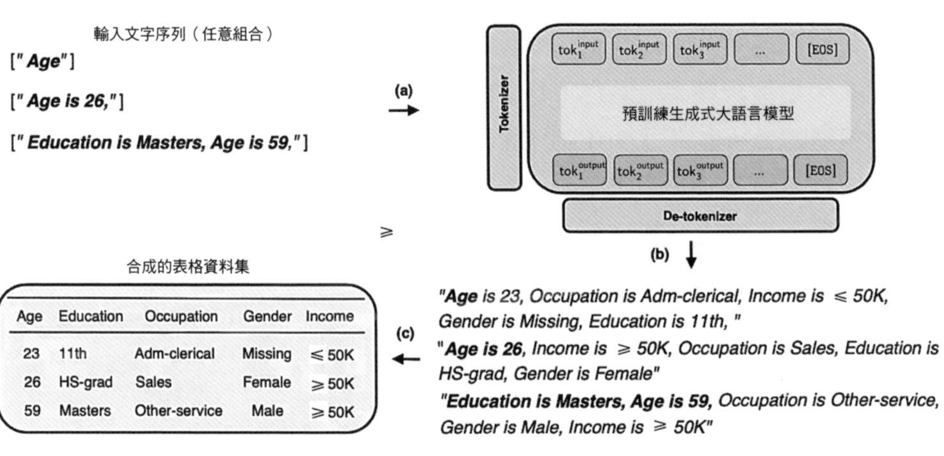

▲ 圖 4-11 將微調後的模型基於一定的策略生成新樣本

② 可以將任何特徵名稱或特徵名稱和值組合。

GReaT 目前已經在 GitHub 上開放原始碼，使用簡單的程式即可透過表格類資料生成更多的樣本資料。

```
from be_great import GReaT
from sklearn.datasets import fetch_california_housing # California Housing dataset
data = fetch_california_housing(as_frame=True).frame
model = GReaT( 大模型 ='distilgpt2', batch_size=32, epochs=25)
model.fit(data)
synthetic_data = model.sample(n_samples=100)
```

針對某個樣本資料，如果你訓練好了 GReaT 模型，那麼當新收集的資料中包含遺漏值時，就可以利用 GReaT 填補遺漏值，具體程式如下。

```
# test_data: pd.DataFrame 格式的樣本資料
# model: GReaT 在原始樣本資料之上訓練的模型
# 下面為了模擬有遺漏值的新資料，對測試資料隨機剔除
import numpy as np
for clm in test_data.columns:
    test_data[clm]=test_data[clm].apply(lambda x: (x if np.random.rand() > 0.5 else
np.nan))
imputed_data = model.impute(test_data, max_length=200)
```

4.3.2 大模型生成監督樣本資料

目前，出現了大量大模型，其中一個原因是有 LLaMA 等開放原始碼模型作為基礎，大模型開發者「站在了巨人的肩膀上」；另外一個重要原因是大多數發佈大模型的公司會使用 GPT-4[①]生成訓練的監督樣本。基於開放原始碼的大模型測試資料集或人工標注的資料（見表 4-1），透過 GPT-4 獲得對應的答案，從而獲得 < 輸入，輸出 > 樣本對，這些樣本對可以作為大模型微調的監督樣本，圖 4-12 是一個具體的例子。

① 主要原因是可以將 GPT-4 舉出的答案作為正確答案。

▼ 表 4-1 大模型的不同能力對應的建模任務和相關的資料集

層次	能力	建模任務	資料集
基礎	語言生成	語言建模	Penn Treebank、WikiText-103、The Pile、LAMBADA
		條件文字生成	WMT（2014/2016/2019/2020/2021/2022）、Flores-101、DiaBLa、CNN/DailyMail、XSum、WikiLingua、OpenDialKG
		程式生成	PPS、HumanEval、MBPP、CodeContest、MTPB、DS-1000、ODEX
	知識利用	封閉 QA	Natural Questions, ARC、TruthfulQA、Web Questions、TriviaQA、PIQA、LC-quad2.0、GrailQA, KQApro、CWQ、MKQA、ScienceQA
		開放 QA	Natural Questions、OpenBookQA、ARC、TriviaQA、Web Questions、MS MARCO、QASC、SQuAD、WikiMovies
		知識補全	WikiFact、FB15k-237、Freebase、WN18RR、WordNet、LAMA、YAGO3-10、YAGO
	複雜推理	知識推理	CSQA、StrategyQA、HotpotQA、ARC、BoolQ、PIQA、SIQA、HellaSwag、WinoGrande、COPA、OpenBookQA、ScienceQA、proScript、ProPara、ExplaGraphs、ProofWriter、EntailmentBank、ProOntoQA
		符號推理	CoinFlip、ReverseList、LastLetter、Boolean Assignment、Parity、Colored Object、Penguins in a Table、Repeat Copy、Object Counting
		數學推理	MATH、GSM8k、SVAMP、MultiArith、ASDiv、MathQA、AQUA-RAT、MAWPS、DROP、NaturalProofs、PISA、miniF2F、ProofNet

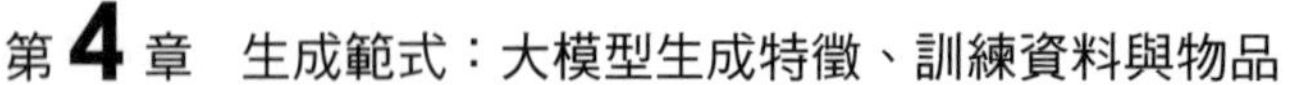

層次	能力	建模任務	資料集
高階	人類對齊	真實性	TruthfulQA、HaluEval
		有幫助	HH-RLHF
		無害性	HH-RLHF、Crows-Pairs、WinoGender、RealToxicityPrompts
	與外部環境互動	家庭場景	VirtualHome、BEHAVIOR、ALFRED、ALFWorld
		網路環境	WebShop、Mind2Web
		開放世界	MineRL、MineDojo
	工具使用	程式執行	GSM8k、TabMWP、Date Understanding
		計算機	GSM8k、MATH、CARP
		模型介面	GPT4Tools、Gorilla
		資料介面	WebQSP、MetaQA、WTQ、WikiSQL、TabFact、Spider

在實際使用時，大模型會透過批次呼叫介面針對開放原始碼的測試資料集（及公司自己累積的資料集）從 GPT-4 獲取結果，獲取結果後還可能需要借助人工抽查、規則策略或（大）模型對傳回結果的置信度評分等手段提升 < 輸入，輸出 > 對的品質，這樣微調的效果才會更好。

▲ 圖 4-12 某個開放原始碼大模型基於問題舉出的回答①

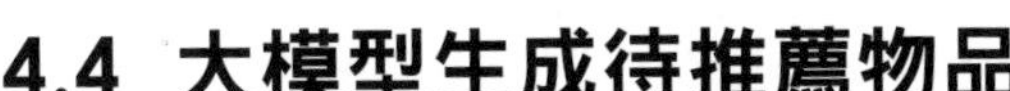

4.4 大模型生成待推薦物品

利用大模型生成的待推薦物品與傳統的推薦系統中的待推薦物品有很大區別。傳統推薦系統中的待推薦物品是已經存在的，而利用大模型生成待推薦的物品是一個從無到有的過程，這正是生成式大模型的主要價值之一。

4.4.1 為使用者生成個性化新聞

下面以 MIND 資料集為例，利用開放原始碼大模型為使用者生成個性化新聞。我們將生成過程拆解為兩個步驟：第一步生成使用者喜歡的新聞類型、標題；第二步基於第一步生成的新聞類型、標題生成滿足使用者個性化需求的新聞。

1. 生成使用者喜歡的新聞類型、標題

下面先舉出程式②，程式中有完整的註釋，對於不熟悉的類別或框架，可以在百度上查詢並學習。

```
import json
import os
import time
import pandas as pd
from langchain.callbacks.manager import CallbackManager
from langchain.callbacks.streaming_stdout import StreamingStdOutCallbackHandler
from langchain_core.prompts import PromptTemplate
from langchain_community.llms import LlamaCpp
from langchain.chains import LLMChain
from tqdm import tqdm
from prompter import MindPrompter, MindColdUser
MIN_INTERVAL = 0
current_path = os.getcwd()
mind_prompter = MindPrompter(current_path + '/data/news.tsv')
user_list = MindColdUser(current_path + '/data/user', mind_prompter).stringify()
```

① 圖中保留了回答中出現的各種問題。

② 程式：src/basic_skills/generative/generate_news/personalized_news_summary_generator.py。

```
# 將新聞讀到 dataframe 中
news_df = pd.read_csv(
    filepath_or_buffer=os.path.join(current_path + '/data/news.tsv'),
    sep='\t',
    header=0,
)
system = """You are asked to capture user's interest based on his/her browsing
history, and generate a piece of news that he/she may be interested. The format of
history is as below:
{input}
You can only generate a piece of news (only one) in the following json format, the
json include 3 keys as follows:
<title>, <abstract>, <category>
where <category> is limited to the following options:
- lifestyle
- health
- news
- sports
- weather
- entertainment
- autos
- travel
- foodanddrink
- tv
- finance
- movies
- video
- music
- kids
- middleeast
- northamerica
- games

<title>, <abstract>, and <category> should be the only keys in the json dict. The news
should be diverse, that is not too similar with the original provided news list. You
are not allowed to response any other words for any explanation or note. JUST GIVE
ME JSON-FORMAT NEWS. Now, the task formally begins. Any other information should not
disturb you."""
save_path = current_path + '/output/personalized_news_summary.log'
```

```
# 下面是呼叫 LLaMA 大模型的語法
callback_manager = CallbackManager([StreamingStdOutCallbackHandler()])
llm = LlamaCpp(
model_path="/Users/liuqiang/Desktop/code/llm/models/gguf/qwen1.5-72b-chat-q5_k_m.gguf
",
    temperature=0.8,
    top_p=0.8,
    n_ctx=6000,
    callback_manager=callback_manager,
    verbose=True,
)
# 先統計出哪些已經計算了，避免後面重複計算
exist_set = set()
with open(save_path, 'r') as f:
    for line in f:
        data = json.loads(line)
        exist_set.add(data['uid'])
# 呼叫大模型迭代計算使用者可能會喜歡的新聞
for uid, content in tqdm(user_list):
    start_time = time.time()
    if uid in exist_set:
        continue
    if not content:
        continue
    try:
        prompt = PromptTemplate(
            input_variables=["input"],
            template=system,
        )
        chain = LLMChain(llm=llm, prompt=prompt)
        enhanced = chain.run(input=content)
        enhanced = enhanced.rstrip('\n')
        with open(save_path, 'a') as f:
            f.write(json.dumps({'uid': uid, 'news': enhanced}) + '\n')
    except Exception as e:
        print(e)
    interval = time.time() - start_time
    if interval <= MIN_INTERVAL:
        time.sleep(MIN_INTERVAL - interval)
```

上面的程式實現了利用使用者的新聞瀏覽歷史生成使用者喜歡的新聞標題和類型。下面是兩個生成的案例，其中 uid 是使用者唯一 ID，news 對應的是使用者喜歡的新聞標題和類型。

```
{"uid": 4, "news":
  {
    "title": "New Study Finds Link Between Gut Health and Mental Well-being",
    "abstract": "A groundbreaking research reveals the intricate relationship between
the health of our gut microbiome and our mental health.",
    "category": "health"
  }
}
{"uid": 7, "news":
  {
    "title": "Jimmy Carter Inspires New Fitness Program for Seniors",
    "abstract": "Following Jimmy Carter's recent recovery from a pelvic fracture, a
new fitness program tailored for seniors has been launched.",
    "category": "health"
  }
}
```

2. 生成個性化的新聞

進一步地，我們還可以利用大模型基於使用者喜歡的新聞標題和類型，為使用者生成個性化的新聞。下面是具體的程式實現[①]。

```
import json
import os
import time
from langchain.callbacks.manager import CallbackManager
from langchain.callbacks.streaming_stdout import StreamingStdOutCallbackHandler
from langchain.chains import LLMChain
from langchain.prompts import PromptTemplate
from langchain_community.llms import LlamaCpp
from tqdm import tqdm
MIN_INTERVAL = 0
```

① 程式：src/basic_skills/generative/generate_news/personalized_news_generator.py。

```
current_path = os.getcwd()
save_path = current_path + '/output/personalized_news.log'
prompt_template = """
you are a news writing expert, The information of the news a user browsed are as
follows:
"news": {news}
The news in the curly braces above are a list of all the news that the user has
browsed, which may contain multiple news articles.
the information in the news stand for the category of news the user likes.
Now please write a new for the user, the news must relevant to the interest of the
user, the news you write must less than 300 words.
"""
# 下面是呼叫 LLaMA 大模型的語法
callback_manager = CallbackManager([StreamingStdOutCallbackHandler()])
llm = LlamaCpp(
    model_path="/Users/liuqiang/Desktop/code/llm/models/gguf/
qwen1.5-72b-chat-q5_k_m.gguf",
    temperature=0.8,
    top_p=0.8,
    n_ctx=6000,
    callback_manager=callback_manager,
    verbose=True,
)
prompt = PromptTemplate(
    input_variables=["news"],
    template=prompt_template,
)
chain = LLMChain(llm=llm, prompt=prompt)
# 先統計出哪些已經計算了，避免後面重複計算
exist_set = set()
with open(save_path, 'r') as f:
    for line in f:
        data = json.loads(line)
        exist_set.add(data['uid'])
# 開啟檔案並建立檔案物件
file = open(current_path + '/output/personalized_news_summary.log', "r")
# 使用 readlines() 方法將檔案內容按行存入清單 lines
lines = file.readlines()
# 關閉檔案
```

```
file.close()
# 輸出檔案內容
for line in tqdm(lines):
    info = eval(line)
    uid = info['uid']
    news = info['news']
    start_time = time.time()
    if uid in exist_set:
        continue
    if not news:
        continue
    try:
        enhanced = chain.run(news)
        enhanced = enhanced.rstrip('\n')
        news_ = {"uid": uid, "news": enhanced}
        with open(save_path, 'a') as f:
            f.write(f'{str(news_)}\n')
    except Exception as e:
        print(e)
    interval = time.time() - start_time
    if interval <= MIN_INTERVAL:
        time.sleep(MIN_INTERVAL - interval)
```

執行上面的程式，就為每個使用者生成了個性化的新聞，下面是一個案例。

```
Title: "Boxing Legend Floyd Mayweather Jr. Announces Collaboration with Luxury Spirits
Brand"

Abstract: Renowned boxing champion Floyd Mayweather Jr. is stepping into the world of
luxury spirits with an exciting collaboration. Mayweather, known for his impeccable
taste and opulent lifestyle, has partnered with a prestigious spirits brand to create
a limited-edition premium whiskey.

The collaboration between Mayweather and the luxury spirits brand aims to fuse the
worlds of sports and luxury. The yet-to-be-named whiskey is reported to have been aged
to perfection, with a rich and complex flavor profile.

Scheduled for release later this year, the limited-edition Floyd Mayweather Jr.
collaboration whiskey is expected to attract considerable attention from luxury
```

```
spirits enthusiasts and sports fans alike.

Category: sports
```

4.4.2 生成個性化的視訊

大模型也能用於視訊生成，參考文獻 [11] 提出了一個基於大模型生成短影音並用於個性化推薦的 AI 生成（AI Generator）內容框架 GeneRec，如圖 4-13 所示。該框架利用大模型對推薦函式庫中已有的視訊進行編輯（Editor）修改或新生成（Creator）短影音。為了讓生成的內容更加個性化，AI 生成內容需要與指令（Instructor）配合，結合使用者指令（User Instruction）與回饋（Feedback），例如使用者瀏覽、點擊、點讚、倒讚等，進行更加個性化的內容創作。

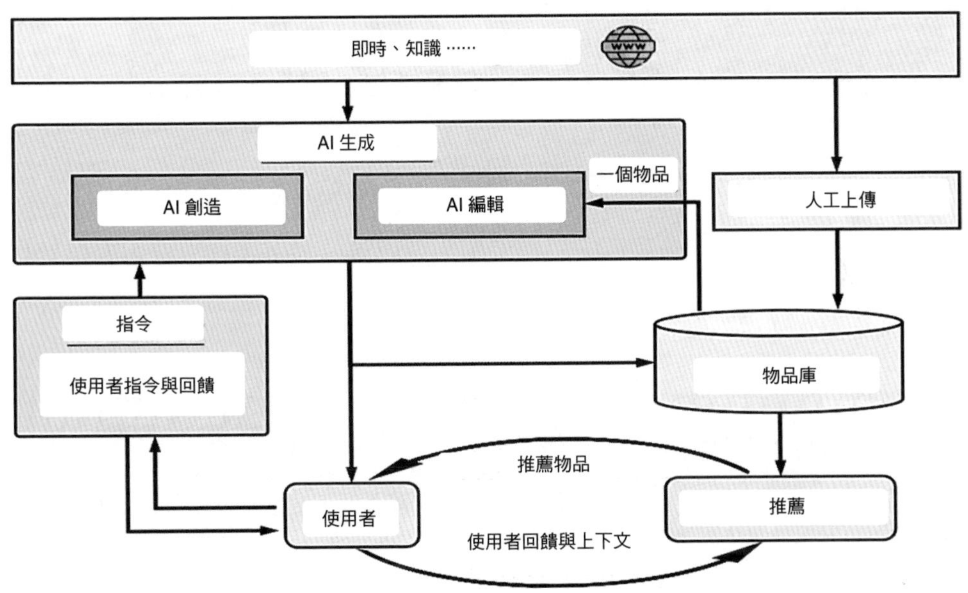

▲ 圖 4-13 AI 生成內容框架 GeneRec

這裡說的生成短影音的範圍比較廣，從零開始創作算生成，基於已有的短影音完成微創作也算生成。下面對 AI 生成內容中涉及的視訊編輯和創造過程說明並提供案例。

考慮到個性化縮圖可能會更進一步地吸引使用者點擊短影音，可以利用大模型完成個性化縮圖選擇和生成任務，為使用者呈現更具吸引力和個性化的微視訊縮圖。

對於縮圖選擇，可以利用使用者過去喜歡的短影音學習使用者對縮圖的偏好特徵，然後透過訓練好的模型針對某個短影音選擇最合適的縮圖，如圖 4-14 所示。

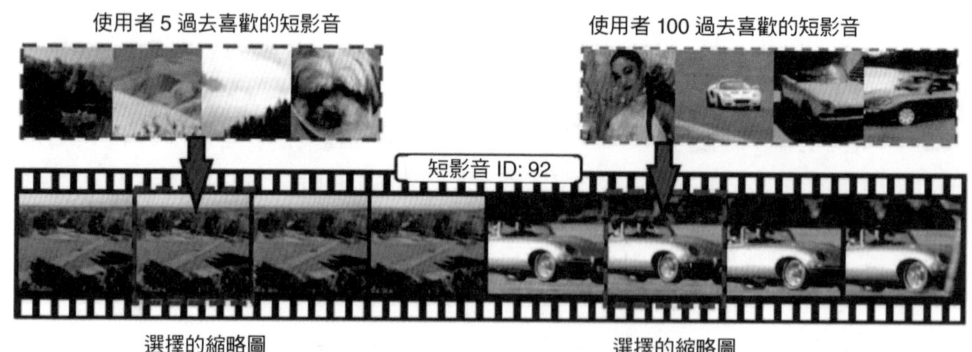

▲ 圖 4-14 縮圖選擇

可以基於使用者過去喜歡的短影音和當前視訊，利用擴散模型等大模型直接生成縮圖，如圖 4-15 所示。

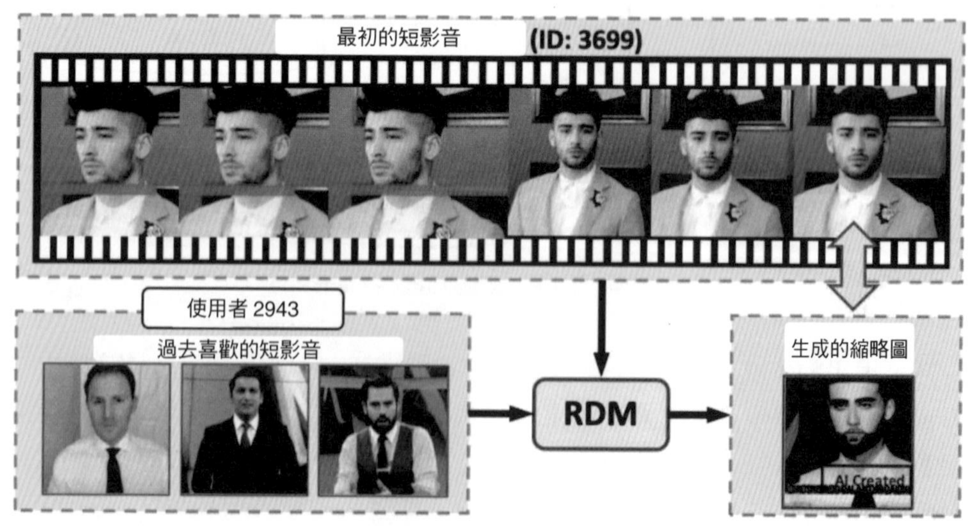

▲ 圖 4-15 縮圖生成

對比較長的短影音（例如超過 3 分鐘），還可以利用 CLIP 技術進行個性化剪輯，如圖 4-16 所示，目標是只推薦使用者喜歡的剪輯部分，這樣可以節省使用者的時間並改善使用者的使用體驗。

上面的縮圖選擇和剪輯基本保持了視訊原樣，原視訊沒有改變。我們還可以對視訊內容進行真正意義上的編輯，包括對視訊風格進行遷移，如圖 4-17 所示；以及對視訊內容進行編輯調整，如圖 4-18 所示。

▲ 圖 4-16 基於使用者歷史偏好對短影音進行個性化剪輯

▲ 圖 4-17 利用 VToonify 技術對視訊風格進行遷移

▲ 圖 4-18 利用 MCVD 技術對對視訊內容進行編輯調整

當然，最複雜的操作方式是從零開始創造一個短影音。Runway 公司發佈的 Gen-2、Pika 以及字節跳動發佈的 MagicVideo-V2，都可以基於文字生成視訊。圖 4-19 是利用 MCVD 技術生成的短影音。

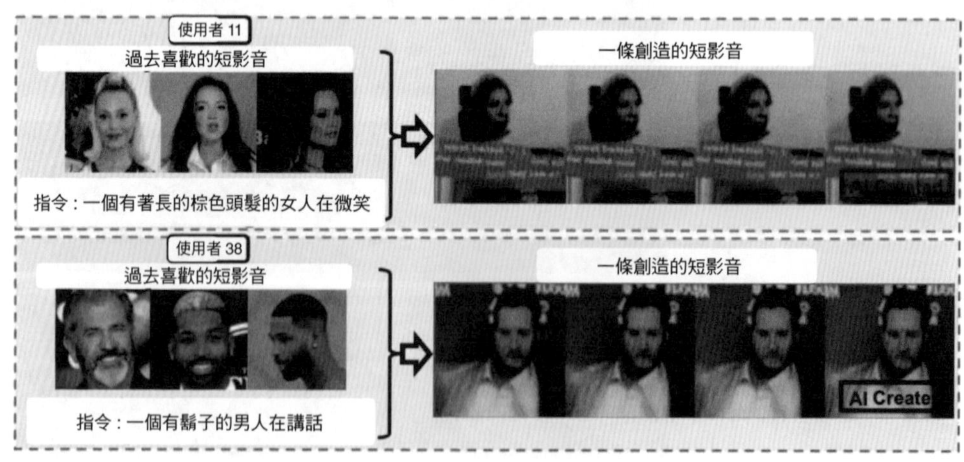

▲ 圖 4-19 利用 MCVD 技術生成的短影音

4.5 總結

本章借助 MIND 資料集和開放原始碼大模型，用程式實現了向量嵌入和生成文字特徵，生成的嵌入特徵可以直接用於推薦，也可以整合到傳統的模型中進行個性化推薦。我們也可以直接利用大模型進行推薦。嵌入方法是推薦系統中非常重要的一種方法，利用大模型進行嵌入表示，可以獲得更好的語義理解，能夠抓住使用者和新聞之間的深層語義關係，最終獲得更好的推薦效果，甚至可以解決冷啟動和跨領域推薦問題。

借助大模型優秀的文字生成能力、概要總結能力，可以生成相關的文字特徵，這些文字特徵可以更進一步地描述新聞和使用者興趣偏好，進而被更進一步地用於後續的推薦業務中。

大模型既可以生成表格類資料，也可以生成用於微調的監督訓練樣本。遊戲中的虛擬空間、智慧汽車中的路況等都可以透過大模型生成，大大提升效率並節省成本。目前，合成資料是大模型最有價值的應用方向之一。

大模型的生成能力可以輔助我們進行創作，本章中利用大模型生成個性化新聞和短影音是兩個容易理解的案例。如果將大模型與機器人、3D 列印等技術結合，那麼具備大模型能力的具身智慧體還能生產出實物商品，例如做一道個性化的美味菜肴，設計一身最適合客戶體型的衣服等。

MEMO

5 預訓練範式：透過大模型預訓練進行推薦

大模型的輸入主要是自監督的巨量文字資訊，主流架構是 Transformer。如果基於推薦系統相關資料預訓練一個大模型，也需要採用類似的想法和步驟。由於推薦系統有其自身的特點，本章重點講解在推薦系統這一背景下，如何預訓練一個解決推薦問題的大模型。

本章首先講解預訓練推薦大模型的一般想法和方法，然後講解兩個案例，最後以 MIND 資料集為例，透過程式實現推薦系統大模型的預訓練。

5.1 預訓練的一般想法和方法

本節整理大模型預訓練的一般想法和方法。

5.1.1 預訓練資料準備

推薦系統的資料主要包括使用者資料、物品資料、場景資料、行為資料。基於推薦系統資料的特性，如果要將推薦系統相關的資料轉為自監督的文字資料供大模型進行預訓練，那麼可以利用使用者行為序列①，行為序列與無監督文字序列非常相似。一般可以採用以下 3 種方式將使用者行為序列轉為適合大模型處理的資料格式。

1. 直接利用使用者行為序列

使用者行為序列對應的 ID（新聞 ID）雖然不是文字（一般是整數或字串），但是可以採用與文字一樣的建模方式處理。可以採用遮罩的方式將行為序列中間一個或多個行為擋住，利用模型預測遮罩部分，處理方式與 BERT 類似。還可以基於前面的行為序列預測後面一個或多個行為序列，類似 GPT 的處理方式。PTUM 模型採用的就是該處理方式。另外，參考文獻 [1] 中提出的 BERT4Rec 也採用了類似方式。

2. 將使用者行為序列 ID 轉為相關的文字

這種方式用新聞的標題等文字資訊替換 ID，採用範本的方式將使用者行為序列轉為文字，下面是一個案例。

使用者過去一段時間點擊過的新聞是 title_1、title_2、……、title_n，使用者下一個點擊的新聞是 title_t。

具體的範本可以非常靈活。按照上面的方式可以將所有使用者行為序列建構為自監督樣本，然後將其作為大模型的輸入。

---

① 按照使用者點擊物品的時間排序。

當然，為了包含更多有價值的資訊，讓大模型預訓練得更加精準，可以將新聞的摘要、標籤等資訊整合到提示詞中，具體實現將在案例中說明。

3. 在使用者行為序列中整合使用者的特徵

除了新聞相關的文字資訊，還可以整合使用者相關的特徵，讓模型更加個性化。下面是一個案例。

使用者 ID：uid 使用者名稱：username

使用者的點擊歷史是：title_1、title_2、……、title_n

使用者下一個點擊的新聞是 title_t

其實第 2 種方式中不同使用者的點擊歷史已經隱含了使用者的個性化資訊（不同使用者點擊的新聞是不一樣的），只不過第 3 種方式將使用者更多的資訊（例如使用者 ID、使用者名稱等）放入預訓練語料，這可以讓大模型更進一步地學習到不同使用者的差異。

5.1.2 大模型架構選擇

推薦系統的實現方式有多種，例如直接推薦、推薦解釋、評分預測、序列推薦等。以前的推薦系統針對不同場景分別訓練模型，在大模型時代，這幾種推薦範式可以用統一的框架實現，大大增加了訓練的資料量，因此可以訓練規模更大的預訓練模型。下面講解的 P5 模型就是這樣實現的。

由於推薦系統涉及的資料規模無法與巨量的文字相比，一般預訓練推薦系統的大模型參數規模相對較小（例如幾千萬到幾億個），所以可以採用較小的 Transformer 架構。具體實現上可以採用遮罩或 GPT，通常將 BERT、GPT1、GPT2、T5 等模型的架構作為基底，再利用上面提到的推薦系統資料集進行預訓練。

關於上面幾個模型的架構細節，你可以自行詳細了解。第 1 章也對 Transformer 架構的基本原理做過詳細介紹，這裡不再細說。圖 5-1 展示了 Transformer 架構及對應的 BERT4Rec 架構。

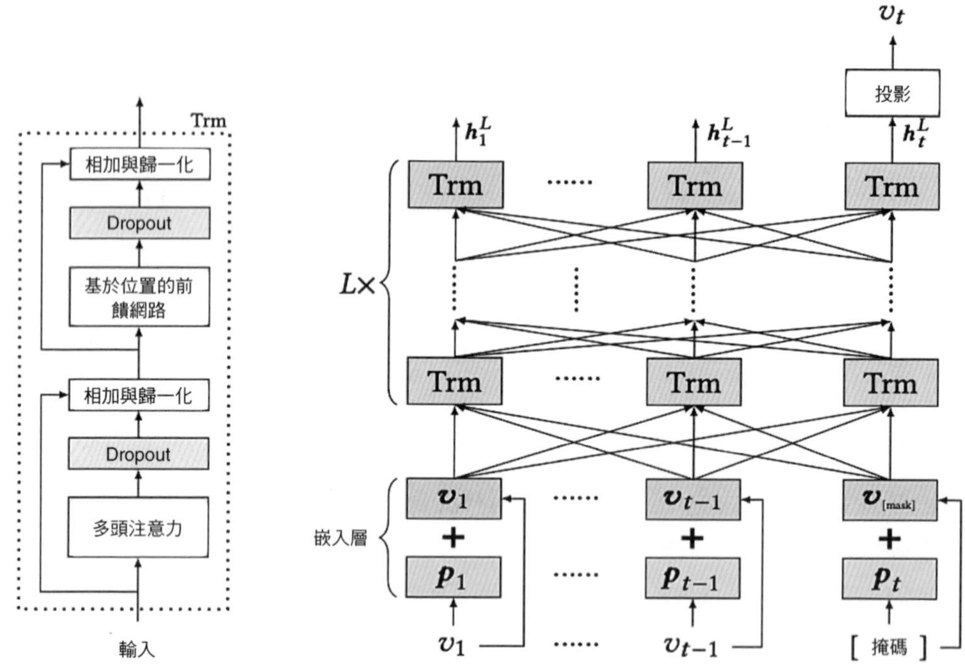

▲ 圖 5-1 Transformer 架構（左）及對應的 BERT4Rec 架構（右）

圖 5-1 右側底部的 $v_1, v_2, \cdots, v_{t-1}$ 是新聞 ID，上面的 $p_1, p_2, \cdots, p_{t-1}, p_t$ 和 $v_1, v_2, \cdots, v_{t-1}, v_{[mask]}$ 分別是 $v_1, v_2, \cdots, v_{t-1}$ 對應的位置嵌入和 ID 嵌入表示。Trm 是 Transformer 層。右上角的投影是分類層或回歸層，可以進行預測（例如預測下一個新聞），我們可以將投影層畫完整，如圖 5-2 所示。

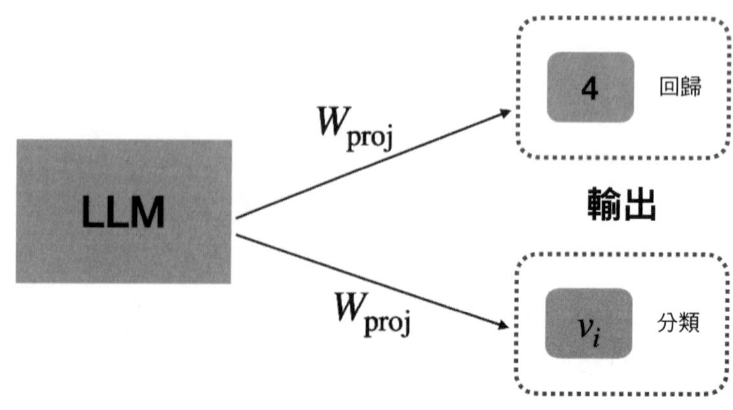

▲ 圖 5-2 透過在 Transformer 中增加一層進行預測

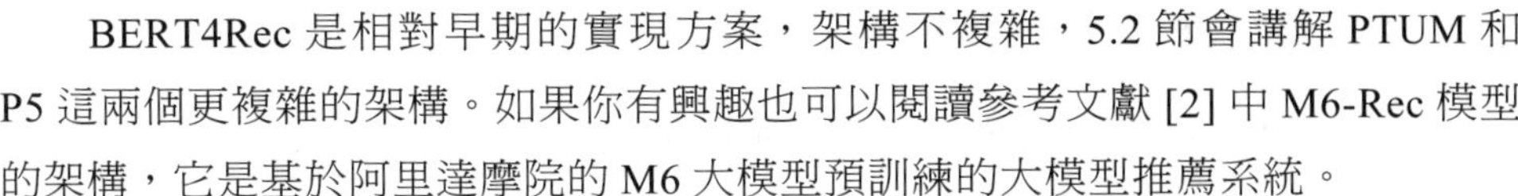

BERT4Rec 是相對早期的實現方案，架構不複雜，5.2 節會講解 PTUM 和 P5 這兩個更複雜的架構。如果你有興趣也可以閱讀參考文獻 [2] 中 M6-Rec 模型的架構，它是基於阿里達摩院的 M6 大模型預訓練的大模型推薦系統。

5.1.3 大模型預訓練

準備好預訓練資料並確定了模型架構後，就可以對模型進行預訓練了。目前有很多開放原始碼的框架提供了預訓練的處理邏輯和對應的工具程式，我們可以非常方便地進行預訓練。下面提供兩種可行的預訓練方案。

1. 基於 Transformers 函式庫的預訓練

Transformers 函式庫是 Hugging Face 社區開放原始碼的非常流行的處理大模型的框架，支援 PyTorch、TensorFlow、JAX 作為後端訓練框架。GitHub 社區中有預訓練大模型的程式 run_clm.py[3]，使用起來非常簡單，下面是一個訓練指令稿。

```
python run_clm.py \
    --model_name_or_path gpt2 \  # 預訓練的模型類型，這裡是 gpt2
    --train_file path_to_train_file \ # 預訓練資料檔案
    --validation_file path_to_validation_file \ # 驗證資料檔案
    --config_overrides="n_embd=768,n_head=12,n_layer=12,n_positions=1024" \ # 模型參數
    --per_device_train_batch_size 8 \ # 預訓練的 batch size
    --per_device_eval_batch_size 8 \ # 驗證的 batch size
    --do_train \ # 需要進行預訓練
    --do_eval \ # 需要進行驗證
    --output_dir /tmp/test-clm # 預訓練好的模型的輸出路徑
```

在筆者的 MacBook Pro（M3 Max 晶片、16 核心 CPU、40 核心 GPU、128 GB 顯示記憶體）上可以利用 MPS（Metal Performance Shaders） backend①進行預訓練，40 分鐘左右可以訓練 3 個 epoch②。上述程式也可以在輝達的 GPU 上執行。

① PyTorch 從 2022 年發佈的 1.12 版本開始就透過 MPS 支援蘋果的 GPU，可以將 MPS 等價為蘋果生態的 CUDA。

② 使用最早版本的 GPT-2，參數量約 1.2 億個。

2. 基於蘋果的 MLX 框架預訓練

如果使用 MacBook，那麼還可以利用蘋果開放原始碼的 MLX 機器學習框架 [4] 進行預訓練①，這個框架是蘋果的機器學習團隊專為 MacBook 打造的，性能高於基於 MPS 的 PyTorch，支援各種大模型的預訓練和微調，並且有相當豐富的程式案例。

利用 MLX 進行預訓練非常簡單，下面是一個具體的案例，你可以從參考文獻 [5] 中獲得更多的案例，其中涵蓋了主流的大模型。

```
python main.py \ # main.py 請參考參考文獻 [9]
--gpu \ # 利用 GPU 進行預訓練
--context_size 1024 \ # 模型 token 的 context size
--num_blocks 12 \ #Transformer 區塊的個數
--dim 768   \  # 嵌入層和隱藏層的維數
--num_heads 12  \ # Transformer 模型的 multi-head 個數
--checkpoint  \ # 需要進行 checkpoint，如果預訓練過程中斷，則可以從 checkpoint 恢復
--num_iters 1000  \ # 預訓練迭代次數
--learning_rate 3e-4  \ # 模型的學習率
--weight_decay 1e-5  \ # 權重 decay 的幅度
--lr_warmup 200  \ #LR linear warmup iterations
--steps_per_report 3 \ # 每 3 步列印 loss
--dataset wikitext2 # 利用 wikitext2 資料預訓練模型
```

上面只是簡單提供了預訓練大模型的兩種可行方案，5.3 節會基於 MIND 資料集，實現一個真正的推薦系統大模型的預訓練。

上面兩個案例在單一機器上預訓練大模型，如果機器配置不高，模型參數較多，預訓練的資料量比較大，預訓練就會非常慢。如果你有多個顯示卡，需要加速預訓練過程，或模型太大，單一卡無法完成訓練，那麼可以使用一些大模型的分散式預訓練的函式庫，比較流行的函式庫包括 ColossalAI、DeepSpeed、Megatron-LM 等，第 9 章會詳細介紹。

① 這是一個類似 PyTorch 的機器學習框架。

5.1.4 大模型推理（用於推薦）

在完成大模型的預訓練後，還需要進行推理才能讓大模型用起來，即為使用者生成個性化的推薦。推理的流程如圖 5-3 所示，輸入資料是使用者行為（可能需要加上新聞文字、使用者文字資訊），輸出是推薦結果，參見 5.1.1 節中的 3 種資料轉換方式。

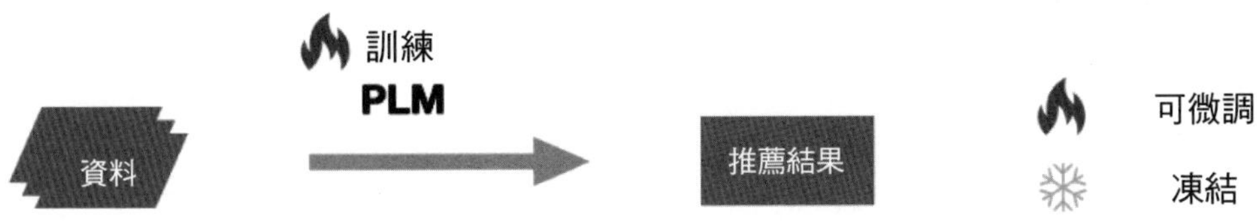

▲ 圖 5-3 利用預訓練的推薦大模型進行推理的流程

在真實使用場景中，推理速度非常重要，目前有很多框架支援大模型的高性能推薦。比較流行的加速大模型推理的框架有 Transformers 函式庫、llama.cpp（使用 C/C++ 推理，可以較好支援 MacBook）、vLLM（基於 CUDA 生態）、TensorRT-LLM（由輝達開放原始碼，支援 CUDA 生態）、MLX 等。這裡以 Transformers 函式庫和 llama.cpp 為例說明如何提升推薦系統大模型的推理速度。

1. 使用 Transformers 函式庫進行推理

Transformers 函式庫已經提供了非常豐富的生態用來對大模型進行推理，如果你預訓練好了大模型，或選定了開放原始碼的大模型，那麼可以直接用 Transformers 函式庫進行推理。下面是一個具體的程式案例，供你參考。

```
from transformers import pipeline
from transformers import AutoTokenizer, AutoModelForCausalLM
from langchain.chains import LLMChain
from langchain_community.llms.huggingface_pipeline import HuggingFacePipeline
from langchain.prompts import PromptTemplate

MODEL = "/Users/liuqiang/Desktop/code/llm/models/Yi-34B-Chat" # 從 Hugging Face 上下載的模
# 型，如果是利用上面步驟預訓練的模型，則填寫預訓練模型的位址

tokenizer = AutoTokenizer.from_pretrained(MODEL)
```

```
model = AutoModelForCausalLM.from_pretrained(MODEL)
pipe = pipeline(
    "text-generation",
    model=model,  # 這是模型的具體儲存路徑
    tokenizer=tokenizer,
    max_length=1024, # 最大輸入長度
    temperature=0.1, # 溫度參數
    top_p=0.9, # top_p 參數
    repetition_penalty=1.15
)

llm = HuggingFacePipeline(pipeline=pipe)
prompt_template = f"""
        '< 指令 > 根據使用者的新聞點擊歷史為使用者進行個性化推薦，答案請使用中文 </ 指令 >\n'
        '< 已知資訊 >{{context}}</ 已知資訊 >\n' # 這裡可以是使用者的新聞點擊歷史
        '< 問題 >{{question}}</ 問題 >\n'  # 例如問使用者下一個會點擊的新聞是什麼
"""

prompt = PromptTemplate(
    input_variables=["context", "question"],
    template=prompt_template,
)

try:
    chain = LLMChain(prompt=prompt, llm=llm)
    res = chain.invoke(input={"context": context, "question": query})
except Exception as e:
    print(e)
print(res)
```

2. 使用 llama.cpp 進行推理

llama.cpp 對 MacBook 支援較好，可以將大模型部分層的參數快取到 GPU 上，因此能充分利用 MacBook 的 GPU，推理速度很快。推理的具體實現方式與上面類似，下面舉出對應的程式案例。

```
from langchain.chains import LLMChain
from langchain_community.llms import LlamaCpp
```

```
MODEL = "/Users/liuqiang/Desktop/code/llm/models/gguf/yi-34b-chat.Q5_K_M.gguf"
# 模型位址，如果利用 llama.cpp 則需要使用 gguf 格式的大模型

n_gpu_layers = 128  # 根據模型和 GPU 大小設置參數
n_batch = 4096  # 可以設置為 1 到 n_ctx 之間的任何值，根據電腦的 RAM 大小設置
llm = LlamaCpp(
    model_path=MODEL,
    n_gpu_layers=n_gpu_layers,
    n_batch=n_batch,
    f16_kv=True,  # 設置為 True, 避免執行過程中遇到問題
    temperature=0.1,
    top_p=0.9,
    n_ctx=10000,
    verbose=True,
    stop=["<|im_end|>"]  # 生成的答案中出現這些詞就停止
)

prompt_template = f"""
        '< 指令 > 根據使用者的新聞點擊歷史為使用者進行個性化推薦，答案請使用中文 </ 指令 >\n'
        '< 已知資訊 >{{context}}</ 已知資訊 >\n' # 這裡可以是使用者的新聞點擊歷史
        '< 問題 >{{question}}</ 問題 >\n'  # 例如問使用者下一個會點擊的新聞是什麼
"""

prompt = PromptTemplate(
    input_variables=["context", "question"],
    template=prompt_template,
)

try:
    chain = LLMChain(prompt=prompt, llm=llm)
    res = chain.invoke(input={"context": context, "question": query})
except Exception as e:
    print(e)
print(res)
```

上面是具體的推理步驟程式，在真實業務場景中還需要將大模型部署成服務，這可以使用 FastAPI 等 Python Web 框架實現，也可以採用開放原始碼的方案，例如 SGLang、Lepton AI。大模型服務的相關內容將在專案實踐部分講解，這裡不再贅述。

5.2 案例講解

本節在 5.1 節的基礎上延伸，重點講解兩個預訓練大模型推薦系統的框架，讓你可以更進一步地了解大模型推薦系統的架構特點及具體的實現方案。

5.2.1 基於 PTUM 架構的預訓練推薦系統

PTUM 是 Pre-Training User Model 的縮寫，指基於使用者行為的預訓練模型，模型的架構比較簡單，如圖 5-4 所示。PTUM 同時利用遮罩行為預測和後續行為預測，可以獲得更好的泛化能力和預測效果，這一想法類似於 1.3.1 節提到的將語言建模過程作為微調的輔助目標，GPT-1 就是這樣預訓練的。下面分別對遮罩行為預測和後續行為預測詳細說明。

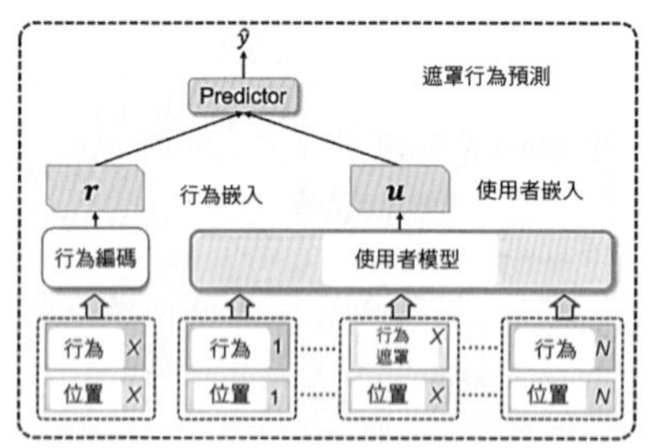

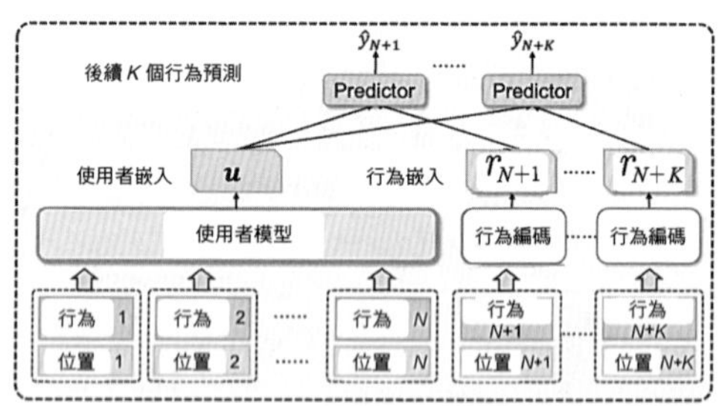

▲ 圖 5-4 PTUM 的架構

1. 遮罩行為預測

遮罩行為預測（Masked Behavior Prediction，MBP）就是將使用者行為序列中間的或多個行為掩蓋起來，然後建模預測掩蓋的行為，類似 BERT 的建模想法。圖 5-4 左側的 Predictor 是預測函式 $\hat{y}=f(\boldsymbol{u},\boldsymbol{r})$，其中，$\boldsymbol{u},\boldsymbol{r}$ 分別是使用者行為和待預測的 item 的嵌入向量，遮罩預測的是左邊行為 X 的機率。f 可以是神經網路模型，也可以是簡單的向量餘弦相似度函式。由於使用者行為的所有可能情況其實就是 item 的數量，因此預測使用者的行為是一個多分類問題，所以可以採用交叉熵損失函式，具體如下。

$$\mathcal{L}_{\text{MBP}} = -\sum_{y \in \mathcal{S}_1} \sum_{i=1}^{P+1} y_i \lg(\widehat{y_i}) \tag{5-1}$$

（5-1）上式中的 S_1 是從使用者行為構造的遮罩預測訓練資料集。y_i 和 $\widehat{y_i}$ 分別是遮罩真實的和預測的 label（item ID）的機率。P 是負樣本的數量，針對每個遮罩訓練資料，從其他使用者行為中隨機取出 P 個 item 作為負樣本。

2. 後續行為預測

後續行為預測（Next K Behaviors Prediction，NBP）基於使用者過去的行為預測接下來的 k 個行為，類似 GPT 系列模型中預測下一個單字出現的機率。圖 5-4 中的 $\hat{y}_{N+1}$ 和 $\hat{y}_{N+K}$ 與上面一樣，是預測的後續 item 的機率，也可以是神經網路模型或向量餘弦函式。損失函式還是採用交叉熵，具體如式（5-2）所示。

$$\mathcal{L}_{\text{NBP}} = -\frac{1}{K} \sum_{y \in \mathcal{S}_2} \sum_{k=1}^{K} \sum_{i=1}^{P+1} y_{i,k} \ln(\hat{y}_{i,k}) \tag{5-2}$$

式（5-2）中的 S_2 是從使用者行為構造的後續行為預測訓練資料集，$y_{i,k}$ 和 $\hat{y}_{i,k}$ 分別是真實的和預測的後續第 k 個行為是 item i 的機率。這裡 P 也是負樣本的個數。

將式（5-1）和式（5-2）結合起來，就獲得了 PTUM 模型的損失函式，如式（5-3）所示，其中 λ 是超參數，代表的是後續行為預測所佔的權重，可以透過交叉驗證的方式取一個合適的值。

$$\mathcal{L} = \mathcal{L}_{\text{MBP}} + \lambda \mathcal{L}_{\text{NBP}} \tag{5-3}$$

圖 5-4 中既有行為嵌入也有位置嵌入，PTUM 對 item 進行嵌入然後建構 Transformer 模型，item 嵌入（編碼）需要結合其他演算法，例如 GRU、BERT 等。PTUM 模型比較簡單，這裡不再贅述，可以透過參考文獻 [6] 了解更多細節。

5.2.2 基於 P5 的預訓練推薦系統

筆者認為，預訓練個性化提示詞預測框架 [7]（Pretrain, Personalized Prompt, and Predict Paradigm，P5）是一個更加通用的、具有大模型精髓的框架，因此重點講解。P5 將五大類推薦問題統一在一個自然語言處理框架下，這樣可以透過預訓練一個模型一次性解決這 5 類問題，這正是大模型泛化能力的最好證明。圖 5-5 舉出 P5 解決的 5 類推薦問題及對應的模型預訓練資料。

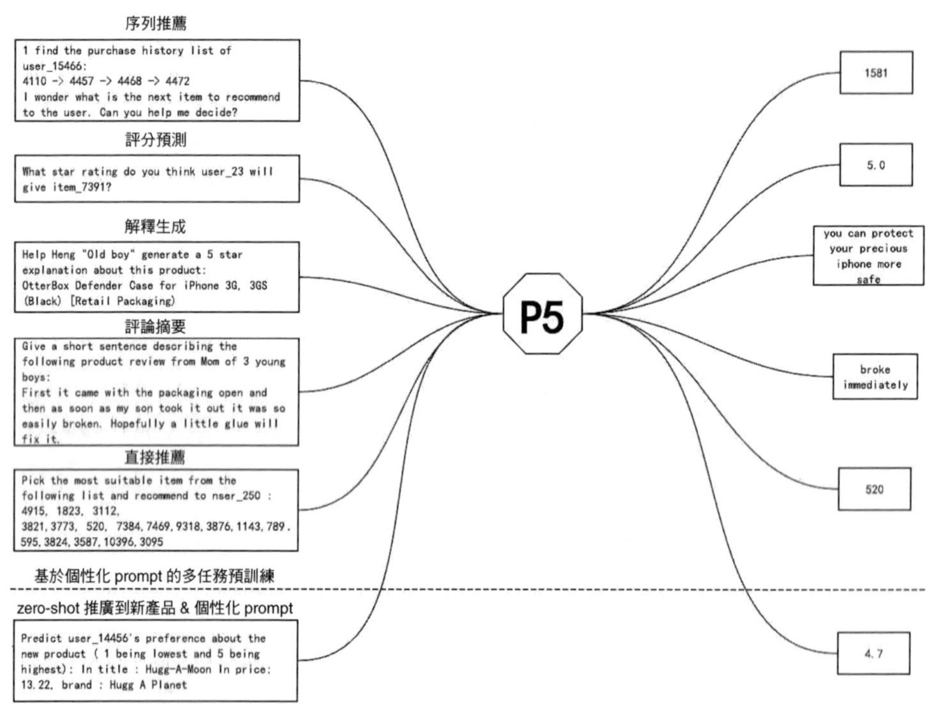

▲ 圖 5-5　P5 解決的 5 類推薦問題及對應的模型預訓練資料

從圖 5-5 中可以看到，P5 不是完全意義上的自監督學習，評分預測、解釋生成等是需要監督資料的，但是序列推薦可以基於使用者行為建構自監督樣本，然後「餵」給大模型。下面詳細說明 P5 預訓練的實現流程。

1. 預訓練資料準備

P5 基於不同的推薦類型建構對應的提示詞範本（Prompt Template），然後對收集的推薦系統原始資料進行處理，最後按照定義好的範本組織資料，這樣就可以建構大模型預訓練需要的訓練語料了。

根據參考文獻 [7] 中設計的個性化提示詞範本，透過原始資料建構輸入 - 目標對，只需將提示詞中的欄位（大括號中的部分）替換為原始資料中的相應資訊，P5 預訓練資料範本如圖 5-6 所示。P5 的 5 類任務的原始資料有三個獨立的來源。具體而言，評分預測、評論摘要、推薦解釋提示詞具有共用的原始資料，見圖 5-6(a)。序列推薦（見圖 5-6(b)）和直接推薦（見 5-6(c)）使用相似的原始資料，但序列推薦需要使用者互動歷史。

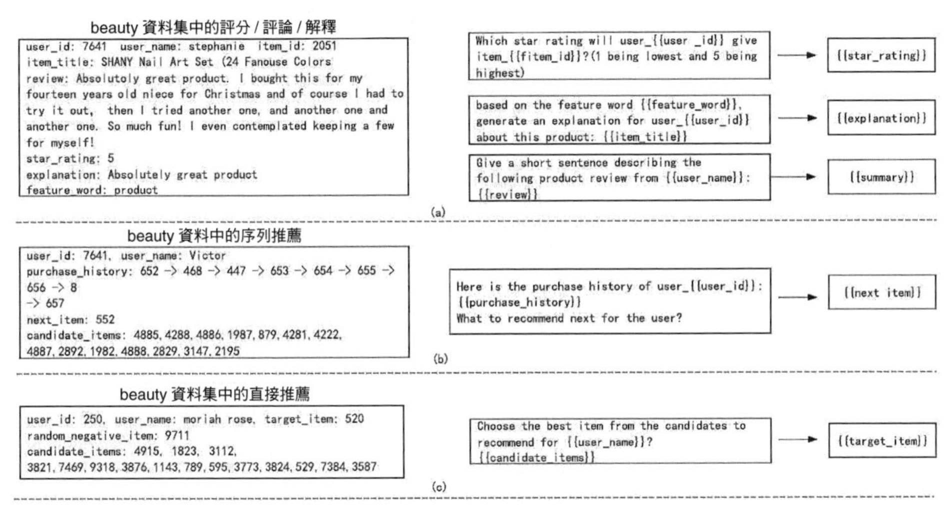

▲ 圖 5-6 P5 預訓練資料範本

圖 5-6 中中間是輸入範本（Input Template），右側是目標範本（Target Template）。舉例來說，最後一行的輸入範本和目標範本分別是：

> 輸入範本：Choose the best item from the candidates to recommend for {{user_name}} ? \n {{candiate_items}}
> 目標範本：{{target_item}}

這裡只舉出了幾個簡單的資料範本範例，參考文獻 [7] 的附錄中有完整的範本。5.3 節的程式案例會採用類似的想法處理，這裡不再贅述。

2. P5 的模型架構

個性化提示詞集便於建立大量可用的預訓練資料，這些資料涵蓋了廣泛的推薦任務。得益於提示詞範本，所有預訓練資料共用統一格式的輸入 - 目標 token 序列，打破了不同任務之間的界限。在大模型「自然語言條件生成」的統一框架下，預訓練多個推薦任務建構的統一資料集可以用於所有的單一推薦任務。

有了預訓練的資料，下面就可以選擇對應的模型架構了。P5 將 T5 作為基底模型，包括一個小型模型和一個基礎模型：6075 萬個參數的 P5-small 和 2328 萬個參數的 P5-base。P5 預訓練模型架構如圖 5-7 所示。

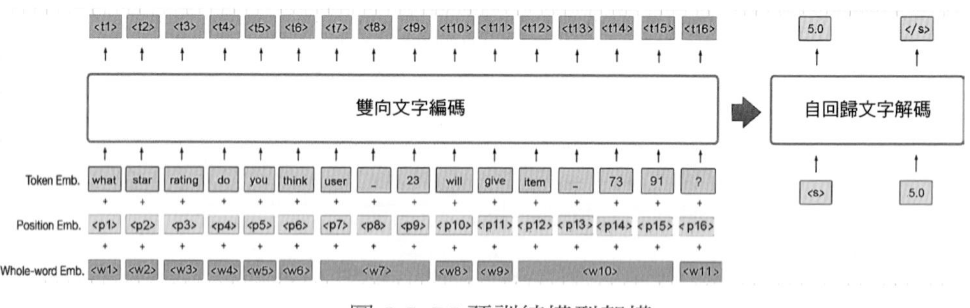

▲ 圖 5-7 P5 預訓練模型架構

對於範例輸入提示詞「what star rating do you think user_23 will give item _7391?」，P5 採用編碼器 - 解碼器框架：先用雙向文字編碼器對輸入進行編碼，再透過文字解碼器自回歸生成答案。

參考文獻 [7] 還透過整詞嵌入（whole-word embedding）***W*** 來指示連續的子詞（sub-word） token 是否來自同一原始詞。舉例來說，如果直接將 ID 編號為 7391 的 item 表示為「item_7391」，則單字將被句子部分標記化（Sentence-Piece Tokenizer）拆分為 4 個獨立的 token（item、_、73、91）。在共用的整詞嵌入 ⟨w10⟩ 的幫助下，可以將 item、_、73、91 作為同一個詞的資訊整合到模型中，因此，P5 可以更進一步地辨識具有個性化資訊的部分。

與特定任務的推薦模型相比，P5 依賴在大規模個性化提示詞集上基於多工提示詞的預訓練，這使得 P5 能夠適應不同的推薦任務，甚至能推廣／泛化到新的個性化提示詞場景。

針對 P5 模型，採用預測下一個詞的框架來建構模型，那麼針對一筆訓練樣本，損失函式可以按照式（5-4）定義，這裡的 X 是輸入文字序列（上面提到的輸入提示詞），$y_{<j}$ 代表生成的第 1~ $j-1$ 個 token，然後預測第 j 個 token y_j，P_θ 就是預測 y_j 的機率。

$$\mathcal{L}_\theta^P = -\sum_{j=1}^{|y|} \ln P_\theta \left(y_j \mid y_{<j}, X \right) \tag{5-4}$$

3. P5 的預訓練

有了資料和模型架構，預訓練就比較容易了。P5 的程式已經在 GitHub 上開放原始碼，Transformers 函式庫提供了模型預訓練的函式和類別。下面簡單說明基於 Transformers 函式庫的預訓練想法。

Transformers 函式庫的 Trainer 類別提供了對大模型進行預訓練的所有封裝，可以進行預訓練。下面是一個簡單的程式範例，供你參考。

```
from transformers import (
    T5Config,
    T5ForConditionalGeneration,
    AutoTokenizer,
    Trainer，
    TrainingArguments
)

config = T5Config.from_pretrained("t5-small")
model = T5ForConditionalGeneration.from_pretrained("t5-small", config=config)
tokenizer = AutoTokenizer.from_pretrained("t5-small")

args=TrainingArguments(
    warmup_steps=warmup_steps, # warmup 步數
    num_train_epochs=epochs, #epochs 數量
```

```
    learning_rate=lr, # 學習率
    optim=optim, # 最佳化方法
    evaluation_strategy="steps",
    save_strategy="steps",
    eval_steps=200,
    save_steps=200,
    output_dir=output_dir,
    save_total_limit=3,
    load_best_model_at_end=True
)

trainer = Trainer(
    model=model, # 對應的模型
    args=training_args, # 預訓練相關參數，例如學習率等
    train_dataset=train_dataset, # 預訓練資料集
    eval_dataset=eval_dataset, # 評估資料集，如果沒有則填 None
    tokenizer=tokenizer, # token
    compute_metrics=compute_metrics # 評估的指標
)
```

將上面的虛擬程式碼完善為一個具體的 Python 指令稿 train.py 後，就可以利用 PyTorch 進行預訓練了。下面是使用 PyTorch 在單機上啟動多個處理程序進行預訓練的範例，單機預訓練耗時較長，可以使用多個 GPU 進行加速。

```
torchrun --standalone \
    --nnodes=1 \ # 節點數量
    --nproc-per-node=4 \ # 啟動 4 個處理程序執行
    train.py \ # 待預訓練的模型指令稿
    --datasets Beauty \ # 具體的資料集
    --epochs 10 \  # epochs 數量
    --batch_size 128 \ #batch 大小
```

上面是簡單的實現想法，5.3 節的程式案例也是基於 T5 架構實現的，會舉出詳細的程式實現過程。

4. P5 的推理

預訓練好的 P5 模型已經在 Hugging Face 社區開放原始碼（模型 ID：makitanikaze/P5），本節使用該模型展示推理過程。

```
# Use a pipeline as a high-level helper
from transformers import pipeline, T5ForConditionalGeneration,AutoTokenizer

model_path = "/Users/liuqiang/Desktop/code/llm4rec/P5/snap/p5_beauty_base"  # 下載後的本
# 地路徑

model = T5ForConditionalGeneration.from_pretrained(model_path)
tokenizer = AutoTokenizer.from_pretrained(model_path)

tokenizer.pad_token_id = (
    0  # unk. we want this to be different from the eos token
)
tokenizer.padding_side = "left"

generate_num = 3

prediction = model.generate(
input_ids=input_ids,
# token 後的，例如 input_ids = tokenizer("Considering {dataset} user_{user_id}
# has interacted with {dataset} items {history} . What is the next recommendation for
the
# user ?", return_tensors="pt").input_ids
    max_length=30, # 最大長度
    num_beams=generate_num, # 集束搜尋大小
    num_return_sequences=generate_num, # 生成的數量
    output_scores=True, #傳回生成的預測得分
    return_dict_in_generate=True,
)

prediction_ids = prediction["sequences"] # 預測的 token 張量

generated_sents = tokenizer.batch_decode(  # 最終生成的結果
    prediction_ids, skip_special_tokens=True
)
```

執行上面程式，最終結果基於 input_ids 中的提示詞預測使用者接下來會購買的商品，如下所示。

```
['Beauty item_1679', 'Beauty item_1668', 'Beauty item_1240', 'Beauty item_1263',...]
```

5.3 基於 MIND 資料集的程式實戰

本節以 MIND 資料集為例，進一步用程式實現大模型的預訓練，一步步教你從零開始預訓練一個自己的推薦系統大模型。

本節依然將 T5 作為基底模型，採用與 P5 類似的框架實現推薦大模型的預訓練。

5.3.1 預訓練資料集準備

我們基於 P5 模型預訓練資料的格式來處理 MIND 資料集。

```
U13740 N55189 N42782 N34694 N45794 N18445 N63302 N10414 N19347 N31801
U73700 N10732 N25792 N7563 N21087 N41087 N5445 N60384 N46616 N52500 N33164 N47289
N24233 N62058 N26378 N49475 N18870
U34670 N45729 N2203 N871 N53880 N41375 N43142 N33013 N29757 N31825 N51891
U79199 N37083 N459 N29499 N38118 N37378 N24691 N27235 N34694 N13137 N35022 N16617
N39762 N35452 N58580 N1258 N24492 N31801 N43142 N14761
```

上面的每行都代表一筆樣本。每行的第一個字串是使用者 ID，後面的字串是使用者點擊過的新聞 ID，按照時間昇冪排列。為了模型更加精準，只保留點擊了 5 筆或更多新聞的使用者。根據這一資料格式，可以使用 MIND 資料集的 behaviors.tsv 資料表，其中的 User ID 欄位是使用者 ID，User Click History 欄位是使用者點擊過的歷史新聞。

1. 生成預訓練的序列資料

可以使用 generate_user_sequence.py 函式將 behaviors.tsv 資料表中的資料轉為上面的格式[①]。

```
import csv
user_sequence_list = []
with open('../data/mind/behaviors.tsv', 'r') as file:
    reader = csv.reader(file, delimiter='\t')
    for row in reader:
        uid = row[1]
        history = row[3]
        if len(history.split(" ")) >= 5:  # 使用者至少要有 5 次歷史點擊行為
            r = uid + " " + history
            user_sequence_list.append(r)

user_sequence_path = "../data/mind/user_sequence.txt"
with open(user_sequence_path, 'a') as file:
    for r in user_sequence_list:
        file.write(r + "\n")
```

經過上述處理，我們就生成了使用者行為序列資料，這個資料在程式倉庫的路徑是 /src/basic_skills/train-llm/data/mind/user_sequence.txt。

2. 生成預訓練資料

為了更進一步地處理模型，需要將使用者行為序列進行編碼，我們為使用者 ID 和新聞 ID 順序編碼，從第一個使用者開始，一個一個為使用者的連續點擊行為分配編號連續的 ID。當一個使用者迭代完成後，進入下一個使用者的迭代過程，為任何尚未被分配的新聞分配一個新的、遞增的 ID。順序編碼方法僅用於預訓練資料，以避免評估階段的潛在資料洩露。

注意：user_sequence.txt 中的每個使用者點擊的最後兩個新聞會被用於測試和驗證，不在預訓練資料考慮之列。生成預訓練資料的具體程式實現以下[②]。

① 程式：/src/basic_skills/train-llm/data-process/generate_user_sequence.py。

② 程式：/src/basic_skills/train-llm/data-process/generate_dataset_train.py。

```
import os
import json
import sys
import fire
sys.path.append('../')
from utils import sequential_indexing, load_prompt_template, check_task_prompt,
get_info_from_prompt
from utils import construct_user_sequence_dict, read_line

def main(data_path: str, item_indexing: str, task: str, dataset: str, prompt_file: str,
sequential_order: str, max_his: int, his_sep: str, his_prefix: int, skip_empty_his:
int):
    file_data = dict()
    file_data['arguments'] = {
        "data_path": data_path, "item_indexing": item_indexing, "task": task,
        "dataset": dataset, "prompt_file": prompt_file, "sequential_order":
sequential_order,"max_his": max_his, "his_sep": his_sep, "his_prefix": his_prefix,
"skip_empty_his": skip_empty_his
    }
    file_data['data'] = []
    tasks = list(task)
    user_sequence = read_line(os.path.join(data_path, dataset, 'user_sequence.txt'))
    user_sequence_dict = construct_user_sequence_dict(user_sequence)
    reindex_user_seq_dict, item_map = sequential_indexing(data_path, dataset,
                                                    user_sequence_dict, sequential_order)
    # get prompt
    prompt = load_prompt_template(prompt_file, tasks)
    info = get_info_from_prompt(prompt)
    check_task_prompt(prompt, tasks)
    # Load training data samples
    training_data_samples = []
    for user in reindex_user_seq_dict:
        items = reindex_user_seq_dict[user][:-2]
        for i in range(len(items)):
            if i == 0:
                if skip_empty_his > 0:
                    continue
            one_sample = dict()
            one_sample['dataset'] = dataset
            one_sample['user_id'] = user
```

```
            if his_prefix > 0:
                one_sample['target'] = 'item_' + items[i]
            else:
                one_sample['target'] = items[i]
            if 'history' in info:
                history = items[:i]
                if max_his > 0:
                    history = history[-max_his:]
                if his_prefix > 0:
                    one_sample['history'] = his_sep.join(["item_" + item_idx for item_
idx in history])
                else:
                    one_sample['history'] = his_sep.join(history)
            training_data_samples.append(one_sample)
    print("load training data")
    print(f'there are {len(training_data_samples)} samples in training data.')
    # construct sentences
    for i in range(len(training_data_samples)):
        one_sample = training_data_samples[i]
        for t in tasks:
            datapoint = {'task': dataset + t, 'data_id': i}
            for pid in prompt[t]['seen']:
                datapoint['instruction'] = prompt[t]['seen'][pid]['Input']
                datapoint['input'] = prompt[t]['seen'][pid]['Input'].format(**one_
sample)
                datapoint['output'] =
prompt[t]['seen'][pid]['Output'].format(**one_sample)
                file_data['data'].append(datapoint.copy())

    print("data constructed")
    print(f"there are {len(file_data['data'])} prompts in training data.")
    # save the data to json file
    output_path = f'{dataset}_{task}_{item_indexing}_train.json'

    with open(os.path.join(data_path, dataset, output_path), 'w') as openfile:
        json.dump(file_data, openfile)

if __name__ == "__main__":
    fire.Fire(main)
```

上面的程式會相依一些輔助函式，這些函式存放在程式倉庫 /src/basic_skills/train-llm/utils.py 中，你可以自己查看，這裡不再贅述。後續程式中沒有定義的函式都在 utils.py 中，不再說明。我們可以透過下面的指令稿執行上面的程式。

```
python generate_dataset_train.py --data_path
/Users/liuqiang/Desktop/code/llm4rec/llm4rec_abc/src/train-llm/data/ \
 --item_indexing  sequential --task sequential,straightforward --dataset mind
--prompt_file ../prompt.txt \
 --sequential_order original --max_his 10 --his_sep ' , ' --his_prefix 1 --skip_empty_
his 1
```

表 5-1 舉出了 generate_dataset_train.py 指令稿的參數說明。注意，這裡透過 Python 的 fire 套件將參數注入程式。

▼ 表 5-1 generate_dataset_train.py 指令稿的參數說明

參數	說明
data_path	資料目錄
item_indexing	新聞的索引方式，這裡採用 sequential，是順序索引
task	推薦任務，提供序列（sequential）推薦和直接（straightforward）推薦兩種方式
dataset	這裡採用 MIND 資料集
prompt_file	提示詞範本檔案
sequential_order	按照使用者在檔案 user_sequence.txt 中本來的順序對使用者索引
max_his	預訓練時使用者的歷史取 max_his，這裡是 10，也就是只用最近的 10 個點擊行為預訓練模型
his_sep	使用者歷史的分隔符號，本節用逗點
his_prefix	歷史中是否加首碼，1 代表是。如果一個新聞的索引是 1023，那麼加首碼就是 item_1023
skip_empty_his	是否跳過歷史為空的使用者，1 代表是，也就是歷史為空的使用者不會用於預訓練模型

運行 generate_dataset_train.py 指令稿會生成 4 個檔案，表 5-2 分別解釋說明。

▼ 表 5-2 執行 generate_dataset_train.py 指令稿後生成的檔案

檔案名稱	解釋說明
item_sequential_indexing_original.txt	新聞 ID 索引，從 1001 開始
user_indexing.txt	使用者 ID 索引，從 1 開始
user_sequence_sequential_indexing_original.txt	將最原始的 user_sequence.txt 資料的使用者 ID 和新聞 ID 替換成索引後的資料
mind_('sequential', 'straightforward')_sequential_train.json	大模型預訓練資料。這個資料是 JSON 格式的 資料格式以下 {"arguments": {"data_path": "/Users/liuqiang/Desktop/code/llm4rec/ llm4rec_abc/src/train-llm/data/", "item_indexing": "sequential", "task": ["sequential", "straightforward"], "dataset": "mind", "prompt_file": "../prompt.txt", "sequential_order": "original", "max_his": 10, "his_sep": " , ", "his_prefix": 1, "skip_empty_his": 1}, "data": [{"task": "mindsequential", "data_id": 0, "instruction": "Considering {dataset} user_{user_id} has interacted with {dataset} items {history} . What is the next recommendation for the user ?", "input": "Considering mind user_1 has interacted with mind items item_1001 . What is the next recommendation for the user ?", "output": "mind item_1002"}]}

表 5-2 最後一行右邊的預訓練資料案例中包含 instruction 和 input。instruction 是提示詞範本，input 是透過將使用者行為資料（user_sequence_sequential_indexing_original.txt）替換為提示詞範本中的參數獲得的。提示詞範本有 22 個，具體如下。

```
(1) sequential; seen; Considering {dataset} user_{user_id} has interacted with
{dataset} items {history} . What is the next recommendation for the user ?; {dataset}
{target}
```

(2) sequential; seen; Here is the purchase history of {dataset} user_{user_id} : {dataset} item {history} . I wonder what is the next recommended item for the user .; {dataset} {target}
(3)sequential; seen; {dataset} user_{user_id} has purchased {dataset} items {history}, predict next possible item to be bought by the user ?; {dataset} {target}
(4) sequential; seen; I find the purchase list of {dataset} user_{user_id} : {dataset} items {history} , I wonder what other itmes does the user need . Can you help me decide ?; {dataset} {target}
(5) sequential; seen; According to what items {dataset} user_{user_id} has purchased : {dataset} items {history} , Can you recommend another item to the user ?; {dataset} {target}
(6) sequential; seen; What would {dataset} user_{user_id} be likely to purchase next after buying {dataset} items {history} ?; {dataset} {target}
(7)sequential; seen; By analyzing the {dataset} user_{user_id} 's purchase of {dataset} items {history} , what is the next item expected to be bought ?; {dataset} {target}
(8) sequential; seen; Can you recommend the next item for {dataset} user_{user_id} , given the user 's purchase of {dataset} items {history} ?; {dataset} {target}
(9) sequential; seen; After buying {dataset} items {history} , what is the next item that could be recommended for {dataset} user_{user_id} ?; {dataset} {target}
(10) sequential; seen; The {dataset} user_{user_id} has bought items : {dataset} items {history} , What else do you think is necessary for the user ?; {dataset} {target}
(11) sequential; unseen; What is the top recommended item for {dataset} user_{user_id} who interacted with {dataset} item {history} ?; {dataset} {target}
(12) straightforward; seen; What should we recommend for {dataset} user_{user_id} ?; {dataset} {target}
(13) straightforward; seen; {dataset} user_{user_id} is looking for some items . Do you have any recommendations ?; {dataset} {target}
(14) straightforward; seen; Do you have any suggested items for {dataset} user_{user_id} ?; {dataset} {target}
(15) straightforward; seen; Which recommendation should we provide to {dataset} user_{user_id} ?; {dataset} {target}
(16) straightforward; seen; How can we assist {dataset} user_{user_id} with a recommendation ?; {dataset} {target}
(17) straightforward; seen; What would be a suitable recommendation for {dataset} user_{user_id} ?; {dataset} {target}
(18) straightforward; seen; What would be a helpful recommendation for {dataset} user_{user_id} ?; {dataset} {target}
(19) straightforward; seen; Can you recommend an item for {dataset} user_{user_id} ?; {dataset} {target}

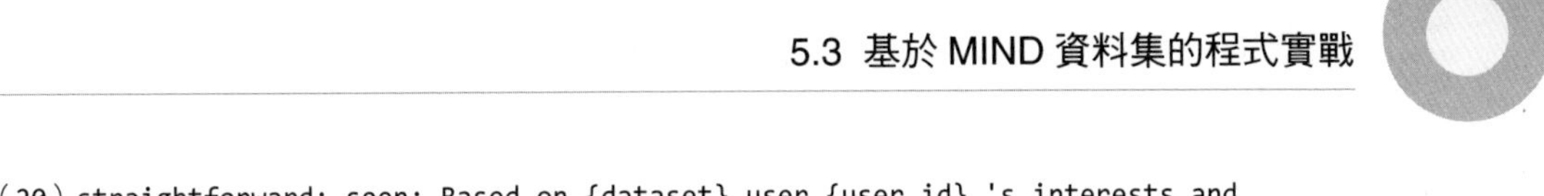


```
(20) straightforward; seen; Based on {dataset} user_{user_id} 's interests and
requirements , what item would you suggest to try ?; {dataset} {target}
(21) straightforward; seen; For {dataset} user_{user_id} , what item stands out as a
top recommendation that they should consider ?; {dataset} {target}
(22)straightforward; unseen; What is the top recommendation for {dataset} user_{user_
id} ?; {dataset} {target}
```

表 5-3 對提示詞中的幾個欄位說明，方便你更進一步地理解提示詞和相關程式。

▼ 表 5-3 提示詞欄位說明

欄位	說明
sequential	sequential 指序列推薦，也就是基於使用者的點擊歷史，預測他接下來要點擊的新聞
straightforward	straightforward 是直接推薦，也就是不告訴模型使用者的點擊歷史，預測使用者下一個要點擊的新聞
seen	從每個任務中，選擇一個提示詞範本作為可見 / 不可見提示詞（在預訓練集中沒有相關範本建構的資料），用於評估模型的零樣本泛化能力
unseen	
dataset	預訓練模型的資料集，這裡採用的是 MIND 資料集，在程式中用小寫
user_id	使用者 ID，具體會根據使用者行為序列填充提示詞，形成預訓練資料
history	使用者過去的新聞點擊歷史
target	預測使用者下一個要點擊的新聞

3. 生成測試、驗證資料

有了上面的訓練資料就可以對推薦大模型進行預訓練了。為了更進一步地評估預訓練好的大模型，還需要生成測試和驗證資料，下面舉出具體的程式實現[①]。

① 程式：/src/train-llm/basic_skills/data-process/generate_dataset_eval.py

```
import os
import json
import sys
import fire
sys.path.append('../')
from utils import sequential_indexing, load_prompt_template, check_task_prompt,
get_info_from_prompt
from utils import construct_user_sequence_dict, read_line, load_test, load_validation

def main(data_path: str, item_indexing: str, task: str, dataset: str, prompt_file: str,
sequential_order: str,
         max_his: int, his_sep: str, his_prefix: int, skip_empty_his: int,
         mode: str, prompt: str):

    file_data = dict()
    file_data['arguments'] = {
        "data_path": data_path, "item_indexing": item_indexing, "task": task,
        "dataset": dataset, "prompt_file": prompt_file, "sequential_order":
sequential_order,
        "max_his": max_his, "his_sep": his_sep, "his_prefix": his_prefix, "skip_empty_
his": skip_empty_his,
        "mode": mode, "prompt": prompt
    }
    file_data['data'] = []
    tasks = list(task)

    user_sequence = read_line(os.path.join(data_path, dataset, 'user_sequence.txt'))
    user_sequence_dict = construct_user_sequence_dict(user_sequence)
    reindex_user_seq_dict, item_map = sequential_indexing(data_path, dataset,
                                                  user_sequence_dict, sequential_order)

    # get prompt
    prompt_ = load_prompt_template(prompt_file, tasks)
    info = get_info_from_prompt(prompt_)
    check_task_prompt(prompt_, tasks)
    # Load data samples
    if mode == 'validation':
        data_samples = load_validation(reindex_user_seq_dict, info, dataset, his_
```

```
prefix, max_his, his_sep)
        prompt_info = prompt.split(':')
        output_path = f'{dataset}_{task}_{item_indexing}_validation_{prompt}.json'
    elif mode == 'test':
        data_samples = load_test(reindex_user_seq_dict, info, dataset, his_prefix,
max_his, his_sep)
        prompt_info = prompt.split(':')
        output_path = f'{dataset}_{task}_{item_indexing}_test_{prompt}.json'
    else:
        raise NotImplementedError
    # construct sentences
    for i in range(len(data_samples)):
        one_sample = data_samples[i]
        for t in tasks:
            datapoint = {'task': dataset + t,
                         'instruction':
prompt_[t][prompt_info[0]][prompt_info[1]]['Input'],
                         'input':
prompt_[t][prompt_info[0]][prompt_info[1]]['Input'].format(**one_sample),
                         'output':
prompt_[t][prompt_info[0]][prompt_info[1]]['Output'].format(**one_sample)}
            file_data['data'].append(datapoint.copy())

    with open(os.path.join(data_path, dataset, output_path), 'w') as openfile:
        json.dump(file_data, openfile)

if __name__ == "__main__":
    fire.Fire(main)
```

程式中出現了兩個新的參數，表 5-4 舉出它們的解釋。

▼ 表 5-4 程式中兩個新參數的解釋

參數	說明
mode	mode 只有兩個設定值：test 和 validation，分別表示生成測試資料和驗證資料
prompt	prompt 的設定值也有兩個：seen:0 和 unseen:0

可以透過以下指令稿生成測試資料。

```
python generate_dataset_eval.py --dataset mind --data_path
/Users/liuqiang/Desktop/code/llm4rec/llm4rec_abc/src/train-llm/data/ \
 --item_indexing  sequential --task sequential,straightforward --prompt_file
/Users/liuqiang/Desktop/code/llm4rec/llm4rec_abc/src/train-llm/prompt.txt \
 --mode validation --prompt seen:0 --sequential_order original --max_his 10 --his_sep
' , ' --his_prefix 1 --skip_empty_his 1

python generate_dataset_eval.py --dataset mind --data_path
/Users/liuqiang/Desktop/code/llm4rec/llm4rec_abc/src/train-llm/data/ \
 --item_indexing  sequential --task sequential,straightforward --prompt_file
/Users/liuqiang/Desktop/code/llm4rec/llm4rec_abc/src/train-llm/prompt.txt \
 --mode test --prompt seen:0 --sequential_order original --max_his 10 --his_sep ' , '
--his_prefix 1 --skip_empty_his 1

python generate_dataset_eval.py --dataset mind --data_path
/Users/liuqiang/Desktop/code/llm4rec/llm4rec_abc/src/train-llm/data/ \
 --item_indexing  sequential --task sequential,straightforward --prompt_file
/Users/liuqiang/Desktop/code/llm4rec/llm4rec_abc/src/train-llm/prompt.txt \
 --mode test --prompt unseen:0 --sequential_order original --max_his 10 --his_sep ' , '
--his_prefix 1 --skip_empty_his 1
```

上面的 3 個指令稿最終會生成 3 個 JSON 格式的資料，其中包括 2 個測試資料、1 個驗證資料（在程式倉庫的 /src/basic_skills/train-llm/data 目錄下）。表 5-5 進行了說明。

▼ 表 5-5 測試資料和驗證資料說明

檔案名稱	解釋說明
mind_('sequential', 'straightforward')_sequential_test_seen:0.json	seen 的測試資料
mind_('sequential', 'straightforward')_sequential_test_unseen:0.json	unseen 的測試資料
mind_('sequential', 'straightforward')_sequential_validation_seen:0.json	seen 的驗證資料

5.3.2 模型預訓練

我們的新聞推薦模型基於 T5 架構，可以直接利用 T5 的基底模型和 config，下面是預訓練的程式[①]。

```
import os
import fire
import transformers
from datasets import load_dataset
from transformers import (
    T5Config,
    T5ForConditionalGeneration,
    AutoTokenizer,
    Trainer,
    TrainingArguments
)

def main(backbone: str, data_path: str, item_indexing: str, task: str, dataset: str,
valid_prompt: str, cutoff: int, model_dir: str, batch_size: int, valid_select: int,
epochs: int, lr: float, warmup_steps: int, gradient_accumulation_steps: int,
logging_steps: int, optim: str, eval_steps: int, save_steps: int, save_total_limit:
int):
    config = T5Config.from_pretrained(backbone)
    model = T5ForConditionalGeneration.from_pretrained(backbone, config=config)
    tokenizer = AutoTokenizer.from_pretrained(backbone)
    train_data_file = os.path.join(data_path, dataset,
                                f'{dataset}_{task}_{item_indexing}_train.json')
    valid_data_file = os.path.join(data_path, dataset,

f'{dataset}_{task}_{item_indexing}_validation_{valid_prompt}.json')
    train_data = load_dataset("json", data_files=train_data_file, field='data')
    valid_data = load_dataset("json", data_files=valid_data_file, field='data')
    def tokenize(prompt, add_eos_token=True):
        result = tokenizer(
            prompt, truncation=True, max_length=cutoff, padding=False, return_
tensors=None,
```

① 程式：/src/basic_skills/train-llm/pre-train/t5_pre-train.py

```
        )
        if (isinstance(result["input_ids"][-1], int) and result["input_ids"][-1] !=
tokenizer.eos_token_id
                and len(result["input_ids"]) < cutoff
                and add_eos_token
        ):
            result["input_ids"].append(tokenizer.eos_token_id)
            result["attention_mask"].append(1)
        elif isinstance(result["input_ids"][-1], list) and add_eos_token:
            for i in range(len(result['input_ids'])):
                if result["input_ids"][i][-1] != tokenizer.eos_token_id and
len(result["input_ids"][i]) < cutoff:
                    result["input_ids"][i].append(tokenizer.eos_token_id)
                    result["attention_mask"][i].append(1)
        result["labels"] = result["input_ids"].copy()
        return result
    def process_func(datapoint):
        encoding = tokenize(datapoint['input'], add_eos_token=True)
        labels = tokenize(datapoint['output'], add_eos_token=True)
        encoding['labels'] = labels['input_ids'].copy()
        # return encoding
        return {**datapoint, **encoding}
    tokenizer.pad_token_id = (
        0  # unk. we want this to be different from the eos token
    )
    tokenizer.padding_side = "left"
    train_set = train_data['train'].shuffle().map(process_func, batched=True)
    valid_set = valid_data['train'].shuffle().map(process_func, batched=True)
    output_dir = os.path.join(model_dir, dataset, item_indexing, backbone)
    trainer = Trainer(
        model=model,
        train_dataset=train_set,
        eval_dataset=valid_set if valid_select > 0 else None,
        args=TrainingArguments(
            per_device_train_batch_size=batch_size,
            gradient_accumulation_steps=gradient_accumulation_steps,
            warmup_steps=warmup_steps,
            num_train_epochs=epochs,
            learning_rate=lr,
```

```
            logging_steps=logging_steps,
            optim=optim,
            evaluation_strategy="steps" if valid_select > 0 else "no",
            save_strategy="steps",
            eval_steps=eval_steps if valid_select > 0 else None,
            save_steps=save_steps,
            output_dir=output_dir,
            save_total_limit=save_total_limit,
            load_best_model_at_end=True if valid_select > 0 else False,
            group_by_length=False,
        ),
        data_collator=transformers.DataCollatorForSeq2Seq(
            tokenizer, pad_to_multiple_of=8, return_tensors="pt", padding=True
        ),
    )
    trainer.train()  # 模型預訓練
    model.save_pretrained(output_dir)  # 儲存預訓練好的模型
    tokenizer.save_pretrained(output_dir)  # 儲存 token

if __name__ == "__main__":
    fire.Fire(main)
```

這裡又出現了一些新的參數，如表 5-6 所示。

▼ 表 5-6 模型預訓練程式參數說明

參數	說明
backbone	基底模型，這裡是 t5-small
valid_prompt	驗證的 prompt，這裡取 seen:0
cutoff	資料超過這個長度就截斷
model_dir	預訓練好的模型儲存路徑
batch_size	每一批次處理多少個資料
valid_select	是否利用驗證集的 loss 選擇模型
epochs	預訓練的資料跑多少輪
lr	模型的學習率

參數	說明
warmup_steps	經過多少步學習率逐步增加到預設的最大值
gradient_accumulation_steps	經過多少步更新梯度
logging_steps	迭代多少次
optim	梯度下降的最佳化器
eval_steps	迭代多少次進行一次評估
save_steps	迭代多少次進行一次 checkout
save_total_limit	一共 checkout 多少次

需要注意的是，資料在放入模型進行預訓練時需要 token 化，這就是上面程式中的 tokenizer 和 process_func 兩個函式所做的工作。有了這些準備工作，就可以將相關參數傳入 Transformers 函式庫的 Trainer 類別中進行預訓練了，可以用下面的指令稿進行預訓練。

```
#!/bin/bash
dir_path="../logs/mind/"

if [ ! -d "$dir_path" ]; then
   mkdir -p "$dir_path"
fi

PYTORCH_ENABLE_MPS_FALLBACK=1 torchrun t5_pre-train.py --item_indexing sequential \
--task sequential,straightforward --dataset mind --epochs 1 --batch_size 1024 \
--backbone t5-small --cutoff 1024 --data_path
/Users/liuqiang/Desktop/code/llm4rec/llm4rec_abc/src/train-llm/data \
--valid_prompt seen:0 --model_dir
/Users/liuqiang/Desktop/code/llm4rec/llm4rec_abc/src/train-llm/models \
--lr 1e-3 --valid_select 1 --warmup_steps 100 --gradient_accumulation_steps 10
--logging_steps 10 --optim 'adamw_torch' \
--eval_steps 200 --save_steps 200 --save_total_limit 3
```

在我的電腦上預訓練以上指令稿需要 24 小時左右，時間非常長（這還是 epochs=1 的時候）。預訓練完成後模型會被儲存起來，預訓練好的模型包含圖

5-8 中的檔案，其中 checkpoint-800 表示在預訓練過程中訓練了 800 個 batch 之後將模型參數儲存下來，以便後續預訓練出現問題時從 checkpoint 開始訓練。以 .json 結尾的是一些模型配置參數，其他兩個比較大的檔案（model.safetensors 和 spiece.model）是模型相關參數。

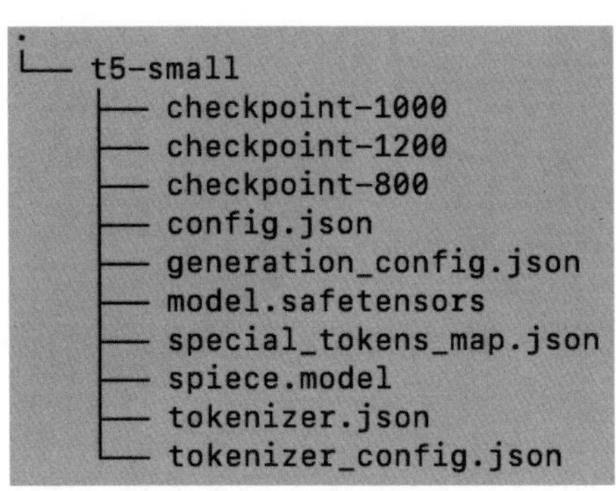

.
└── t5-small
├── checkpoint-1000
├── checkpoint-1200
├── checkpoint-800
├── config.json
├── generation_config.json
├── model.safetensors
├── special_tokens_map.json
├── spiece.model
├── tokenizer.json
└── tokenizer_config.json

▲ 圖 5-8 預訓練後的模型檔案

5.3.3 模型推理與驗證

按照上面的步驟預訓練好模型後，就可以利用測試資料集進行驗證了，透過推理來檢驗推薦模型的效果。下面是具體的程式實現①。

```
import os
import logging
import fire
from torch.utils.data import DataLoader
from datasets import load_dataset
from tqdm import tqdm
import sys
import numpy as np
from transformers import (
    T5ForConditionalGeneration,
    AutoTokenizer,
)
sys.path.append('../')
```

① 程式：/src/basic_skills/train-llm/evaluation/t5_evaluate.py。

```
from utils import EvaluationDataset, evaluation_results, get_metrics_results


def main(log_dir: str, checkpoint_path: str, data_path: str, item_indexing: str, task:
str,
         dataset: str, cutoff: int, test_prompt: str, eval_batch_size: int, metrics:
str):
    # setup
    log_file = os.path.join(log_dir, dataset,
                            checkpoint_path.replace('.', '').replace('/', '_') + '.log')
    for handler in logging.root.handlers[:]:
        logging.root.removeHandler(handler)
    logging.basicConfig(filename=log_file, level=logging.INFO,
                        format='%(asctime)s - %(name)s - %(levelname)s - %(message)s')
    logging.getLogger().addHandler(logging.StreamHandler(sys.stdout))
    model = T5ForConditionalGeneration.from_pretrained(checkpoint_path)
    tokenizer = AutoTokenizer.from_pretrained(checkpoint_path)
    tokenizer.pad_token_id = (
        0  # unk. we want this to be different from the eos token
    )
    tokenizer.padding_side = "left"
    # load test data
    test_data_file = os.path.join(data_path, dataset,
f'{dataset}_{task}_{item_indexing}_test_{test_prompt}.json')
    logging.info("test_data_file=" + test_data_file)
    test_data = load_dataset("json", data_files=test_data_file, field='data')
    model.eval()
    metrics = list(metrics)
    generate_num = max([int(m.split('@')[1]) for m in metrics])
    task_list = np.unique(test_data['train']['task'])
    for t in task_list:
        logging.info(f'testing on {t}')
        subset_data = test_data.filter(lambda example: example['task'] == t)
        dataset = EvaluationDataset(subset_data['train'], tokenizer, cutoff)
        dataloader = DataLoader(dataset, batch_size=eval_batch_size, shuffle=False)
        test_total = 0
        metrics_res = np.array([0.0] * len(metrics))
        for batch in tqdm(dataloader):
            """
            下面是一個 batch 的案例：
```

```
            {'input_ids': tensor([[    3, 21419, 12587,  ...,     0,     0,     0],
            ...,
            [    3, 21419, 12587,  ...,     0,     0,     0]]),

            'attention_mask': tensor([[1, 1, 1,  ..., 0, 0, 0],
            ...,
            [1, 1, 1,  ..., 0, 0, 0]]),

            'label': tensor([[12587,  2118,   834, 22504,  2577,     1,     0],
            [12587,  2118,   834, 19993,  4867,     1,     0],
            ...,
            [12587,  2118,   834, 19993,  5062,     1,     0]])}
        """
        prediction = model.generate(  # 大模型生成函式
            input_ids=batch["input_ids"],  # torch.LongTensor of shape
(batch_size, sequence_length)
            attention_mask=batch["attention_mask"],  # torch.FloatTensor of shape
(batch_size, sequence_length)
            max_length=30,
            num_beams=generate_num,
            num_return_sequences=generate_num,
            output_scores=True,
            return_dict_in_generate=True,
        )
        output_ids = batch['label']
        prediction_ids = prediction["sequences"]  # 利用大模型進行預測輸出的向量化表示，
需要解碼
        prediction_scores = prediction["sequences_scores"]
        gold_sents = tokenizer.batch_decode(  # 使用者真實的點擊記錄
            output_ids, skip_special_tokens=True
        )
        generated_sents = tokenizer.batch_decode(  # 大模型預測的點擊記錄
            prediction_ids, skip_special_tokens=True
        )  # 生成的推薦清單結構：['mind item_1679', 'mind item_2837',
'mind item_1089', 'mind item_1240']
        rel_results = evaluation_results(generated_sents, gold_sents,
prediction_scores, generate_num)
        test_total += len(rel_results)
        metrics_res += get_metrics_results(rel_results, metrics)
```

```
        metrics_res /= test_total
        for i in range(len(metrics)):
            logging.info(f'{metrics[i]}: {metrics_res[i]}')

if __name__ == "__main__":
    fire.Fire(main)
```

表 5-7 對新出現的參數舉出了解釋。

▼ 表 5-7　模型推理程式參數說明

參數	說明
test_prompt	取 seen:0 和 unseen:0 兩個值，分別用前面講到的 mind_('sequential', 'straightforward')_ sequential_test_seen:0.json 和 mind_('sequential', 'straightforward')_sequential_test_unseen:0.json 資料集驗證模型的效果
eval_batch_size	多少步計算一次評估指標
metrics	評估模型效果的度量指標，本章使用 hit@5、hit@10、ndcg@5 和 ndcg@10 這 4 個度量指標（utils.py 中有具體的程式）

下面的兩個指令稿分別評估表 5-7 中的兩個測試資料集。

```
python t5_evaluate.py --dataset mind --task sequential,straightforward --item_indexing
sequential --backbone t5-small \
--checkpoint_path /Users/liuqiang/Desktop/code/llm4rec/llm4rec_abc/src/train-llm/
models/mind/sequential/t5-small \
 --test_prompt seen:0 --log_dir '../logs' \
--data_path /Users/liuqiang/Desktop/code/llm4rec/llm4rec_abc/src/train-llm/data \
--cutoff 1024 --eval_batch_size 32 --metrics hit@5,hit@10,ndcg@5,ndcg@10

python t5_evaluate.py --dataset mind --task sequential,straightforward --item_indexing
sequential --backbone t5-small \
--checkpoint_path
/Users/liuqiang/Desktop/code/llm4rec/llm4rec_abc/src/train-llm/models/mind/sequential/
t5-small \
--test_prompt unseen:0 --log_dir '../logs' \
--data_path /Users/liuqiang/Desktop/code/llm4rec/llm4rec_abc/src/train-llm/data \
--cutoff 1024 --eval_batch_size 32 --metrics hit@5,hit@10,ndcg@5,ndcg@10
```

執行上面的程式，獲得的評估指標如下。

```
hit@5: 0.010642579260385459
hit@10: 0.020659124446630595
ndcg@5: 0.006426694605760801
ndcg@10: 0.00965856762769121
```

5.4 總結

本章講解利用大模型預訓練原生的推薦系統的想法和方法。由於推薦系統自身的資料特性[①]，其在資料準備和模型選擇上與常規的大模型有一定差異。

為了套用大模型的自監督學習進行預訓練，可以採用 3 種方式準備資料：直接利用使用者行為序列、將使用者行為序列 ID 轉為相關的文字、在使用者行為序列中整合使用者的特徵。推薦系統的資料量與網際網路上的巨量的文字還有很大差距，因此在架構選擇上可以採用 BERT、GPT1、GPT2、T5 等參數規模在幾百萬個到幾億個的大模型。

利用大模型進行預訓練的方式雖然實現了個性化推薦，但由於模型規模較小，其泛化能力、指令跟隨能力較弱，筆者更推薦基於已經預訓練好的更大的大模型，利用推薦系統相關資料進行微調的方案，這是下一章的主題。

本章還講解了開放原始碼的預訓練推薦系統框架 PTUM 和 P5 的核心思想。其中，P5 的適用範圍更廣，能夠極佳地利用大模型進行自然語言推理，並且可以泛化到多個推薦場景，希望你可以認真閱讀 P5 相關的參考資料，真正掌握該框架的本質。

P5 預訓練的模型框架是基於 T5 的，P5 最大的創新在於將 5 類推薦任務統一在同一個範式下，這樣可以獲得以下兩個好處。

① 包含使用者行為序列和使用者自身的特性。

一是可以獲得更多的預訓練資料。大量的預訓練資料正是讓大模型強大的基礎，P5 將 5 類資料一起使用，彌補了推薦系統資料量少的不足，因此可以將比 BERT 更大的模型（T5）作為基底模型。

二是將 5 類推薦任務統一在一個範式下，一次預訓練就可以解決 5 類問題，因此 P5 是一個多工學習系統。同時，由於 P5 是基於大模型進行預訓練的，因此還具備了大模型的 zero-shot 的能力。

本章基於 T5 框架，參考 OpenP5 的程式實現，一步步展示了如何從零開始預訓練一個專為推薦系統服務的大模型。透過對本章的學習，你一定能深刻體會到一個大模型從資料處理到預訓練完成再到推理評估的不易。我們一起見證了一個較小的大模型的誕生過程，這是一件非常有成就感的事情！

6

微調範式：微調大模型進行個性化推薦

本章講解如何基於微調大模型的範式進行個性化推薦，也就是基於推薦系統獨特的資料（例如使用者行為資料）建構訓練樣本，然後透過監督學習的想法對已有的開放原始碼大模型進行參數微調，讓大模型更加調配個性化推薦任務。

本章聚焦微調大模型，6.1 節介紹微調的價值、步驟、方法、困難與挑戰；6.2 節透過兩個具體的案例說明如何微調大模型進行個性化推薦，以便你更進一步地理解微調大模型進行推薦的演算法原理和應用場景；6.3 節基於 MIND 資料集舉出實現微調的程式。

6.1 微調的方法

微調大模型是大模型中非常重要的主題，也有一定的技術門檻，本節先整理微調相關的基礎知識，方便你進行後續案例和程式的學習。

6.1.1 微調的價值

大模型是基於巨量文字資料進行預訓練的，當訓練資料和模型規模足夠大具備湧現能力時，大模型就具備了一定的解決下游任務的能力，例如文字生成、摘要、翻譯，甚至是程式、邏輯推理。但推薦系統是特定的下游任務，與文字類的、大模型比較擅長的任務差距較大。很多學者直接用 ChatGPT 等大模型進行推薦，發現推薦效果可能不一定有經過特定訓練的深度學習推薦模型好。

基於上面的背景，透過將推薦系統特定的領域知識建構成監督學習樣本對，再灌輸給大模型進行微調就可以讓大模型更進一步地調配推薦這一特定場景。很多學者的研究成果也表明，利用推薦資料進行微調的大模型會表現出非常好的效果，甚至超過了基於傳統深度學習演算法精調的模型。大模型具有很好的泛化能力和知識遷移能力，微調少量的監督樣本就可以激發出大模型強大的個性化推薦能力。

使用推薦系統特定領域的資料對大模型進行微調，是讓大模型推薦系統取得更加好的效果，並且超越傳統推薦演算法的可行方法。

6.1.2 微調的步驟

圖 6-1 展示了微調大模型架構。微調主要基於推薦系統相關的資料（人物誌、物品畫像、使用者行為、場景資料）、候選集（Candidate）、gen_data[①]，或其他用於大模型微調的資料，建構監督學習訓練樣本，再選擇合適的開放原始碼的基底大模型和合適的目標函式進行微調，包括以下 5 個步驟。

① 基於前面提到的資料，利用演算法生成的中間輔助資料，是可選的。

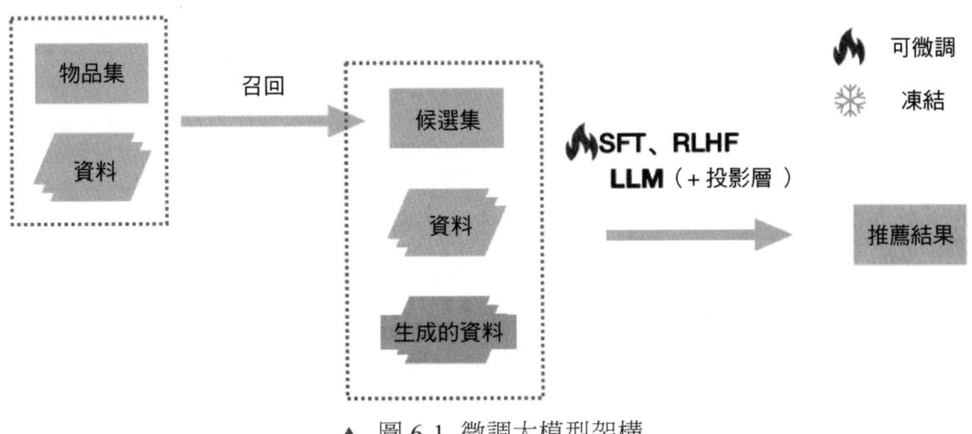

▲ 圖 6-1 微調大模型架構

1. 選擇合適的基底大模型

一般選擇參數量比較小的開放原始碼大模型（例如 7B、13B 等）作為基底大模型，模型太大了訓練成本、推理成本很高，並且推理速度可能較慢，達不到企業級推薦的要求。合適的大模型可以是 LLaMA 7B 或 LLaMA 13B，它們是 Meta 開放原始碼的大模型，也是目前業內熱門的開放原始碼大模型，表 6-1 舉出了一些範例，供你參考。對於中文語境，可以選擇幾個基於中文語料預訓練的大模型。

▼ 表 6-1 可以微調的大模型範例

開放原始碼大模型	可選參數量	開放原始碼公司	說明
LLaMA	7B、13B	Meta	目前熱門的開放原始碼大模型，參考連結 6-1
Gemma	2B、7B	Google	Google DeepMind 開放原始碼的大模型，參考連結 6-2
Flan-T5	3B、11B	Google	參考連結 6-3
BLOOMZ	7B	BigScience	參考連結 6-4
Alpaca	7B	Stanford	參考連結 6-5

續表

開放原始碼大模型	可選參數量	開放原始碼公司	說明
BELLE	7B、13B	鏈家網	基於中文微調 LLaMA 而成，參考連結 6-6
ChatGLM2	6B	智譜 AI	雙語對話語言模型，參考連結 6-7
Baichuan2	7B、13B	百川智慧	參考連結 6-8
Yi	6B	零一萬物	參考連結 6-9
Qwen、Qwen1.5	0.5B、1.8B、4B、7B、14B	阿里－通義千問	參考連結 6-10

2. 根據監督樣本選擇合適的目標函式

一般微調是監督機器學習問題，需要一個比較明確的目標函式進行最佳化，透過最佳化更新模型參數，最終讓模型調配下游的推薦系統任務。具體需要根據提供的監督訓練樣本 < 輸入，輸出 > 的形式而定，輸入一般是文字（這是大模型決定的），而輸出可以是文字，也可以是預測的評分或分類任務等。所以根據這兩個不同目標可以有兩類目標函式。

另外，最好選擇經過指令微調（Instruction Tune）的大模型，這類大模型可以更進一步地（效果更好、泛化能力更強）基於人類自然語言輸入獲得輸出結果。如果基底大模型沒有經過指令微調，也可以在進行推薦資料微調之前進行一輪指令微調。

< 文字，文字 > 監督樣本的目標函式

一般來說，微調大模型有兩種方式：一種是對全部模型參數進行微調；另一種是微調部分（額外增加的）參數。當然，第二種方式對模型參數改動更小（也可能不改動，而是增加新的參數），微調速度會更快，並且能夠保證模型在之前的任務上的效果不變[①]。下面我們分別介紹這兩種微調方法對應的目標函式。

① 反例是，經過推薦系統資料微調，模型在之前表現很好的任務上的效果反而變差了。

- 微調整個模型。

如果微調整個模型，則可以採用式（6-1）所示的目標函式，這裡的(x,y)是輸入 - 輸出監督樣本對，都是文字的形式，Z 是所有輸入 - 輸出監督樣本對組成的訓練集，y_t 是 y 的第 t 個 token，$y_{<t}$ 是 y_t 之前的所有 token，Φ 是大模型的所有參數。$P_\Phi\left(y_t x, y_{<t}\right)$ 是條件機率，也就是基於輸入樣本 x 和 y_t 之前的所有 token $y_{<t}$ 預測的 y_t 的機率。

$$\max_{\Phi} \sum_{(x,y)\in Z} \sum_{t=1}^{|y|} \ln\left(P_{\Phi}(y_t \mid x, y_{<t})\right) \tag{6-1}$$

- 微調額外參數（LoRA）。

在微調大模型時，直接對模型進行微調是計算密集型且耗時的。考慮到這些因素，我們可以採用輕量級的微調策略來執行指令調優。輕量級微調的核心前提是，大模型擁有大量的參數，並且它們的資訊集中在較低的內在維度上。通俗一點說，就是模型在推理時，只有部分參數真正起作用。

因此，我們可以透過僅微調一小部分參數來實現與整個模型相當的性能。具體來說，使用 LoRA 微調（見圖 1-17）凍結預訓練的模型參數，將可訓練的秩分解矩陣引入 Transformer 架構的每一層，以進行輕量級微調。因此，透過最佳化秩分解矩陣，我們可以有效地合併補充資訊，同時將原始參數保持在凍結狀態。基於 LoRA 的參數有效的微調方法的目標函式如式 6-2 所示。Θ 就是 LoRA 對應的參數，其他參數的意義與式（6-1）中的相同。

$$\max_{\Theta} \sum_{(x,y)\in Z} \sum_{t=1}^{|y|} \ln\left(P_{\Phi+\Theta}(y_t \mid x, y_{<t})\right) \tag{6-2}$$

< 文字，回歸／分類 label> 監督樣本的目標函式

如果希望最終的大模型推薦系統預測物品的評分或進行分類，那麼可以在大模型頂部增加一個投影層（Projection Layer），這樣就可以對大模型進行分類或回歸訓練了，如圖 6-2 所示。下面分別說明兩者的目標函式。

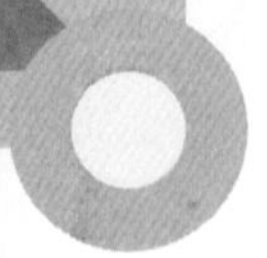

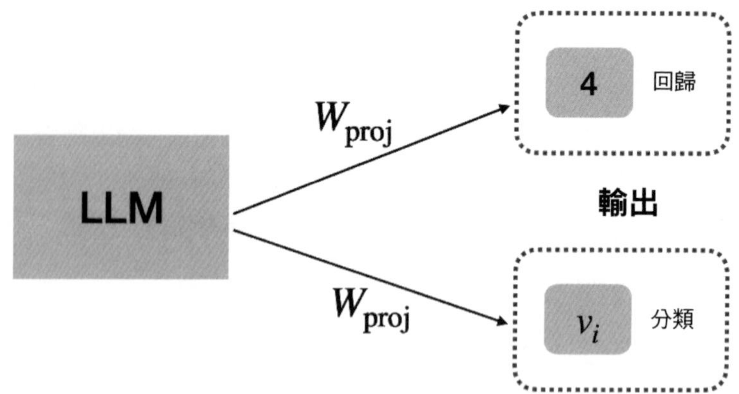

▲ 圖 6-2 在大模型頂部增加一個投影層，進行分類或回歸訓練

- 回歸問題。

如果將最終的推薦問題看成回歸問題，就是預測使用者對推薦物品的評分（0 ～ 10 分），那麼最終的目標函式可以是式（6-3）所示的樣子。其中，N 是訓練樣本的個數，$logits_{\text{dec}} = W_{\text{proj}} \times h_{\text{dec}}$ ，h_{dec} 是大模型的最後一層 Transformer 的輸出，$logits_{\text{dec}}^{i}$ 是第 i 個分量。對於回歸問題，使用 MSE（Mean Squared Error）損失函式。

$$L_{\text{regression}} = \frac{1}{|N|}\sum_{i=1}^{N}\left(\text{logits}_{\text{dec}}^{i} - r^{i}\right)^{2} \tag{6-3}$$

- 分類問題。

如果將最終的推薦問題看成分類問題，那麼可以是 0-1 分類①，也可以是多分類②，這時目標函式可以是式（6-4）所示的樣子。這裡的 r^i 是第 i 個訓練樣本的真實 label，其他參數與上面的類似，這裡不再說明。對於分類問題，使用交叉熵（Cross Entropy）損失函式。

① 預測使用者是否點擊某個物品。

② 例如預測 10 個使用者可能舉出的分值，甚至會預測使用者會選擇某個物品，這時分類的數量就是物品的數量，這是一個超大規模的多分類問題。

$$L_{\text{cross\_entropy}} = -\sum_{i=1}^{N} r^i \ln\left(\text{logits}_{\text{dec}}^i\right) \tag{6-4}$$

從式（6-3）和式（6-4）可以看出，可以將大模型的參數固定不變，只最佳化投影矩陣 W_{proj} 對應的參數，相比大模型的參數量少了幾個量級，這時候的微調方法是一種參數有效的方法。

3. 準備微調的監督資料

監督資料指形如 (x, y) 的訓練樣本，x, y 一般是文字形式的，將使用者行為、使用者興趣偏好、候選集等資訊透過一定的範本組成指令輸入（Instruction Input），例如：

你需要預測使用者對電影的評分，評分從 1 分到 5 分。下面會給你使用者的評分歷史記錄，格式是：電影標題，類型，評分。

例子：

問題：

使用者的評分歷史是：

movie_1，喜劇，4

movie_2，愛情、科幻，5

……

現在使用者看了電影「movie_*k*, 動作」，你給該電影評分，只需要評分就可以，不需要做解釋。

答案：4

問題：

使用者的評分歷史是：

movie_6，倫理，5

movie_8，恐怖、科幻，3

……

現在使用者看了電影「movie_j, 愛情」，你給該電影評分，只需要評分就可以，不需要做解釋。

上面的 x 可以看作評分預測問題，也可以看作 1~5 的分類問題，基於上面的 x ，可行的訓練樣本的 label y 可以如下。

> 3

如果是前面提到的 < 文字，文字 > 訓練樣本，那麼 y 可能是一段文字。按照這裡提供的範例，基於推薦的訓練樣本資料可以建構這樣的樣本對。一般大模型具備很好的泛化能力，只需要幾十上百對訓練樣本就可以獲得比較好的效果。

4. 基於監督資料進行微調訓練

有了上面的準備工作，這一步就是具體的微調了。這時需要基於選擇的模型、目標函式、準備的訓練資料，進行適當的訓練程式開發，然後選擇適當的硬體平臺（GPU 配置）進行微調訓練，訓練過程中使用的技巧也與訓練大模型類似，這裡不再贅述。具體的微調實現方法在 6.3 節中詳細說明。

5. 測試與推斷

當模型微調好後，可以在測試集中測試一下模型的效果，一般的測試評估指標與常規的推薦系統類似，例如 RMSE、MAE、MAP、精準度、召回率、NDCG、ROC、AUC 等。這時通常還需要與一些傳統的模型（例如矩陣分解、協作過濾、深度學習、SASRec 等）進行對比，看看基於大模型的推薦模型是否表現出更好的效果。除此以外，資源使用情況、推斷的速度等也是在現實場景中需要考慮的重要指標。

如果大模型推薦系統的效果滿足業務要求，就可以將大模型推薦系統部署成推斷服務，供業務使用。線上上場景使用時，需要做 A/B 測試，並且要時刻關注線上指標的變化情況，關注大模型推薦系統是否穩定。

6.1.3 微調的方法

目前有兩種主流的大模型微調方法：指令微調和 RLHF（基於人類回饋的增強學習），本節重點講解 RLHF。RLHF 在某些推薦系統場景中會用到，6.2 節會舉出 RLHF 在應徵職務推薦中的應用案例。

基於 PPO 演算法的 RLHF 方法是將大模型與人類的價值觀（有用、真實、無害）對齊的主要方法，這也是 ChatGPT 採用的方法，圖 6-3 舉出了 RLHF 方法的三個步驟。步驟 1：監督微調（SFT）；步驟 2：訓練獎勵模型（RM）；步驟 3：使用 PPO 演算法針對獎勵模型最佳化策略。步驟 1、2 中人工標注的資料會用於訓練模型，在步驟 2 中，A、B、C、D 是基於模型生成的樣本，由標注人員進行排名。下面對這 3 個步驟進行詳細說明。

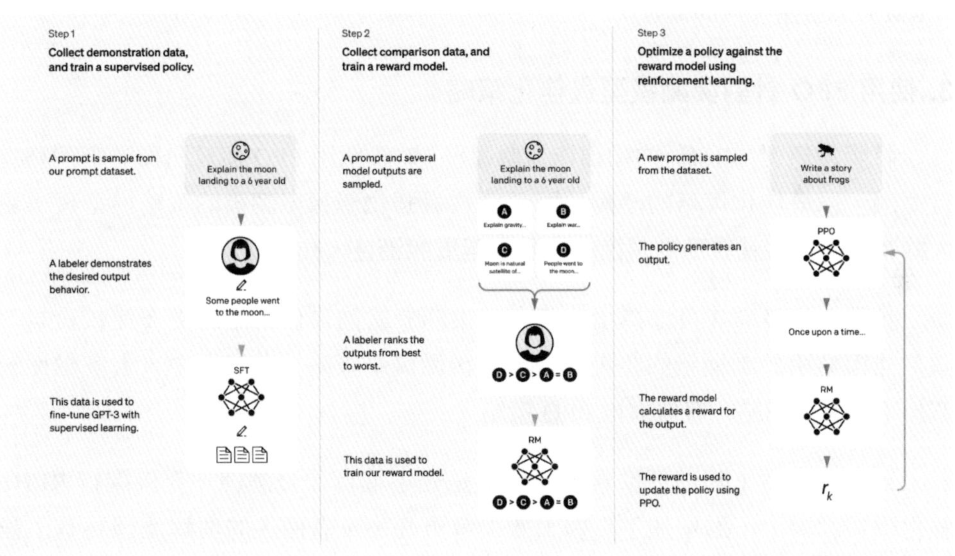

▲ 圖 6-3 RLHF 方法的三個步驟 (圖片來源：OpenAI)

1. 收集範例資料，訓練監督策略

在步驟 1 中，標注人員基於輸入大模型（例如 GPT-3）的提示詞和人類標準建構範例（Demonstration），也就是監督的輸入 - 輸出樣本對。一般來說，提示詞範例要多樣，要覆蓋到模型建構者希望解決的領域中的各種情況，從而增強模型的泛化能力。有了訓練的樣本對，再利用監督學習對需要使用 RLHF 演算法進行微調的大模型（例如 GPT-3）進行微調。

2. 收集比較資料，訓練獎勵模型

基於待對齊的模型（例如 GPT-3），收集模型多個輸出之間的待比較的資料集，然後標注人員對同一個提示詞的多個輸出進行排序，排序的依據依然是有幫助、真實性、無害性。最後，利用剛剛標注的排序資料訓練一個獎勵模型（RM）來預測人類的偏好。

3. 使用 PPO 針對獎勵模型最佳化策略

將獎勵模型（RM）的輸出作為獎勵[①]，然後使用 PPO 演算法對監督策略進行微調，以最佳化策略的獎勵，即希望透過強化學習微調後的策略模型獲得更大的獎勵值，這就表示微調後的策略模型更能滿足人類的需求。

步驟 2 和步驟 3 可以持續迭代，收集關於當前最佳策略的更多比較資料，該最佳策略用於訓練新的獎勵模型，然後微調新策略。在實踐中，大多數比較資料來自監督策略，有些來自 PPO 策略。

另外，步驟 1 其實可以省去，也就是先不進行監督訓練，直接進行 RLHF 強化學習訓練，不過增加了步驟 1 效果會更好，筆者個人的理解是因為有了監督微調，大模型可以更進一步地對齊人類需求，RLHF 進一步與人類的需求進行對齊。

① 純量值，例如在 0 ～ 1 之間，越靠近 1 表示模型的輸出越符合人類的需求。

6.1.4 微調的困難與挑戰

要獲得好的微調效果需要克服很多困難，這主要表現在以下 3 方面。

1. 選擇基底模型

目前比較流行的開放原始碼大模型是 LLaMA，在英文場景中效果比較好，如果推薦場景是中文的，那麼如何選擇基底大模型就是一個難題。雖然也有一些開放原始碼的選擇，但是它們用於推薦的效果還需要探索②。

2. 訓練樣本

需要多少訓練樣本、如何建構簡單學習的樣本範本、如何構造訓練樣本、選擇哪些目標函式都需要結合實際情況進行權衡，可能也需要進行大量的測試。我們在後面的案例中也會提到一些針對具體場景進行微調的方法。另外，你可以參考相關的論文，將論文提供的方法及參數作為參考的起始條件。

3. 選擇微調方法

前面提到了微調整個模型、基於 LoRA 微調、增加投影層微調 3 種方法，需要結合具體的推薦場景及模型選擇。這裡建議你採用基於 LoRA 微調、增加投影層微調等參數有效的方法，整體資源使用率更高，嘗試的成本和代價也更小。

本節的基礎知識非常重要，希望你極佳地掌握它們，也可以參考相關論文加深對基礎知識的理解和記憶。

6.2 案例講解

本節是對上一節知識的鞏固，希望本節的案例能夠幫助你更進一步地了解在具體場景下如何對大模型進行微調以適應推薦業務。

② 目前，關於大模型推薦系統的論文基本是基於英文資料的，中文大模型推薦系統相關的材料較少。

第一個案例建構了一個將大模型應用於推薦系統的微調框架 TALLRec[1]，該框架可以應用在各類推薦任務（例如電影推薦、圖書推薦等）中。第二個案例是基於網際網路應徵的具體業務問題，即在應徵場景中如何為求職者推送工作描述（Job Description，JD），這個案例創新性地利用 RLHF 框架對模型進行微調，以更加精準的方法為候選集進行 JD 推薦。

6.2.1 TALLRec 微調框架

TALLRec（Tuning framework for Aligning LLMs with Recommendation）是一個非常通用的利用大模型對齊推薦系統場景的微調框架。主要用於調整大模型的語言處理任務和推薦任務之間的巨大差距，希望透過微調大模型的方式對齊推薦系統，讓大模型具備更加優秀的推薦能力。下面重點講解 TALLRec 框架的核心架構和實現原理。

1. TALLRec 框架介紹

利用 TALLRec 框架進行微調主要有兩個步驟：第一步是將推薦資料集範本化為適合大模型微調的指令資料（Instruction Data）格式；第二步是採用 LoRA 等參數有效的微調方法進行微調。微調好的模型可以用於效果評估測試，如果效果及格就可以部署供業務使用了。

TALLRec 框架如圖 6-4 所示。左上角展示了一個序列推薦的例子，利用使用者的互動歷史預測他對下一個物品的興趣偏好。右上角展示了將該場景中的推薦資料組織為用於推薦微調（rec-tuning）的指令資料（Instruction Data）的方法。底部展示了 TALLRec 框架的流程，透過 LoRA 來提高 TALLRec 框架的效率。

2. 指令資料範本的建構

指令微調（Instruction Tuning）是大模型微調過程的重要組成部分，這是一種被廣泛使用的技術，透過微調多樣化的人工標注的指令和回饋來提升大模型的能力。微調過程會賦予模型強大的泛化能力，從而使其能夠熟練處理新任務，並確保其對新場景和新挑戰的適應性。一般來說，指令微調的範本建構有四個步驟，圖 6-5 是一個翻譯任務的指令微調中的微調資料的範例。

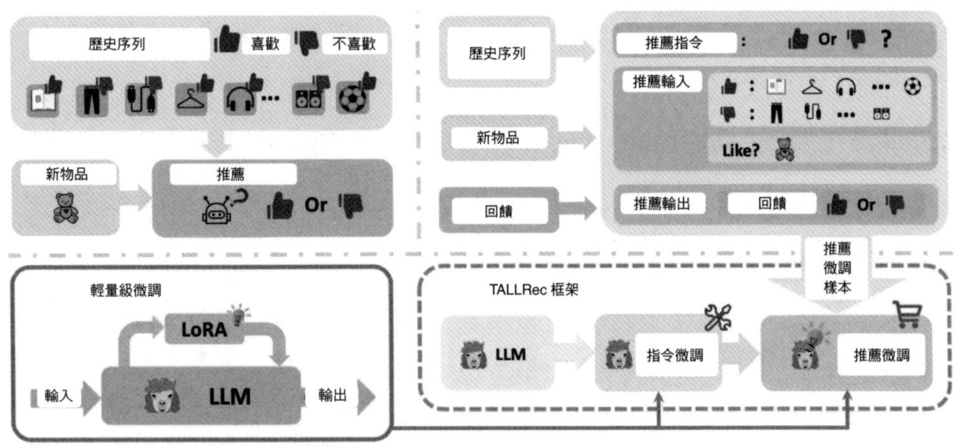

▲ 圖 6-4 TALLRec 框架

Instruction Input	
Task Instruction:	Translate from English to Chinese.
Task Input:	Who am I ?
Instruction Output	
Task Output:	我是誰？

▲ 圖 6-5 翻譯任務的指令微調中的微調資料的範例

步驟 1：辨識任務，並使用自然語言說清指令，以有效完成任務。

該指令可以包含任務的明確定義，以及該問題的具體解決方案。這些描述被稱為任務指令（Task Instruction）。圖 6-5 中為「將英文翻譯為中文」。

步驟 2：利用自然語言說明任務的輸入和輸出，即「Task Input」和「Task Output」。

圖 6-5 中的「Task Input」和「Task Output」分別是「Who am I ？」和「我是誰？」。

步驟 3：將「Task Instruction」和「Task Input」組裝成「Instruction Input」，將「Task Output」作為對應的「Instruction Output」。

每個用於微調的樣本都需要完成這一過程，這些樣本即灌入大模型進行微調的樣本對。

步驟 4：基於「Instruction Input」和「Instruction Output」樣本對微調大模型。

可以對模型的所有參數進行微調，也可以微調很少一部分參數，舉例來說，TALLRec 就是基於 LoRA 進行局部參數微調的。

針對推薦系統，上面的定義也是適用的，只不過將 Task 改為 Rec（Recommendation），結合上面的介紹和圖 6-4 右上角的範例，表 6-2 舉出一個推薦任務的 Instruction Input 和 Instruction Output 範例。

▼ 表 6-2 推薦任務的 Instruction Input 和 Instruction Output 範例

指令類型	指令內容	說明
Instruction Input	Rec Instruction：預測使用者是否喜歡物品 Rec Input：該使用者過去的興趣歷史是 {(item_1, 1), (item_2, 0), ..., (item_k, 1)}，那麼使用者是否會喜歡新的 item_j 呢？	這裡的 1 和 0 分別代表使用者喜歡和不喜歡 item_1、item_2 等是待推薦物品的標題等文字資訊
Instruction Output	Rec Output：(item_j, 1)	

3. 二次微調原理

TALLRec 框架的底層大模型是開放原始碼的 LLaMA-7B，對其進行了兩次微調：指令微調（Instruction Tuning）和推薦微調（rec-tuning）。

步驟 1：指令微調。

指令微調使用一般的監督微調資料，目的是讓大模型可以更進一步地理解人類的輸入指令，按照人類的輸入輸出進行回饋，讓大模型具備更好的泛化能力，這裡用到的微調資料是參考文獻 [3] 中提供的 self-Instruction 資料，用到的目標函式是 6.1 節式（6-1）中所示的函式，對整個模型進行微調。

步驟 2：推薦微調。

這一步驟是步驟 1 之後，利用推薦樣本進行進一步微調，讓大模型調配推薦業務場景，獲得更好的推薦效果。我們可以基於表 6-2 建構 Rec Input——User Preference: item1, $item_4$, ..., $item_n$. User Unpreference: $item_2$, $item_3$, ..., $item_{n-1}$. Whether the user will enjoy the target movie/book: $item_{n+1}$①和 Rec Output——Yes./No.（喜歡 / 不喜歡），然後根據表 6-2 中的 Rec Input 和 Rec Output 建構出 Instruction Input 和 Instruction Output。這樣就可以對大模型進行微調了，在推薦這一步，我們利用 LoRA 進行參數有效的微調，以減少微調的參數量，具體的目標函式是式（6-2）中所示的函式。

由於微調過程使用了 LoRA，參數量至多只有原來的千分之一，所以微調非常高效，論文作者是在一台輝達 RTX 3090（24GB）GPU 上進行微調的。

經過步驟 1 微調後的大模型已經具備比較好的泛化能力了，步驟 2 只需要不到 100 個微調樣本就可以獲得非常好的效果。同時，該模型具備很好的泛化能力，舉例來說，基於電影推薦場景進行微調後可以直接應用於圖書推薦，並且效果很不錯。

6.2.2 GIRL：基於人類回饋的微調框架

GIRL[2]（Generative job Recommendation paradigm based on LLM） 框 架是一種生成式推薦框架，利用大模型生成個性化的 JD，然後基於求職者簡歷（CV）、個性化 JD，以及 RLHF 微調大模型為求職者進行 JD 推薦，筆者認這個實現方案非常好，將 RLHF 用得恰到好處。下面對 GIRL 框架進行詳細介紹。

1. 三種 JD 推薦範式

傳統的 JD 推薦將求職者 CV 和函式庫中所有有效的 JD 進行匹配，按照 JD 和求職者 CV 的匹配得分進行排序，將最適合求職者的 JD 推薦給求職者（當然還要考慮地域、薪資等因素），這就是圖 6-6 中常規的工作推薦範式，這裡不再贅述。

① 這裡將使用者喜歡的和不喜歡的物品分開列出了。

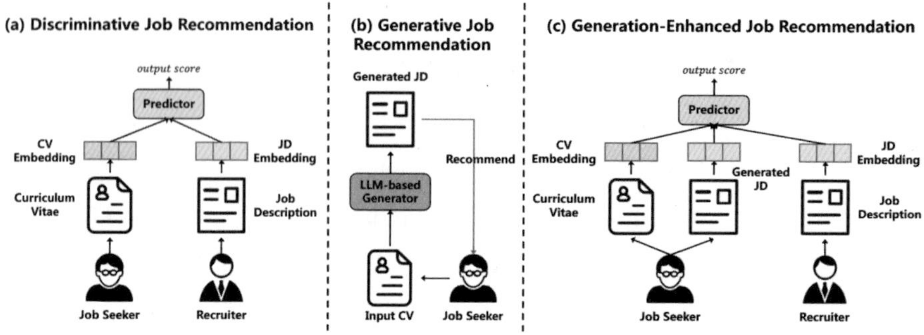

▲ 圖 6-6 3 種不同的 JD 推薦範式

GIRL 提出了兩種新的 JD 推薦範式：生成式工作推薦和生成增強的工作推薦，如圖 6-6 所示。生成式工作推薦利用大模型為求職者生成個性化的 JD，生成的 JD 是最匹配求職者興趣的；生成增強的工作推薦借助生成的 JD 為求職者進行個性化 JD 推薦，推薦的工作更匹配。這兩種范式生成的 JD 至少有兩個價值：一是為求職者提供就業指導，並且具備可解釋性（自然語言形式）；二是作為資料供推薦系統進行更加精準的推薦（生成增強的工作推薦）。

2. 基於 RLHF 的 JD 生成框架

透過大模型為求職者生成個性化的 JD 在 GIRL 框架中是透過 RLHF 演算法實現的。GIRL 框架是分為三步實現的，如圖 6-7 所示。

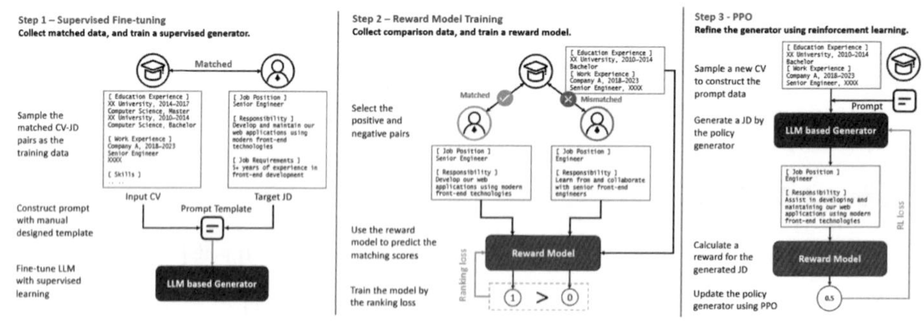

▲ 圖 6-7 GIRL 框架

步驟 1：監督微調

這一步的目的是透過微調樣本來指導大模型基於使用者的 CV 生成合適的 JD。微調樣本是匹配的（CV，JD）對，見圖 6-6(a)，應徵者投了這個 JD，發佈 JD 的 HR 也約了求職者面試，也就是求職者和應徵企業互相滿意的（CV，JD）對。

為了讓大模型進行簡單微調，需要建構監督微調的提示詞範本，這非常容易，因為 CV 和 JD 都是自然語言的形式，圖 6-8 就是一個可行的範本，其中，Human、Assistant 是角色；Please generate ... following candidate: 是指令（Instruction），指導大模型要做什麼；Basic information ... and Netty. 就是使用者的 CV，即 Task Input；而 Job title: ... is preferred. 就是 JD，即 Task Output（在模型推斷時，這部分留白就可以了）。

針對每一個匹配的（CV，JD）對，按照圖 6-8 建構監督樣本，然後採用監督微調的方法訓練大模型 $(\mathcal{G}: C \rightarrow J')$ ，大模型的作用是基於給定的 CV 生成 JD，具體的目標函式參見 6.1.2 節，這裡不再贅述。

步驟 2：獎勵模型訓練

訓練獎勵函式 $(\mathcal{U}: (C, J) \rightarrow \mathbf{R})$ 也需要建構樣本對。在 6.1.3 節，利用大模型對提示詞生成多個答案，然後根據生成答案是否滿足人類目標排序，再訓練獎勵函式，見圖 6-6(b)。GIRL 採用比較巧妙的方式實現類似的功能。具體是怎麼做的呢？

從使用者行為資料（包括求職者的行為和 HR／獵頭的行為）中收集匹配和不匹配的（CV，JD）對[①]，將與 CV 匹配的工作集合記為 J^+ ，與 CV 不匹配的工作集合記為 J^- ，透過最佳化下面的成對排序（Pairwise Ranking）損失函式來訓練獎勵模型（ σ 是 Sigmoid 啟動函式，C 即 CV）。

① 匹配的前面提到了，不匹配的意思是求職者投遞了簡歷，但是 HR／獵頭覺得不合適。

Human:
Please generate a suitable job description based on the curriculum vitae of the following candidate:
Basic information: The candidate is a 27-years-old female with a bachelor's degree, majoring in Software Engineering.
personal profile:
1. She has research experience in the field of data processing and computation, and development experience with components from Elasticsearch, HBase, and the Flink ecosystem.
2. She possesses strong abilities to read and understand source code, having reviewed core source code from JDK8, Spring, Spring Boot, and Netty.

Assistant:
Job title: Java Developer
Salary range: 25k-40k* 15
Responsibilities:
1. Support the recommendation product and algorithm departments, rapidly iterate daily needs, complete effect analysis and continuous improvements.
2. Track industry developments, timely introduction of external advanced experiences.

Qualifications:
1. Proficient in Java programming basics, with excellent abilities and experience in Java IO and multithreading programming.
2. In-depth understanding of JVM, JVM tuning experience, and experience with distributed systems and concurrency scenarios are preferred.
3. Proficient in applying mainstream development frameworks and open-source systems such as Spring Boot, MyBatis, MySQL, Redis, ES, Kafka, etc.
4. Good stress resistance, communication, learning, collaboration skills, and a strong sense of responsibility.
5. Prior experience in recommendation/search engineering development in Internet companies is preferred.

▲ 圖 6-8 監督訓練的提示詞範本

$$\mathcal{L}_{\mathrm{rmt}} = \ln \sigma\left(\mathcal{U}(C, J^{+}) - \mathcal{U}(C, J^{-})\right)$$

這種方法使獎勵模型能夠根據 HR 的回饋來捕捉市場偏好。此外，我們可以使用獎勵模型來預測求職者與職務描述之間的匹配度，從而提前驗證推薦演算法的適用性。

步驟 3：PPO

這一步驟基於 PPO 演算法，對齊大模型與利用獎勵模型衡量的 HR 的偏好，見圖 6-6 的 (c)，這讓大模型不僅考慮了求職者的需求，也考慮了這些職務實際的市場受歡迎度[①]。

① 你可以想一想，為什麼這麼說？

具體的訓練過程是：首先收集一批步驟 1、2 中沒有使用過的 CV②，基於步驟 1 的大模型($\mathcal{G}$: C → J') 生成對應的 JD 集合，然後利用步驟 2 中的$\mathcal{U}$: $(C,J) \to \mathbf{R}$ 生成（CV，JD）對對應的獎勵值，最後透過 PPO 方法更新$\mathcal{G}$和$\mathcal{U}$的參數 [2]。

3. 生成加強的 JD 推薦方案

基於 RLHF 對 JD 生成的大模型($\mathcal{G}$: C → J') 進行微調後，可以生成更匹配使用者期望和市場需求的 JD，那麼自然可以想到，我們可以將生成的 JD 作為輔助資訊，把($\mathcal{G}$: C → J') 看作一個特徵取出大模型，提升 JD 推薦的效果，見圖 6-6(c)，下面詳細說明。

假設某個使用者的簡歷是C，待推薦的 JD 是J，J'是大模型$\mathcal{G}$利用簡歷C生成的 JD，推薦系統要計算C與J的匹配度得分，我們可以採用以下方式計算。

我們可以利用編碼器對C進行編碼：$C = \text{Encoder}(C)$③。同樣的方法可以對J'、J分別進行編碼：$j' = Encoder(J')$、$j = Encoder(J)$。然後利用 MLP 神經網路計算簡歷C與 JD J的匹配得分，公式如下。

$$\text{score} = \text{MLP}([c; j; j']) \tag{6-5}$$

計算出了使用者的簡歷C與任何一個 JD J的匹配得分，向使用者推薦的過程是：先召回該使用者的所有簡歷，透過式（6-5）計算每份簡歷與使用者的簡歷C的匹配得分，然後按照得分降冪排列，選擇 Top N 作為最終給使用者的 JD 推薦。

② 你可以考慮一下為什麼要用沒有使用過的 CV ？

③ Encoder 可以是一個複雜的深度學習編碼器，例如 BERT。

6.3 基於 MIND 資料集實現微調

本節基於 MIND 資料集實現一個微調推薦系統，採用參數有效的微調方案——LoRA 微調，另外，採用 TALLRec 推薦系統架構實現微調推薦系統的架構。

6.3.1 微調資料準備

我們基於 MIND 資料集建構微調資料[①]。LoRA 微調需要的資料格式如下。

```
{"instruction": "Given the user's preference and unpreference, identify whether the
user will like the target movie by answering \"Yes.\" or \"No.\".",
"input": "User Preference: \"Opinion: Colin Kaepernick is about to get what he deserves:
a chance\"\nUser Unpreference:
\"Browns apologize to Mason Rudolph, call Myles Garrett's actions 'unacceptable'\",
\"I've been writing about tiny homes for a year and finally spent 2 nights in a 300-
foot home to see what it's all about here's how it went\",\"The Kardashians Face
Backlash Over 'Insensitive' Family Food Fight in KUWTK Clip\",
  \"THEN AND NOW: What all your favorite '90s stars are doing today\",\"Report: Police
investigating woman's death after Redskins' player Montae Nicholson took her to
hospital\",\"U.S. Troops Will Die If They Remain in Syria,
  Bashar Al-Assad Warns\",\"3 Indiana judges suspended after a night of drinking
turned into a White Castle brawl\",
  \"Cows swept away by Hurricane Dorian found alive   but how?\",\"Surviving Santa
Clarita school shooting victims on road to recovery: Latest\",\"The Unlikely Star of
My Family's Thanksgiving Table\",\"Meghan Markle and Hillary Clinton Secretly Spent
the Afternoon Together at Frogmore Cottage\",\"Former North Carolina State, NBA player
Anthony Grundy    dies in stabbing, police say\",\"85 Thanksgiving Recipes You Can
Make Ahead\",\"Survivor Contestants Missy Byrd and Elizabeth Beisel Apologize For
Their Actions\",\"Pete Davidson, Kaia Gerber Are Dating, Trying to Stay 'Low Profile'\"
,\"There's a place in the US where its been over 80 degrees since March\",\"Taylor
Swift Rep Hits Back at Big Machine, Claims She's Actually Owed $7.9 Million in Unpaid
Royalties\",\"The most talked about movie moments of the 2010s\",
   \"Belichick mocks social media in comments on Garrett incident\",\"13 Reasons
Why's Christian Navarro Slams Disney for Casting 'the White Guy' in The Little
```

① 基於 MIND small 的 Validation 資料微調，這個資料集雖然相對較小，微調也很慢。

```
Mermaid\"\nWhether the user will like the targe news \"66 Cool Tech Gifts
    Anyone Would Be Thrilled to Receive\"?",
    "output": "No."},
...
]
```

將使用者行為資料和新聞資料轉為上述格式，因此需要讀取 news.tsv 和 behaviors.tsv 兩個檔案。news.tsv 檔案中有 News ID 和對應的新聞標題，而 behaviors.tsv 檔案中有一個 Impression News 欄位，裡面包含使用者點擊和沒有點擊的新聞，剛好可以用於建構微調資料：Impression News 欄位最後一個 News ID 用於預測使用者是否喜歡該新聞，前面的欄位可以作為使用者過去的行為偏好（參見上面範例中的 input）。利用這兩個檔案，我們可以建構具體的微調資料集，下面是詳細的程式實現②。

```
import csv
import json
from enum import Enum

class Action(Enum):
    YES = "Yes."
    NO = "No."

instruction = ("Given the user's preference and unpreference, identify whether the
user will like the target movie by "
               "answering \"Yes.\" or \"No.\".")
news_dict = {}  # 從 news.tsv 獲取每個新聞 ID 到標題的映射字典
with open('../data/mind/news.tsv', 'r') as file:
    reader = csv.reader(file, delimiter='\t')
    for row in reader:
        news_id = row[0]
        news_title = row[3]
        news_dict[news_id] = news_title
data_list = []  # 利用 behaviors.tsv 檔案獲取使用者喜歡和不喜歡的新聞
with open('../data/mind/behaviors.tsv', 'r') as file:
    reader = csv.reader(file, delimiter='\t')
```

② 程式：src/basic_skills/finetune-llm/data-process/generate_json_data.py。

```
    for row in reader:
        impression = row[4]
        impre_list = impression.split(" ")
        if len(impre_list) >= 5:  # 使用者至少要有 5 個曝光歷史
            preference = []
            unpreference = []
            for impre in impre_list[:-1]:  # 將前面的新聞作為微調資料，最後一個新聞作為預測
                [impre_id, action] = impre.split("-")
                title = news_dict[impre_id]
                if int(action) == 1:
                    preference.append("\""+title+"\"")
                else:
                    unpreference.append("\""+title+"\"")
            input = "User Preference: " + ','.join(preference) + "\n" + "User 
Unpreference: " + ','.join(unpreference)
            [impre_id, action] = impre_list[-1].split("-")
            output = Action.YES.value if int(action) == 1 else Action.NO.value
            input = input + "\n" + "Whether the user will like the targe news " + "\"" 
+ news_dict[impre_id] + "\"?"
            res_dic = {
                "instruction": instruction,
                "input": input,
                "output": output
            }
            data_list.append(res_dic)
res = json.dumps(data_list)
user_sequence_path = "../data/mind/train.json"  # 將生成的微調資料儲存起來
with open(user_sequence_path, 'a') as file:
    file.write(res)
```

上述程式會在 data 目錄下生成一個 JSON 微調資料 train.json。下一節我們利用這個微調資料對推薦模型進行微調。

6.3.2 模型微調

本節利用 6.3.1 節建構的微調資料，基於 LoRA 對 chinese-alpaca-2-7B 模型[①]進行微調，微調好的模型具備個性化推薦能力。本節提供 4 個微調實現方案，其中重點介紹基於 alpaca-lora 框架的微調。

1. 基於 alpaca-lora 框架的微調

參考文獻 [1] 提供了一種方便的微調方法，本節採用參考文獻 [4] 中的程式實現微調②。

```
import os
import sys
from typing import List
import fire
import torch
from datasets import load_dataset
from peft import (
    LoraConfig,
    get_peft_model,
    get_peft_model_state_dict,
    prepare_model_for_int8_training,
    set_peft_model_state_dict,
)
from transformers import LlamaForCausalLM, LlamaTokenizer, Trainer,
DataCollatorForSeq2Seq, TrainingArguments
from utils.prompter import Prompter

def train(
    # model/data params
    base_model: str = "",  # the only required argument
    data_path: str = "./data/mind/train.json",
    output_dir: str = "./lora-alpaca",
    # training hyperparams
    batch_size: int = 128,
    micro_batch_size: int = 4,
    num_epochs: int = 3,
    learning_rate: float = 3e-4,
    cutoff_len: int = 256,
    val_set_size: int = 2000,
    # lora hyperparams
    lora_r: int = 8,
```

① 這是一個利用中文語料對 Stanford' s Alpaca 模型進行微調得到的模型。

② 程式：/src/basic_skills/finetune-llm/finetune_tallrec.py。

```
    lora_alpha: int = 16,
    lora_dropout: float = 0.05,
    lora_target_modules: List[str] = ["q_proj", "v_proj"],
    # llm hyperparams
    train_on_inputs: bool = True,  # if False, masks out inputs in loss
    add_eos_token: bool = False,
    group_by_length: bool = False,  # faster, but produces an odd training loss curve
    # wandb params
    wandb_project: str = "",
    wandb_run_name: str = "",
    wandb_watch: str = "",  # options: false | gradients | all
    wandb_log_model: str = "",  # options: false | true
    resume_from_checkpoint: str = None,  # either training checkpoint or final adapter
    prompt_template_name: str = "alpaca",  # The prompt template to use, will default
to alpaca.
):
    assert (
        base_model
    ), "Please specify a --base_model, e.g. --base_model='huggyllama/llama-7b'"
    gradient_accumulation_steps = batch_size // micro_batch_size
    prompter = Prompter(prompt_template_name)
    device_map = "auto"
    world_size = int(os.environ.get("WORLD_SIZE", 1))
    ddp = world_size != 1
    if ddp:
        device_map = {"": int(os.environ.get("LOCAL_RANK") or 0)}
        gradient_accumulation_steps = gradient_accumulation_steps // world_size
    # Check if parameter passed or if set within environ
    use_wandb = len(wandb_project) > 0 or (
        "WANDB_PROJECT" in os.environ and len(os.environ["WANDB_PROJECT"]) > 0
    )
    # Only overwrite environ if wandb param passed
    if len(wandb_project) > 0:
        os.environ["WANDB_PROJECT"] = wandb_project
    if len(wandb_watch) > 0:
        os.environ["WANDB_WATCH"] = wandb_watch
    if len(wandb_log_model) > 0:
        os.environ["WANDB_LOG_MODEL"] = wandb_log_model
    model = LlamaForCausalLM.from_pretrained(
```

```
    base_model,
    torch_dtype=torch.float16,
    device_map=device_map,
)
tokenizer = LlamaTokenizer.from_pretrained(base_model)
tokenizer.pad_token_id = (
    0  # unk. we want this to be different from the eos token
)
tokenizer.padding_side = "left"  # Allow batched inference
def tokenize(prompt, add_eos_token=True):
    # there's probably a way to do this with the tokenizer settings
    # but again, gotta move fast
    result = tokenizer(
        prompt,
        truncation=True,
        max_length=cutoff_len,
        padding=False,
        return_tensors=None,
    )
    if (
        result["input_ids"][-1] != tokenizer.eos_token_id
        and len(result["input_ids"]) < cutoff_len
        and add_eos_token
    ):
        result["input_ids"].append(tokenizer.eos_token_id)
        result["attention_mask"].append(1)
    result["labels"] = result["input_ids"].copy()
    return result
def generate_and_tokenize_prompt(data_point):
    full_prompt = prompter.generate_prompt(
        data_point["instruction"],
        data_point["input"],
        data_point["output"],
    )
    tokenized_full_prompt = tokenize(full_prompt)
    if not train_on_inputs:
        user_prompt = prompter.generate_prompt(
            data_point["instruction"], data_point["input"]
        )
```

```
            tokenized_user_prompt = tokenize(
                user_prompt, add_eos_token=add_eos_token
            )
            user_prompt_len = len(tokenized_user_prompt["input_ids"])
            if add_eos_token:
                user_prompt_len -= 1
            tokenized_full_prompt["labels"] = [
                -100
            ] * user_prompt_len + tokenized_full_prompt["labels"][
                user_prompt_len:
            ]  # could be sped up, probably
        return tokenized_full_prompt
    model = prepare_model_for_int8_training(model)
    config = LoraConfig(
        r=lora_r,
        lora_alpha=lora_alpha,
        target_modules=lora_target_modules,
        lora_dropout=lora_dropout,
        bias="none",
        task_type="CAUSAL_LM",
    )
    model = get_peft_model(model, config)
if data_path.endswith(".json") or data_path.endswith(".jsonl"):
# 資料是 JSON 或 JSONL 格式的
        data = load_dataset("json", data_files=data_path)
    else:
        data = load_dataset(data_path)
    if resume_from_checkpoint:
        # Check the available weights and load them
        checkpoint_name = os.path.join(
            resume_from_checkpoint, "pytorch_model.bin"
        )  # Full checkpoint
        if not os.path.exists(checkpoint_name):
            checkpoint_name = os.path.join(
                resume_from_checkpoint, "adapter_model.bin"
            )  # only LoRA model - LoRA config above has to fit
            resume_from_checkpoint = (
                False  # So the trainer won't try loading its state
            )
```

```
        # The two files above have a different name depending on how they were saved, but
# are actually the same.
        if os.path.exists(checkpoint_name):
            print(f"Restarting from {checkpoint_name}")
            adapters_weights = torch.load(checkpoint_name)
            set_peft_model_state_dict(model, adapters_weights)
        else:
            print(f"Checkpoint {checkpoint_name} not found")
model.print_trainable_parameters()  # Be more transparent about the % of trainable
# params.
    if val_set_size > 0:
        train_val = data["train"].train_test_split(
            test_size=val_set_size, shuffle=True, seed=42
        )
        train_data = (
            train_val["train"].shuffle().map(generate_and_tokenize_prompt)
        )
        val_data = (
            train_val["test"].shuffle().map(generate_and_tokenize_prompt)
        )
    else:
        train_data = data["train"].shuffle().map(generate_and_tokenize_prompt)
        val_data = None
    if not ddp and torch.cuda.device_count() > 1:
        # keeps Trainer from trying its own DataParallelism when more than 1 gpu is
available
        model.is_parallelizable = True
        model.model_parallel = True
    trainer = Trainer(
        model=model,
        train_dataset=train_data,
        eval_dataset=val_data,
        args=TrainingArguments(
            per_device_train_batch_size=micro_batch_size,
            gradient_accumulation_steps=gradient_accumulation_steps,
            warmup_steps=100,
            num_train_epochs=num_epochs,
            learning_rate=learning_rate,
            logging_steps=10,
```

```
            optim="adamw_torch",
            evaluation_strategy="steps" if val_set_size > 0 else "no",
            save_strategy="steps",
            eval_steps=200 if val_set_size > 0 else None,
            save_steps=200,
            output_dir=output_dir,
            save_total_limit=3,
            load_best_model_at_end=True if val_set_size > 0 else False,
            ddp_find_unused_parameters=False if ddp else None,
            group_by_length=group_by_length,
            report_to="wandb" if use_wandb else None,
            run_name=wandb_run_name if use_wandb else None,
        ),
        data_collator=DataCollatorForSeq2Seq(
            tokenizer, pad_to_multiple_of=8, return_tensors="pt", padding=True
        ),
    )
    model.config.use_cache = False

    trainer.train(resume_from_checkpoint=resume_from_checkpoint)
    model.save_pretrained(output_dir)

if __name__ == "__main__":
    fire.Fire(train)
```

上述程式首先匯入基底模型，然後基於 LoraConfig 建構 peft 模型，最後利用 Transformers 函式庫中的 Trainer 類別進行微調，當模型微調好後進行儲存。下面的指令稿可以對模型進行微調。

```
python finetune_tallrec.py \
    --base_model '/Users/liuqiang/Desktop/code/llm/models/chinese-alpaca-2-7b' \
# 基底模型的儲存路徑
    --data_path '/Users/liuqiang/Desktop/code/llm4rec/llm4rec_abc/src/finetune-llm/
data/mind/train.json' \ # 微調資料路徑
    --output_dir './lora-weights' \ # 微調好的 LoRA 參數的儲存路徑
    --batch_size 128 \ # 批大小
    --micro_batch_size 4 \ # 微批大小
    --num_epochs 3 \ #epochs 數
    --learning_rate 1e-4 \ # 學習率
```

```
    --cutoff_len 512 \ # cutoff 長度
    --val_set_size 2000 \ # 驗證資料筆數
    --lora_r 8 \ # LoRA 中的秩（rank），它決定了低秩矩陣的大小
    --lora_alpha 16 \ # LoRA 適應的學習率縮放因數。這個參數影響了低秩矩陣的更新速度
    --lora_dropout 0.05 \ # Dropout 的比例
    --lora_target_modules '[q_proj,v_proj]' \ # 對基底模型中的哪些模組進行微調
    --train_on_inputs \ #if False, masks out inputs in loss
    --group_by_length #是否將微調資料集中長度大致相同的樣本分在一組，以最大限度地減少所應用
的填充並 # 提高效率。僅在應用動態填充時有用
```

上面的訓練指令稿速度是非常慢的，在筆者的電腦上需要訓練 4 天左右才能完成。你可以適當調整參數①，加快微調速度。

2. 其他微調方法

下面介紹 3 個利用其他框架進行微調的方法，有興趣的讀者可以詳細閱讀原始程式。

- 基於蘋果的 MLX 框架進行微調。

MLX 是蘋果開放原始碼的基於 mac 系列晶片的大模型框架，在 mlx-examples（在 GitHub 上搜尋 mlx-examples）專案中的 lora 目錄下，有相關的微調參考程式實現，下面是一個參考程式範例。

```
python lora.py --model /Users/liuqiang/Desktop/code/llm/models/Yi-34B-Chat \ # 基底模型
         --train \ # 進行微調
         --iters 1000 \ # 微調迭代次數
         --adapter-file ./adapters.npz \ # 微調後的輸出檔案，採用 .npz 格式儲存
         --data data/ # 微調對應的資料
```

- 基於 LLaMA-Factory 框架進行微調。

LLaMA-Factory（在 GitHub 上搜尋 LLaMA-Factory）是一個開放原始碼好用的微調框架，目前開放原始碼的主流模型都支援用 LLaMA-Factory 進行微調，它支援 LoRA、QLoRA 等微調方法，支援預訓練、監督微調、PPO 等範式，如圖 6-9 所示。

① 思考一下，可以調整哪些參數？

方法	全部參數	部分參數	LoRA	QLoRA
預訓練	☑	☑	☑	☑
監督微調	☑	☑	☑	☑
獎勵模型	☑	☑	☑	☑
PPO 訓練	☑	☑	☑	☑
DPO 訓練	☑	☑	☑	☑

▲ 圖 6-9 LLaMA-Factory 支援的微調類型

LLaMA-Factory 的好用性非常好，還提供微調介面，便於操作。圖 6-10 就是 LLaMA-Factory 的微調介面，具體的參數可以透過填寫或下拉的方式補充完善，然後就可以直接進行微調了。

如果不採用 Web UI 方式進行微調，還可以直接利用指令稿進行微調，下面是一個可行的微調指令稿，供你參考。

```
CUDA_VISIBLE_DEVICES=0 python src/train_bash.py \
--stage sft \ # 監督微調
--do_train \ # 進行微調
--model_name_or_path /Users/liuqiang/Desktop/code/llm/models/chinese-alpaca-2-7b \
# 基底模型
--dataset movie_train \ # 微調的資料集
--template default \ # 提示詞範本
--finetuning_type lora \ # 微調方法
--lora_target q_proj,v_proj \ # 對基底模型中的哪些模組進行微調
--output_dir ./lora-alpaca \ # 微調後的儲存目錄
--overwrite_cache \ # 是否重寫快取
--per_device_train_batch_size 4 \ # 每個 device 的 batch size
--gradient_accumulation_steps 4 \ # 多少步開始累積梯度
--lr_scheduler_type cosine \
--logging_steps 10 \ # 間隔多少步輸出日誌
--save_steps 1000 \  #間隔多少步儲存中間過程
--learning_rate 5e-5 \ # 學習率
--num_train_epochs 3.0 \ # epochs 數
--plot_loss # 列印出損失
```

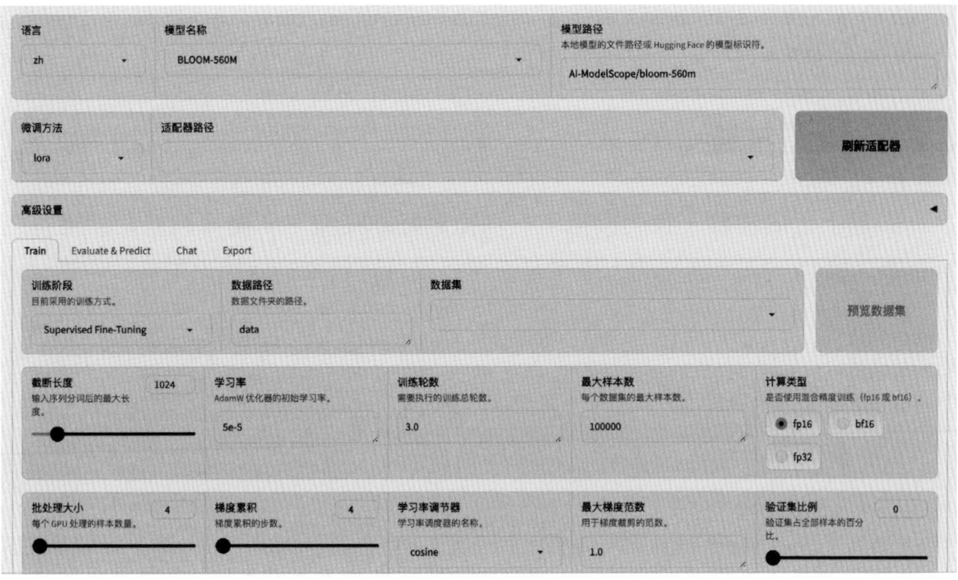

▲ 圖 6-10 LLaMA-Factory 的微調介面 (編按：本圖例為簡體中文介面)

- 基於 Transformers 函式庫的 run_clm.py 進行微調。

Transformers 函式庫中也提供了進行微調的程式案例 run_clm.py，你可以利用該方法進行微調，下面是一個具體的微調案例。

```
PYTORCH_MPS_HIGH_WATERMARK_RATIO=0.0 python run_clm.py \
    --model_name_or_path /Users/liuqiang/Desktop/code/llm/models/chinese-alpaca-2-7b \
    # 基底模型
    --train_file /Users/liuqiang/Desktop/code/llm4rec/TALLRec/data/movie/train.json \
    # 微調的資料
    --validation_file
/Users/liuqiang/Desktop/code/llm4rec/TALLRec/data/movie/valid.json \ # 驗證資料
    --per_device_train_batch_size 8 \ # 每個 device 的訓練集 batch size
    --per_device_eval_batch_size 4 \ # 每個 device 的驗證集 batch size
    --do_train \ # 是否需要進行訓練
    --do_eval \ # 是否需要進行評估
    --optim adamw_torch \ # 最佳化方法
    --learning_rate 5e-5 \ # 學習率
    --num_train_epochs 3.0 \ # epochs 數量
    --low_cpu_mem_usage true \ # 是否採用低 CPU、記憶體進行微調
```

```
--output_dir /Users/liuqiang/Desktop/code/llm4rec/TALLRec/lora-alpaca-transformers
# 微調好的模型輸出路徑
```

6.3.3 模型推斷

透過上面的步驟，我們獲得了一個微調好的模型，下面我們基於該模型進行推斷，檢驗一下微調後的大模型是否可以進行更精準的個性化推薦①。

```
import os
import sys
import fire
import gradio as gr
import torch
import transformers
from peft import PeftModel
from transformers import GenerationConfig, LlamaForCausa大模型 , LlamaTokenizer
from utils.callbacks import Iteratorize, Stream
from utils.prompter import Prompter

def main(
    load_8bit: bool = False,
    base_model: str = "",
    lora_weights: str = "tloen/alpaca-lora-7b",
    prompt_template: str = "",  # The prompt template to use, will default to alpaca.
    server_name: str = "0.0.0.0",  # Allows to listen on all interfaces by providing
'0.
    share_gradio: bool = False,
):
    base_model = base_model or os.environ.get("BASE_MODEL", "")
    assert (
        base_model
    ), "Please specify a --base_model, e.g. --base_model='huggyllama/llama-7b'"
    prompter = Prompter(prompt_template)
    tokenizer = LlamaTokenizer.from_pretrained(base_model)
    device_map = "mps"
    model = LlamaForCausalLM.from_pretrained(
```

① 程式：/src/basic_skills/finetune-llm/infer_tallrec.py。

```
        base_model,
        device_map=device_map,
        torch_dtype=torch.float16,
    )
    model = PeftModel.from_pretrained(
        model,
        lora_weights,
        device_map=device_map,
        torch_dtype=torch.float16,
    )
    # unwind broken decapoda-research config
    model.config.pad_token_id = tokenizer.pad_token_id = 0  # unk
    model.config.bos_token_id = 1
    model.config.eos_token_id = 2
    if not load_8bit:
        model.half()  # seems to fix bugs for some users.
    model.eval()
    if torch.__version__ >= "2" and sys.platform != "win32":
        model = torch.compile(model)
    def evaluate(
        instruction,
        input=None,
        temperature=0.1,
        top_p=0.75,
        top_k=40,
        num_beams=4,
        max_new_tokens=128,
        stream_output=False,
        **kwargs,
    ):
        prompt = prompter.generate_prompt(instruction, input)
        inputs = tokenizer(prompt, return_tensors="pt")
        input_ids = inputs["input_ids"].to(device_map)
        generation_config = GenerationConfig(
            temperature=temperature,
            top_p=top_p,
            top_k=top_k,
            num_beams=num_beams,
            **kwargs,
```

```
        )
        generate_params = {
            "input_ids": input_ids,
            "generation_config": generation_config,
            "return_dict_in_generate": True,
            "output_scores": True,
            "max_new_tokens": max_new_tokens,
        }
        if stream_output:
            # Stream the reply 1 token at a time.
            # This is based on the trick of using 'stopping_criteria' to create an
iterator,
            # from https://github.com/oobabooga/text-generation-webui/
blob/ad37f396fc8bcbab90e11ecf17c56c97bfbd4a9c/modules/text_generation.py#L216-L243.
            def generate_with_callback(callback=None, **kwargs):
                kwargs.setdefault(
                    "stopping_criteria", transformers.StoppingCriteriaList()
                )
                kwargs["stopping_criteria"].append(
                    Stream(callback_func=callback)
                )
                with torch.no_grad():
                    model.generate(**kwargs)
            def generate_with_streaming(**kwargs):
                return Iteratorize(
                    generate_with_callback, kwargs, callback=None
                )
            with generate_with_streaming(**generate_params) as generator:
                for output in generator:
                    # new_tokens = len(output) - len(input_ids[0])
                    decoded_output = tokenizer.decode(output)
                    if output[-1] in [tokenizer.eos_token_id]:
                        break
                    yield prompter.get_response(decoded_output)
            return  # early return for stream_output
        # Without streaming
        with torch.no_grad():
            generation_output = model.generate(
                input_ids=input_ids,
                generation_config=generation_config,
```

```
            return_dict_in_generate=True,
            output_scores=True,
            max_new_tokens=max_new_tokens,
        )
    s = generation_output.sequences[0]
    output = tokenizer.decode(s)
    yield prompter.get_response(output)
gr.Interface(
    fn=evaluate,
    inputs=[
        gr.components.Textbox(
            lines=2,
            label="Instruction",
            placeholder="Tell me about alpacas.",
        ),
        gr.components.Textbox(lines=2, label="Input", placeholder="none"),
        gr.components.Slider(
            minimum=0, maximum=1, value=0.1, label="Temperature"
        ),
        gr.components.Slider(
            minimum=0, maximum=1, value=0.75, label="Top p"
        ),
        gr.components.Slider(
            minimum=0, maximum=100, step=1, value=40, label="Top k"
        ),
        gr.components.Slider(
            minimum=1, maximum=4, step=1, value=4, label="Beams"
        ),
        gr.components.Slider(
            minimum=1, maximum=2000, step=1, value=128, label="Max tokens"
        ),
        gr.components.Checkbox(label="Stream output"),
    ],
    outputs=[
        gr.components.Textbox(
            lines=5,
            label="Output",
        )
    ],
    title="🦙🌲 Alpaca-LoRA",
```

```
        description="Alpaca-LoRA is a 7B-parameter LLaMA model finetuned to follow
instructions. It is trained on the [Stanford
Alpaca](https://github.com/tatsu-lab/stanford_alpaca) dataset and makes use of the
Huggingface LLaMA implementation. For more information, please visit [the project's
website](https://github.com/tloen/alpaca-lora).",  # noqa: E501
    ).queue().launch(server_name=server_name, share=share_gradio)


if __name__ == "__main__":
    fire.Fire(main)
```

我們可以透過以下指令稿執行上面的程式。上面的程式用到了 Gradio 這個框架，這是一個將機器學習能力方便地封裝成 Web 服務的 Python 框架，透過封裝可以提升模型的好用性。程式執行後會出現圖 6-11 所示的介面，你可以直接在這個介面上進行推理驗證，還可以調整參數。

```
python infer_tallrec.py \
    --base_model '/Users/liuqiang/Desktop/code/llm/models/chinese-alpaca-2-7b' \
    # 基底模型路徑
    --lora_weights
'/Users/liuqiang/Desktop/code/llm4rec/llm4rec_abc/src/finetune-llm/lora-weights'
    # LoRA 微調的參數路徑
```

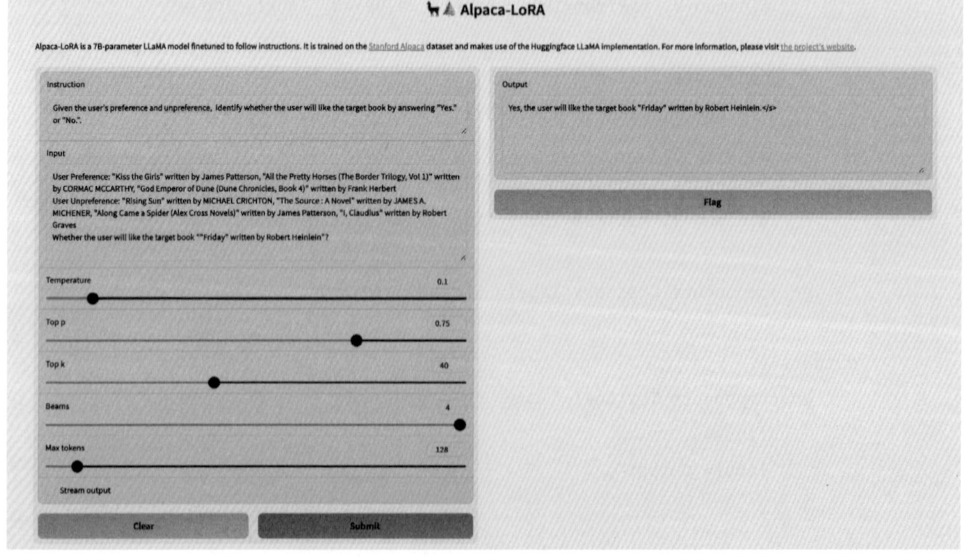

▲ 圖 6-11 微調後的模型互動介面

6.4 總結

本章講解了利用推薦系統資料微調大模型的基礎知識，包括微調的價值、微調的步驟、微調的方法和微調過程面臨的問題和挑戰，重點是微調的步驟和微調的方法。關於微調的步驟，需要重點了解基於不同的監督樣本的形式選擇合適的微調目標函式的方法，以及如何建構監督樣本。關於微調的方法，需要掌握監督微調方法和基於 PPO 的 RLHF 方法。

在本章講解的兩個案例中，TALLRec 是一個非常通用的微調框架，包含待推薦物品的標題等文字資訊的推薦系統都可以透過該框架實現。而 GIRL 巧妙地利用了 RLHF 對應徵場景下的推薦進行微調，這個方法也具有一定的通用性，好友推薦、旅遊目的地推薦、司機推薦等都可以採用該方法實現。TALLRec 和 GIRL 這兩個微調大模型應用於推薦系統的方法剛好覆蓋了本章的兩種微調範式，希望你結合這兩個案例，好好掌握監督微調和 RLHF 微調這兩種方法。

本章基於 MIND 資料集從零開始微調了一個個性化推薦系統。現在開放原始碼工具非常多，整體來說，微調還是比較簡單的。我們最缺少的就是高品質的微調資料和算力。

微調是大模型中非常重要的主題，也是大模型應用於垂直行業非常重要的實現方案。在具體業務場景中，我們需要基於開放原始碼模型和垂直資料建構特定領域的專有小模型。對資料安全要求比較高的公司通常不會考慮透過呼叫第三方 API 來使用大模型，這時，微調是一種非常好的解決方案。只要微調資料足夠優質，微調後的專用模型的效果就可能優於 ChatGPT 等通用大模型。

MEMO

7 直接推薦範式：利用大模型的上下文學習進行推薦

前面的章節講解了如何透過預訓練和微調大模型進行個性化推薦，這類模型是高度調配推薦業務場景的，有很不錯的效果。不管是預訓練還是微調，考慮到成本，一般對參數較少的大模型（<10B）進行處理。當模型的參數規模足夠大時，大模型本身就具備非常強的泛化能力。我們自然想到，直接用這類大模型是否可以進行個性化推薦呢？答案是肯定的，透過上下文學習能力，大模型可以在使用者提供背景知識的前提下，實現下游的推薦任務。

只需要利用特定提示詞激發大模型的推薦能力就可以將其用於下游推薦任務，而無須進行任何微調，我們將這種能力叫作上下文學習，上下文學習分為 zero-shot 和 few-shot。

zero-shot 直接利用預訓練好的大模型，透過設計特定的提示詞和範本讓大模型完成推薦任務。這類推薦的效果主要由大模型自身的通用能力及提示詞的獨特設計決定，一般使用 ChatGPT、GPT-4 等超大的大模型，這類模型的通用能力更好。few-shot 為預訓練好的大模型提供幾個範例樣本，指導大模型進行推薦，啟動大模型的個性化推薦能力。另外，可以將推薦任務分解為簡單的子任務，讓大模型逐一解決，這就是大模型的思維鏈能力。

基於上面的思考，本章講解如何直接使用大模型進行個性化推薦。先講解基本的實現原理和相關案例，讓你對該技術有初步的了解，再提供程式案例指導你實踐大模型個性化推薦。

7.1 上下文學習推薦基本原理

直接推薦范式是本章的核心框架和指導原則，該範式可以將完成預訓練的大模型直接用於個性化推薦，這裡的預訓練大模型一般是通用的大模型（例如 ChatGPT、Claude、Kimi 等），而非專門為推薦任務預訓練或微調的大模型。

這裡再提一下，在經典的召回→排序推薦範式下，直接推薦范式其實是利用大模型進行排序，即對召回的候選物品進行排序（如圖 3-6 所示）。召回階段使用其他推薦演算法，例如傳統的基於標籤的召回等。下面詳細說明直接推薦範式的基本原理。

大模型基於巨量文字資料預訓練，巨量文字本身就壓縮了各個領域的基礎知識，這些知識可以被激發出來用於個性化推薦。只不過，我們需要使用特定的提示詞才能激發大模型的個性化推薦能力。

推薦系統的提示詞需要使用特定的範本（Template）才能更進一步地激發大模型的推薦能力。推薦系統與其他下游任務最大的不同是個性化，因此提示詞也應該是個性化的，個性化提示詞包括不同使用者和物品的個性化欄位。舉例來說，使用者的偏好可以透過商品 ID 或使用者描述（如偏好、畫像等）表示。此外，個性化提示詞的預期模型輸出也應該根據輸入物品的變化而變化，這表示使用者對不同物品的偏好變化。這些物品欄位可以由商品 ID 或包含詳細描述的物品中繼資料表示。下面是一個具體的例子，供你參考。

```
I've watched the following movies in the past in order:

['0. Mirror, The (Zerkalo)', '1. The 39 Steps', '2. Sanjuro', '3. Trouble in
Paradise']\n\n Then if I ask you to recommend a new movie to me according to my
watching history, you should recommend Shampoo and now that I've just watched Shampoo,
there are 20 candidate movies that I can watch next:\n ['0. Manon of the Spring (Manon
des sources)', '1. Air Bud', '2. Citizen Kane', '3. Grand Hotel', '4. A Very Brady
Sequel', '5. 42 Up', '6. Message to Love: The Isle of Wight Festival', '7. Screamers',
'8. The Believers', '9. Hamlet', '10. Cliffhanger', '11. Three Wishes', '12.
Nekromantik', '13. Dangerous Minds', '14. The Prophecy', '15. Howling II: Your Sister
Is a Werewolf', '16. World of Apu, The (Apur Sansar)', '17. The Breakfast Club', '18.
Hoop Dreams', '19. Eddie']

Please rank these 20 movies by measuring the possibilities that I would like to watch
next most, according to my watching history. Please think step by step.\n Please show
me your ranking results with order numbers. Split your output with line break. You
MUST rank the given candidate movies. You can not generate movies that are not in the
given candidate list.
```

下面用具體的案例進一步說明上下文學習能力是如何用於推薦的。

7.2 案例講解

上下文學習推薦的基本原理相當簡單，主要是利用大模型的推理能力及適當的提示詞激發大模型的推薦能力，下面舉出 3 個具體的案例。

7.2.1 LLMRank 實現案例

參考文獻 [1] 將推薦看作一個有條件的排序任務（想法與貝氏估計相似），將使用者的歷史互動序列$\mathcal{H} = \{i_1, i_2, \cdots, i_n\}$（按照互動時間昇冪排列）作為初始條件，大模型推薦系統的任務是對召回的候選集$\mathcal{C} = \{i_j\}_{j=1}^{m}$進行排序，將使用者最喜歡的物品排在前面。這就是一個典型的利用大模型進行推薦排序的方案。

對每個使用者，首先構造兩個自然語言模式（Pattern）：一個是使用者歷史互動序列$\mathcal{H} = \{i_1, i_2, \cdots, i_n\}$，一個是取出候選集$\mathcal{C}$，然後將這兩個模式輸入一個自然語言範本$T$中形成最終的指令。透過這種方式，期望大模型理解指令並按照指令的建議輸出推薦排序結果。利用大模型進行 zero-shot 排序推薦的技術方案如圖 7-1 所示。接下來，描述詳細的指令設計過程。

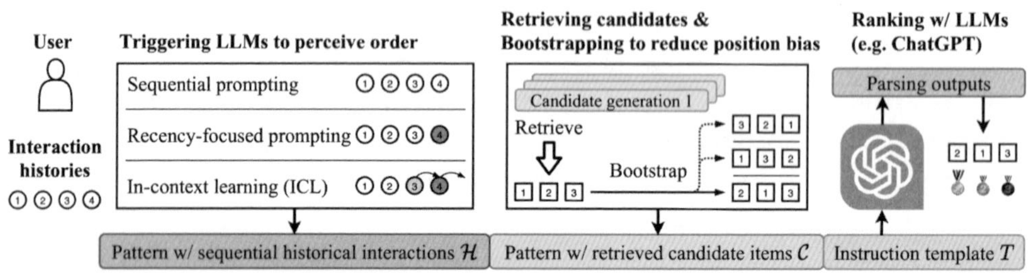

▲ 圖 7-1 利用大模型進行 zero-shot 排序推薦的技術方案

使用者歷史互動序列：為了研究大模型是否可以從使用者歷史互動中捕捉使用者偏好，我們將順序歷史互動序列$\mathcal{H} = \{i_1, i_2, \cdots, i_n\}$作為大模型的輸入包含在指令中。為了使大模型能夠意識到歷史互動的順序性，有以下三種建構指令的方法。

- **順序提示詞**：按時間順序排列歷史互動。例如：「I've watched the following movies in the past in order: '0. Multiplicity', '1. Jurassic Park', …」。
- **關注最近行為的提示詞**：除了順序的互動記錄，還可以增加一句話來強調最近的互動。例如：「I've watched the following movies in the past in order: '0.Multiplicity','1. JurassicPark',… Note that my most recently watched movie is Dead Presidents. …」。

- **上下文學習提示詞**：上下文學習是大模型解決各種任務的一種效果突出的提示方法，它在提示中舉出範例樣本（可能帶有任務描述），並指示大模型解決特定任務。對於個性化推薦任務，簡單地引入其他使用者的範例可能引入雜訊，因為不同的使用者通常具有不同的偏好。我們透過調整上下文學習提示詞，增補輸入互動序列來引入範例樣本。例如：「If I've watched the following movies in the past in order: '0. Multiplicity', '1. Jurassic Park', …, then you should recommend Dead Presidents to me and now that I've watched Dead Presidents, then …」。

取出候選集：要排序的候選物品通常由幾個候選生成模型生成（多路召回）。為了用大模型對這些候選物品排序，先按順序排列候選物品 |C|。例如：「Now there are 20 candidate movies that I can watch next: '0. Sister Act', '1. Sunset Blvd', …」。按照經典的候選生成方法，候選物品沒有特定的順序。我們將不同候選集的生成模型的召回結果放到一個集合中，並隨機排序。我們考慮了一個相對較小的候選集，並對 20 個候選物品（m=20）進行排序。實驗表明，大模型對提示詞中範例的順序很敏感。因此，我們在提示詞中為候選物品生成了不同的順序（採用的方法為自舉法——bootstrapping），這使我們能夠進一步驗證大模型的排序結果是否受到候選集排列順序的影響，即位置偏差（這種位置偏差確實存在），以及如何透過自舉法減少位置偏差。

使用大模型進行排序。為了將大模型作為排序模型，我們最終將上述模式整合到指令範本 T 中。一個可行的範例指令範本可以是「[pattern that contains sequential historical interactions H(conditions)] [pattern that contains retrieved candidate items C(candidates)] Please rank these movies by measuring the possibilities that I would like to watch next most, according to my watching history. You MUST rank the given candidate movies. You cannot generate movies that are not in the given candidate list.」。

解析大模型的輸出。透過將指令輸入大模型，我們可以獲得大模型的排序結果以供推薦。請注意，大模型的輸出仍然是自然語言文字，我們使用啟發式文字匹配方法解析輸出，並將推薦結果與候選集進行匹配。具體來說，當物品的文字較短且能夠區分時，如電影標題，可以在大模型輸出和候選物品的文字

之間直接執行高效的子串匹配演算法（如 KMP）。我們還可以為每個候選物品分配一個索引，並指示大模型直接輸出排序後的物品索引。儘管提示詞中包含了候選物品，但大模型仍可能生成候選集之外的物品。而對於 GPT-3.5，出現這種誤差的機率非常小，約為 3%。在這種情況下，我們可以提示大模型並讓它重新輸出，也可以簡單地將其視為不正確的輸出而忽略。

參考文獻 [1] 在兩個公開資料集上進行實驗，獲得了幾個關鍵發現，這些發現可以用於指導如何將大模型作為推薦系統的排序模型。

- 大模型可以利用使用者歷史互動進行個性化排序，但很難感知給定的歷史互動的順序。
- 透過專門設計的提示詞，如關注最近提示詞和上下文學習提示詞，可以觸發大模型感知歷史互動的順序，從而改善排序效果。
- 大模型優於 zero-shot 推薦方法，特別是在多個不同的召回排序演算法生成的候選集上有較好的表現。
- 大模型在排序時存在位置偏見和熱門偏見，這可以透過提示詞或自舉法來緩解。

7.2.2 多工實現案例

參考文獻 [2] 設計了一組提示詞，並評估了 ChatGPT 在 5 項推薦任務（評分預測、序列推薦、直接推薦、解釋生成和評論摘要）中的表現。與傳統的推薦方法不同，整個評估過程中不會微調 ChatGPT，只依靠提示詞將推薦任務轉為自然語言任務。此外，論文還探索了使用 few-shot 提示詞注入包含使用者潛在興趣的互動資訊，以幫助 ChatGPT 更進一步地了解使用者的需求和興趣。在 Amazon Beauty 資料集上的實驗結果表明，ChatGPT 在某些任務中獲得了較好的效果，在其他任務中也能夠達到基準線水準。

使用 ChatGPT 執行 5 項推薦任務並評估其推薦性能的工作流程如圖 7-2 所示，包括 3 個步驟。首先，根據推薦任務的具體特徵構造不同的提示詞。其次，這些提示詞被用作 ChatGPT 的輸入，期望 ChatGPT 根據提示詞中指定的要求生成推薦結果。最後，最佳化（Refinement）模組對 ChatGPT 的輸出進行檢查和最佳化，將最佳化後的結果作為最終推薦結果傳回給使用者。下面分別對這 3 個步驟加以說明。

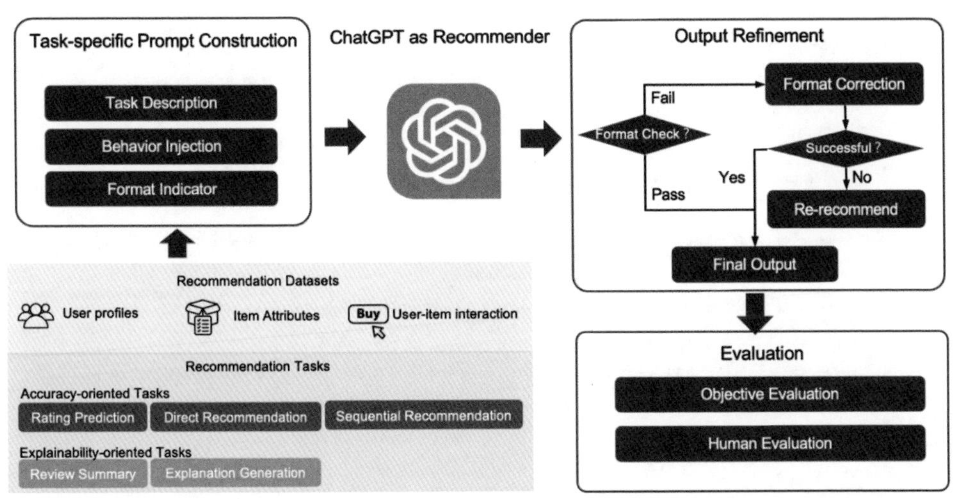

▲ 圖 7-2 使用 ChatGPT 執行 5 項推薦任務並評估其推薦性能的工作流程

步驟 1：構造特定任務的提示詞。

透過設計針對不同任務的提示詞來激發 ChatGPT 的推薦能力。每個提示詞包含 3 部分：任務描述（Task Description）、行為注入（Behavior Injection）和格式指示符號（Format Indicator）。任務描述用於使推薦任務適應自然語言處理任務。行為注入旨在評估 few-shot 提示詞的影響，它整合了使用者 - 物品互動，以幫助 ChatGPT 更有效地確定使用者偏好和需求。格式指示符號用於約束輸出格式，使推薦結果更易於理解和評估。圖 7-3 是一些提示詞的範例。

評分預測
zero-shot
How will user rate this product_title: "SHANY Nail Art Set (24 Famous Colors Nail Art Polish, Nail Art Decoration)" , and product_category: Beauty? (1 being lowest and 5 being highest) Attention! Just give me back the exact number a result , and you don't need a lot of text.
few-shot
Here is user rating history:
1. Bundle Monster 100 PC 3D Designs Nail Art Nailart Manicure Fimo Canes Sticks Rods Stickers Gel Tips, 5.0;
2. Winstonia's Double Ended Nail Art Marbling Dotting Tool Pen Set w/ 10 Different Sizes 5 Colors - Manicure Pedicure, 5.0;
3. Nail Art Jumbo Stamp Stamping Manicure Image Plate 2 Tropical Holiday by Cheeky&reg, 5.0 ;
4.Nail Art Jumbo Stamp Stamping Manicure Image Plate 6 Happy Holidays by Cheeky&reg, 5.0;
Based on above rating history, please predict user's rating for the product: "SHANY Nail Art Set (24 Famouse Colors Nail Art Polish, Nail Art Decoration)", (1 being lowest and5 being highest,The output should be like: (x stars, xx%), do not explain the reason.)
序列推薦
zero-shot
Requirements: you must choose 10 items for recommendation and sort them in order of priority, from highest to lowest. Output format: a python list. Do not explain the reason or include any other words.
The user has interacted with the following items in chronological order: ['Better Living Classic Two Chamber Dispenser, White', 'Andre Silhouettes Shampoo Cape, Metallic Black', ……, 'John Frieda JFHA5 Hot Air Brush, 1.5 inch'].Please recommend the next item that the user might interact with.
few-shot
Requirements: you must choose 10 items for recommendation and sort them in order of priority, from highest to lowest. Output format: a python list. Do not explain the reason or include any other words.
Given the user's interaction history in chronological order: ['Avalon Biotin B-Complex Thickening Conditioner, 14 Ounce', 'Conair 1600 Watt Folding Handle Hair Dryer', ……, 'RoC Multi-Correxion 4-Zone Daily Moisturizer, SPF 30, 1.7 Ounce'], the next interacted item is ['Le Edge Full Body Exfoliator - Pink']. Now, if the interaction history is updated to ['Avalon Biotin B-Complex Thickening Conditioner, 14 Ounce', 'Conair 1600 Watt Folding Handle Hair Dryer',……, 'RoC Multi-Correxion 4-Zone Daily Moisturizer, SPF 30, 1.7 Ounce', 'Le Edge Full Body Exfoliator - Pink'] and the user is likely to interact again, recommend the next item.
直接推薦
zero-shot
Requirements: you must choose 10 items for recommendation and sort them in order of priority, from highest to lowest. Output format: a python list. Do not explain the reason or include any other words.
The user has interacted with the following items (in no particular order): [""Skin Obsession Jessner's Chemical Peel Kit Anti-aging and Anti-acne Skin Care Treatment"", 'Xtreme Brite Brightening Gel 1oz.',……, 'Reviva - Light Skin Peel, 1.5 oz cream']. From the candidates listed below, choose the top 10 items to recommend to the user and rank them in order of priority from highest to lowest. Candidates: ['Rogaine for Women Hair Regrowth Treatment 3- 2 ounce bottles', 'Best Age Spot Remover', ……""L'Oreal Kids Extra Gentle 2-in-1 Shampoo With a Burst of Cherry Almond, 9.0 Fluid Ounce""].
few-shot
Requirements: you must choose 10 items for recommendation and sort them in order of priority, from highest to lowest. Output format: a python list. Do not explain the reason or include any other words.
The user has interacted with the following items (in no particular order): ['Maybelline New York Eye Studio Lasting Drama Gel Eyeliner, Eggplant 956, 0.106 Ounce', ""L'Oreal Paris Healthy Look Hair Color, 8.5 Blonde/White Chocolate"", ……, 'Duo Lash Adhesive, Clear, 0.25 Ounce']. Given that the user has interacted with 'WAWO 15 Color Professionl Makeup Eyeshadow Camouflage Facial Concealer Neutral Palette' from a pool of candidates: ['MASH Bamboo Reusable Cuticle Pushers Remover / Manicure Pedicure Stick', 'Urban Decay All Nighter Long-Lasting Makeup Setting Spray 4 oz', ……,'Classic Cotton Balls Jumbo Size, 100 Count'], please recommend the best item from a new candidate pool, ['Neutrogena Ultra Sheer Sunscreen SPF 45 Twin Pack 6.0 Ounce', 'Blinc Eyeliner Pencil - Black', ……,'Skin MD Natural + SPF15 combines the benefits of a shielding lotion and a sunscreen lotion']. Note that the candidates in the new pool are not ordered in any particular way.

▲ 圖 7-3 亞馬遜 Beauty 資料集上基於準確性任務的提示詞範例

步驟 2：利用 ChatGPT 生成推薦結果。

將上面構造好的提示詞以自然語言對話的形式注入 ChatGPT（例如呼叫 ChatGPT 的 API）獲得最終的輸出（推薦結果）。由於這一步完全基於 ChatGPT 的自然語言生成能力，屬於黑盒，這裡不詳細說明。

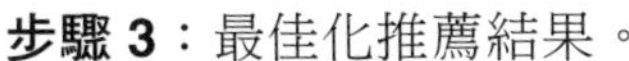

步驟 3：最佳化推薦結果。

為了確保生成結果的多樣性，ChatGPT 在其回應生成過程中加入了一定程度的隨機性，這可能導致對相同輸入產生不同的輸出。因此，當使用 ChatGPT 進行推薦時，這種隨機性有時會導致評估推薦物品困難。雖然提示詞結構中的格式指示符號可以部分緩解這一問題，但在實際使用中，它仍然不能保證預期的輸出格式。因此，需要輸出最佳化模組來檢查 ChatGPT 的輸出格式。如果輸出通過了格式檢查，它將直接用作最終推薦結果，如果沒有透過，則會根據預先定義的規則對其進行校正。如果校正成功，則校正後的結果將用作最終輸出，如果沒有校正成功，則將相應的提示詞輸入 ChatGPT 以進行重新推薦，直到滿足格式要求。

需要注意的是，在評估 ChatGPT 的輸出時，不同的任務有不同的格式要求。舉例來說，對於評分預測，只需要輸出特定的分數，而對於序列推薦或直接推薦，需要輸出推薦物品的清單。特別是對於序列推薦，一次將資料集中的所有物品注入 ChatGPT 是一項挑戰（因為 ChatGPT 對輸入的 token 數量有限制）。因此，ChatGPT 的輸出可能與資料集中的候選集不匹配（輸出的物品不在資料集中）。為了解決這個問題，參考文獻 [2] 中引入了一種基於相似性的文字匹配方法，將 ChatGPT 的預測映射回原始資料集。儘管這種方法可能不能完全反映 ChatGPT 的能力，但它仍然可以間接展示 ChatGPT 在序列推薦中的潛力。

7.2.3 NIR 實現案例

參考文獻 [3] 提出了 NIR（Next-Item Recommendation）框架，透過將大模型推薦任務拆解為 3 個步驟來實現個性化的推薦，這 3 個步驟都利用提示詞激發大模型的推理能力。圖 7-4 是具體的步驟說明，下面詳細說明。

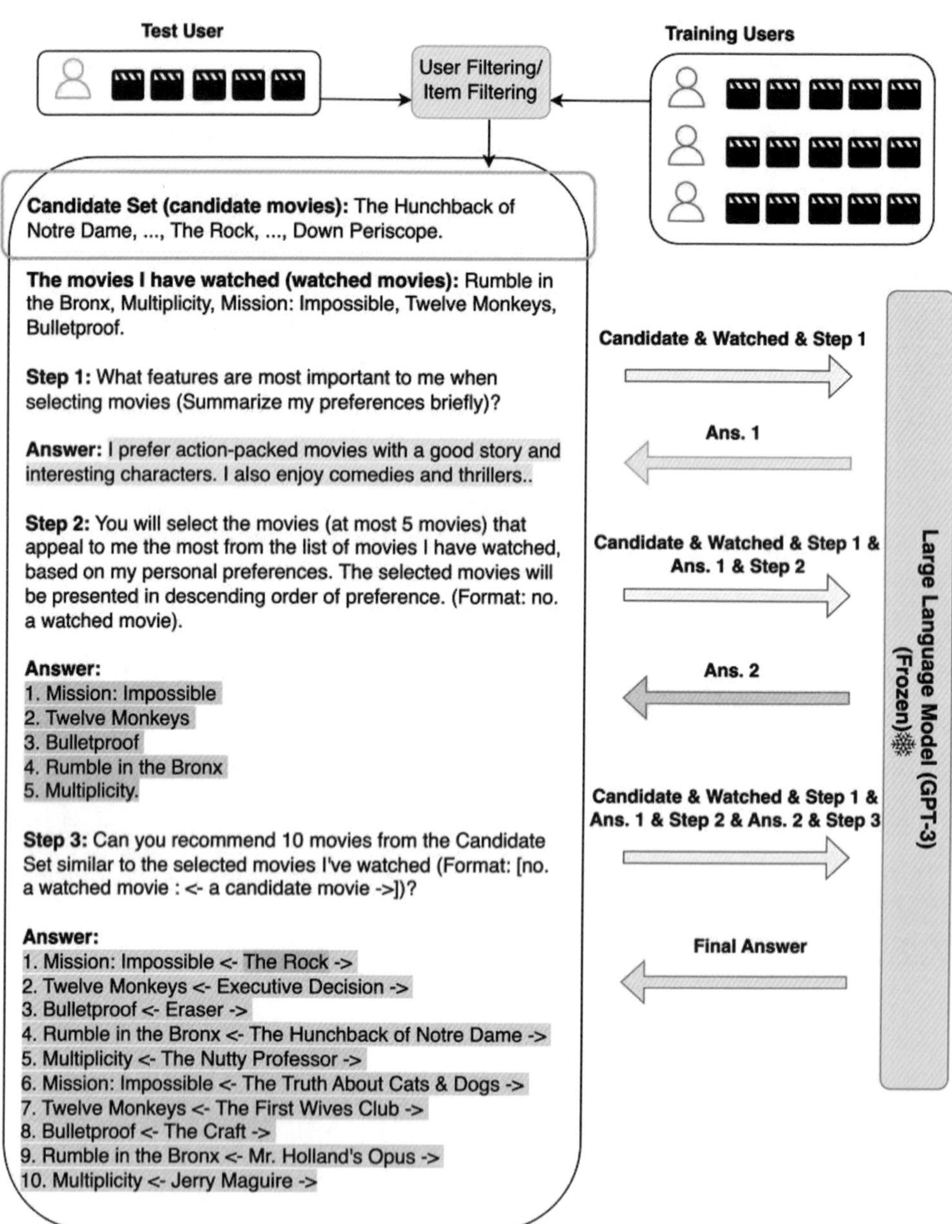
Test User
Training Users
User Filtering/
Item Filtering
Candidate Set (candidate movies): The Hunchback of Notre Dame, ..., The Rock, ..., Down Periscope.
The movies I have watched (watched movies): Rumble in the Bronx, Multiplicity, Mission: Impossible, Twelve Monkeys, Bulletproof.
Step 1: What features are most important to me when selecting movies (Summarize my preferences briefly)?
Answer: I prefer action-packed movies with a good story and interesting characters. I also enjoy comedies and thrillers..
Step 2: You will select the movies (at most 5 movies) that appeal to me the most from the list of movies I have watched, based on my personal preferences. The selected movies will be presented in descending order of preference. (Format: no. a watched movie).
Answer:
1. Mission: Impossible
2. Twelve Monkeys
3. Bulletproof
4. Rumble in the Bronx
5. Multiplicity.
Step 3: Can you recommend 10 movies from the Candidate Set similar to the selected movies I've watched (Format: [no. a watched movie : <- a candidate movie ->])?
Answer:
1. Mission: Impossible <- The Rock ->
2. Twelve Monkeys <- Executive Decision ->
3. Bulletproof <- Eraser ->
4. Rumble in the Bronx <- The Hunchback of Notre Dame ->
5. Multiplicity <- The Nutty Professor ->
6. Mission: Impossible <- The Truth About Cats & Dogs ->
7. Twelve Monkeys <- The First Wives Club ->
8. Bulletproof <- The Craft ->
9. Rumble in the Bronx <- Mr. Holland's Opus ->
10. Multiplicity <- Jerry Maguire ->
Candidate & Watched & Step 1
Ans. 1
Candidate & Watched & Step 1 & Ans. 1 & Step 2
Ans. 2
Candidate & Watched & Step 1 & Ans. 1 & Step 2 & Ans. 2 & Step 3
Final Answer
Large Language Model (GPT-3) (Frozen)

▲ 圖 7-4 上下文學習推薦的步驟

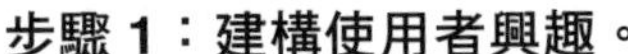

步驟 1：建構使用者興趣。

為了捕捉使用者的偏好，利用大模型（GPT-3）從使用者看過的電影中總結出使用者的偏好特徵。如圖 7-4「回答 1」左側文字所示，大模型傳回的答案總結了目標使用者的偏好。具體的提示詞如下。

```
What features are most important to me when selecting movies (Summarize my preferences briefly)?
```

步驟 2：代表性電影生成。

利用大模型基於使用者看過的電影並結合第一步中總結的使用者偏好，找出 5 部對使用者最有吸引力的電影（這 5 部電影是從使用者看過的電影中找出的），如圖 7-4 中「回答 2」左側的文字所示。具體的提示詞如下。

```
You will select the movies (at most 5 movies) that appeal to me the most from the list of movies I have watched, based on my personal preferences. The selected movies will be presented in descending order of preference. (Format: no. a watched movie).
```

步驟 3：電影推薦。

讓大模型從候選集中選擇與步驟 2 中代表使用者興趣的電影最相似的 10 部電影作為最終的推薦，如圖 7-4 中「最終回答」左側文字所示。具體的提示詞如下。

```
Can you recommend 10 movies from the Candidate Set similar to the selected movies I've watched (Format: [no. a watched movie : <- a candidate movie >])?
```

7.3 上下文學習推薦程式實現

本節基於 MIND 資料集實現大模型上下文學習推薦，希望你透過程式更進一步地學習、了解如何直接利用大模型的泛化能力解決推薦問題。

上下文學習推薦相比預訓練大模型進行推薦和微調大模型進行推薦來說，實現上更簡單、更容易上手，也希望你可以熟練掌握這部分的基礎知識。

本節首先準備相關資料，然後架設一個大模型本機服務，這個本機服務類似於 ChatGPT 的 API，方便我們直接利用大模型實現上下文學習推薦，最後舉出上下文學習單步驟推薦和多步驟推薦兩個程式案例。

7.3.1 資料準備

使用 MIND 資料集建構本節的測試資料①。大模型上下文學習推薦需要的資料格式如下。

```
{"instruction": "You are a recommendation system expert who provides personalized
recommendations to users based on the background information provided.",
"input": "I've browsed the following news in the past in order:
[''Wheel Of Fortune' Guest Delivers Hilarious, Off The Rails Introduction','Hard Rock
Hotel New Orleans collapse: Former site engineer weighs in','Felicity Huffman begins
prison sentence for college admissions scam','Outer Banks storms unearth old shipwreck
from 'Graveyard of the Atlantic'','Tiffany's is selling a holiday advent calendar
for $112,000','This restored 1968 Winnebago is beyond adorable','Lori Loughlin Is
'Absolutely Terrified' After Being Hit With New Charge','Bruce Willis brought Demi
Moore to tears after reading her book','Celebrity kids then and now: See how they've
grown','Felicity Huffman Smiles as She Begins Community Service Following Prison
Release','Queen Elizabeth Finally Had Her Dream Photoshoot, Thanks to Royal Dresser
Angela Kelly','Hundreds of thousands of people in California are downriver of a dam
that 'could fail'','Alexandria Ocasio-Cortez 'sincerely' apologizes for blocking ex-
Brooklyn politician on Twitter, settles federal lawsuit','The Rock's Gnarly Palm Is a
Testament to Life Without Lifting Gloves']
Then if I ask you to recommend a new news to me according to my browsing history, you
should recommend 'Donald Trump Jr. reflects on explosive 'View' chat: 'I don't think
they like me much anymore'' and now that I've just browsed 'Donald Trump Jr. reflects
on explosive 'View' chat: 'I don't think they like me much anymore'', there are 22
candidate news that I can browse next:
1. 'Browns apologize to Mason Rudolph, call Myles Garrett's actions 'unacceptable'',
2. 'I've been writing about tiny homes for a year and finally spent 2 nights in a 300-
foot home to see what it's all about   here's how it went',
3. 'Opinion: Colin Kaepernick is about to get what he deserves: a chance',
4. 'The Kardashians Face Backlash Over 'Insensitive' Family Food Fight in KUWTK Clip',
```

① 基於 MIND small 的 Validation 資料微調，該資料相對較小。

```
5. 'THEN AND NOW: What all your favorite '90s stars are doing today',6. 'Report:
Police investigating woman's death after Redskins' player Montae Nicholson took her to
hospital',
7. 'U.S. Troops Will Die If They Remain in Syria, Bashar Al-Assad Warns',
8. '3 Indiana judges suspended after a night of drinking turned into a White Castle
brawl',
9. 'Cows swept away by Hurricane Dorian found alive   but how?',
10. 'Surviving Santa Clarita school shooting victims on road to recovery: Latest',
11. 'The Unlikely Star of My Family's Thanksgiving Table',
12. 'Meghan Markle and Hillary Clinton Secretly Spent the Afternoon Together at
Frogmore Cottage',
13. 'Former North Carolina State, NBA player Anthony Grundy dies in stabbing, police
say',
14. '85 Thanksgiving Recipes You Can Make Ahead',
15. 'Survivor Contestants Missy Byrd and Elizabeth Beisel Apologize For Their Actions',
16. 'Pete Davidson, Kaia Gerber Are Dating, Trying to Stay 'Low Profile'',
17. 'There's a place in the US where its been over 80 degrees since March',
18. 'Taylor Swift Rep Hits Back at Big Machine, Claims She's Actually Owed $7.9 Million
in Unpaid Royalties',
19. 'The most talked about movie moments of the 2010s',
20. 'Belichick mocks social media in comments on Garrett incident',
21. '13 Reasons Why's Christian Navarro Slams Disney for Casting 'the White Guy' in
The Little Mermaid',
22. '66 Cool Tech Gifts Anyone Would Be Thrilled to Receive'
Please select some news that I would like to browse next according to my browsing
history.
Please think step by step.
Please show me your results. Split your output with line break. You MUST select
from the given candidate news. You can not generate news that are not in the given
candidate list."
"output": "['Opinion: Colin Kaepernick is about to get what he deserves: a chance'\n]"
}
```

要將使用者行為資料和新聞資料轉為上述格式，就需要讀取 news.tsv 和 behaviors.tsv 兩個檔案。news.tsv 檔案中有 News ID 和對應的新聞標題，而 behaviors.tsv 檔案中有兩個欄位：User Click History 和 Impression News，前者包含使用者的觀看歷史，後者包含使用者點擊和沒有點擊的新聞。

利用 news.tsv 和 behaviors.tsv 這兩個檔案就可以建構具體的測試資料[①]集：User Click History 作為提示詞中使用者的點擊歷史，Impression News 作為待大模型排序的候選集（標籤為 1 代表使用者喜歡，為 0 代表使用者不喜歡），下面是詳細的程式實現[②]。

```
import csv
import json
instruction = ("You are a recommendation system expert who provides personalized
recommendations to users based on "
               "the background information provided.")
news_dict = {}  # 從 news.tsv 獲取每個新聞 ID 到標題的映射字典。
with open('../data/mind/news.tsv', 'r') as file:
    reader = csv.reader(file, delimiter='\t')
    for row in reader:
        news_id = row[0]
        news_title = row[3]
        news_dict[news_id] = news_title
data_list = []  # 利用 behaviors.tsv 檔案獲取使用者喜歡和不喜歡的新聞
with open('../data/mind/behaviors.tsv', 'r') as file:
    reader = csv.reader(file, delimiter='\t')
    for row in reader:
        history = row[3]
        impression = row[4]
        history_list = history.split(" ")
        impre_list = impression.split(" ")
        if len(history_list) >= 6 and len(impre_list) >= 10:
# 至少要有 6 個使用者點擊歷史和 10 個曝光歷史
            his = []
            for news_id in history_list[:-1]:
                title = news_dict[news_id]
                his.append(title)
            last_view_id = history_list[-1]
            last_view = news_dict[last_view_id]
            preference = []
            candidate = []
```

① 上下文學習推薦不需要訓練，所以是測試資料。

② 程式：src/basic_skills/icl-rec/data-process/generate_json_data.py。

```
            index = 1
            for impre in impre_list:
                [impre_id, action] = impre.split("-")
                title = news_dict[impre_id]
                candidate.append(str(index) + ". " + title + "\n")
                index = index + 1
                if int(action) == 1:
                    preference.append(title + "\n")
            input = ("I've browsed the following news in the past in order:\n\n" +
                     "[" + ','.join(his) + "]" + "\n\n" +
                     "Then if I ask you to recommend a new news to me according to my 
browsing history, "
                     "you should recommend " + last_view + " and now that I've just 
browsed " + last_view + ", " +
                     "there are " + str(len(impre_list)) + " candidate news that I can 
browse next:" +
                     ','.join(candidate) + "\n\n" +
                     "Please select some news that I would like to browse next 
according to my browsing history. " +
                     "Please think step by step.\n\n" +
                     "Please show me your results. Split your output with line break. 
You MUST select from the given " +
                     "candidate news. You can not generate news that are not in the 
given candidate list."
                     )
            output = "[\n" + ','.join(preference) + "\n]"
            res_dic = {
                "instruction": instruction,
                "input": input,
                "output": output
            }
            data_list.append(res_dic)
res = json.dumps(data_list)
user_sequence_path = "../data/mind/validation.json"  # 將生成的訓練資料儲存起來
with open(user_sequence_path, 'a') as file:
    file.write(res)
```

執行上述程式，就會在 /data/mind/ 目錄下生成一個 test.json 檔案，這個檔案就是我們後面需要用到的資料集。

7.3.2 程式實現

有了上面的資料，就可以開始嘗試進行上下文學習推薦了。本節的程式需要借助本地的大模型服務，這裡使用 Ollama 框架 [4]（也可以嘗試使用 FastChat、lite 等大模型 [5,6]）。

如果使用 macOS 或 Windows 作業系統，則可以直接下載 Ollama 安裝套件。Ollama 的具體架設過程詳見參考文獻 [4]。

1. 單步驟上下文學習推薦

根據 7.1 節的想法利用大模型實現上下文學習個性化推薦，這裡提供 4 種不同框架（Ollama、LlamaCpp、Transformers 函式庫、MLX）的實現方案，具體程式以下①。

```
import requests
import torch
import fire
import json
from datasets import load_dataset
from langchain.chains import LLMChain
from mlx_lm import load, generate
from langchain_community.llms import LlamaCpp
from langchain.prompts.chat import ChatPromptTemplate
from langchain.callbacks import StreamingStdOutCallbackHandler
from transformers import AutoTokenizer, AutoModelForCausalLM, pipeline, TextStreamer
from langchain_community.llms.huggingface_pipeline import HuggingFacePipeline

def icl_rec(
        model_path: str = "",
        icl_type: str = "",
        instruction: str = "",
        prompt: str = "",
        temperature: float = 0.1,
        top_p: float = 0.95,
        ctx: int = 13000,
```

① 程式：src/basic_skills/icl-rec/icl_rec.py。

```
):
    """
    針對單一提示詞為使用者進行上下文學習推薦
    :param model_path: 模型位址
    :param icl_type: 上下文學習推理類型，我們實現了 Ollama、LlamaCpp、Transformers 函式庫、
MLX 4 類別上下文學習推理方法
    :param instruction: 指令
    :param prompt: 提示詞
    :param temperature: 大模型溫度係數
    :param top_p: 大模型的 top_p
    :param ctx: 輸出 token 長度
    :return: 無
    """
    if not instruction:
        instruction = ("You are a recommendation system expert who provides personalized
recommendations to users "
                       "based on the background information provided.")
    assert (
        icl_type
    ), "Please specify a --icl_type, e.g. --icl_type='ollama'"

    if icl_type == "ollama":
        """
        利用 Ollama 框架將大模型封裝成類似於符合 ChatGPT 介面規範的 API 服務，直接透過呼叫介面
實現大模型上下文學習推薦
        """
        url = "http://localhost:11434/api/chat"  # Ollama 的 API 位址
        data = {
            "model": "yi:34b-chat",  # Ollama 安裝的模型名稱
            "options": {
                "temperature": temperature,
                "top_p": top_p,
                "num_ctx": ctx,
                "num_gpu": 128,
            },
            "messages": [
                {
                    "role": "user",
                    "content": prompt
```

```
                }
            ]
        }
        response = requests.post(url=url, json=data, stream=True)
        for chunk in response.iter_content(chunk_size=None, decode_unicode=True):
            j = json.loads(chunk.decode('utf-8'))
            print(j['message']['content'], end="")

    elif icl_type == "llamacpp": # 利用 LlamaCpp 實現上下文學習推薦
        if not model_path:
            model_path =
"/Users/liuqiang/Desktop/code/llm/models/gguf/ qwen1.5-72b-chat-q5_k_m.gguf"
        callback = StreamingStdOutCallbackHandler()
        n_gpu_layers = 128  # 根據模型和 GPU 顯示記憶體池大小更改此值
        n_batch = 4096  # 可以設置為 1 至 n_ctx 之間的任何值，根據電腦的 RAM 大小設置
        llm = LlamaCpp(
            streaming=True,
            model_path=model_path,
            n_gpu_layers=n_gpu_layers,
            n_batch=n_batch,
            f16_kv=True,  # 設置為 True, 避免執行中遇到問題
            temperature=temperature,
            top_p=top_p,
            n_ctx=ctx,
            callbacks=[callback],
            verbose=True
        )
        system = [("system", instruction),
                  ("user", "{input}")]
        template = ChatPromptTemplate.from_messages(system)
        chain = LLMChain(prompt=template, llm=llm)
        chain.invoke({"input": prompt})

    elif icl_type == "transformers":  # 利用 Transformers 函式庫實現上下文學習推薦
        if not model_path:
            model_path = "/Users/liuqiang/Desktop/code/llm/models/Yi-34B-Chat"
        tokenizer = AutoTokenizer.from_pretrained(model_path)
        model = AutoModelForCausalLM.from_pretrained(
            model_path,
            trust_remote_code=True,
```

```
            torch_dtype=torch.float16,
            device_map="mps",
            use_cache=True,
        )
        model = model.to("mps")
        streamer = TextStreamer(tokenizer)
        pipe = pipeline(
            "text-generation",
            model=model,
            tokenizer=tokenizer,
            streamer=streamer,
            max_length=13000,
            temperature=temperature,
            pad_token_id=tokenizer.eos_token_id,
            top_p=top_p,
            repetition_penalty=1.15,
            do_sample=False,
        )
        llm = HuggingFacePipeline(pipeline=pipe)
        system = [("system", instruction),
                  ("user", "{input}")]
        template = ChatPromptTemplate.from_messages(system)
        chain = LLMChain(prompt=template, llm=llm)
        chain.invoke({"input": prompt})

    elif icl_type == "mlx":  # 利用蘋果開放原始碼的 MLX 框架實現上下文學習推薦
        if not model_path:
            model_path = "/Users/liuqiang/Desktop/code/llm/models/Yi-34B-Chat"
        model, tokenizer = load(model_path)
        response = generate(
            model,
            tokenizer,
            prompt=prompt,
            temp=temperature,
            max_tokens=10000,
            verbose=True
        )
        print(response)

    else:
```

```
        raise NotImplementedError("the case not implemented!")

def icl_rec_batch(validation_path: str, icl_type='llamacpp'):
    """
        針對從訓練資料建構的測試集，逐筆為使用者進行上下文學習推理，你可以跟真實值（output）
對比，查看大模型上下文學習推薦的效果
    """
    data = load_dataset("json", data_files=validation_path)
    index = 1
    for row in data['train']:
        instruction = row['instruction']
        input = row['input']
        output = row['output']
        print(str(index) + " : --------------------")
        icl_rec(icl_type=icl_type, instruction=instruction, prompt=input)
        print(output)

if __name__ == "__main__":
    # fire.Fire(icl_rec)
    icl_rec(icl_type="llamacpp")
    # icl_rec_batch("./data/mind/test.json")
```

執行上述程式，得到以下結果①。

```
考慮到使用者的瀏覽歷史，我將選擇那些與使用者之前瀏覽的新聞主題相關或相似的新聞。
1. 'Browns apologize to Mason Rudolph, call Myles Garrett's actions 'unacceptable'','
- 這筆新聞與體育相關，特別是橄欖球，這與使用者瀏覽的「Donald Trump Jr. reflects on explosive
'View' chat」這筆新聞的主題不同，但可能吸引喜歡體育新聞的使用者。

2. 'Opinion: Colin Kaepernick is about to get what he deserves: a chance','
- 這筆新聞與體育和政治相關，特別是關於 NFL 球員 Colin Kaepernick 的爭議。這與使用者瀏覽的
「Donald Trump Jr. reflects on explosive 'View' chat」這筆新聞的主題有相似之處，都是關於公
共人物的爭議。
根據使用者的瀏覽歷史，我推薦使用者瀏覽這兩筆新聞，因為它們與使用者之前瀏覽的新聞主題相關或相似。
```

① 這裡使用 qwen1.5-72B-chat-q5_k_m，舉出的推薦是中英文混雜的，但是明顯可以看到是有邏輯推理的。

2. 多步驟上下文學習推薦

根據 7.2 節中 NIR 實現案例的想法讓大模型舉出個性化的推薦，這裡還是使用 Ollama 框架實現。

下面是筆者在執行 ollama run qwen:72B-chat-v1.5-q5_K_M 後與大模型的互動式對話，其中，「>>>」後面是筆者舉出的提示詞，「⋮」後面是大模型舉出的回答。

```
>>> The news I have browsed:
...         [''Wheel Of Fortune' Guest Delivers Hilarious, Off The Rails Introduction','
Hard Rock Hotel New Orleans collapse: Former site engineer weighs in','Felicity Huffman
begins prison sentence for college adm
... issions scam','Outer Banks storms unearth old shipwreck from 'Graveyard of the
Atlantic'','Tiffany's is selling a holiday advent calendar for $112,000','This restored
1968 Winnebago is beyond adorable','Lori Lo
... ughlin Is 'Absolutely Terrified' After Being Hit With New Charge','Bruce Willis
brought Demi Moore to tears after reading her book','Celebrity kids then and now: See
how they've grown','Felicity Huffman Smiles
... as She Begins Community Service Following Prison Release','Queen Elizabeth Finally
Had Her Dream Photoshoot, Thanks to Royal Dresser Angela Kelly','Hundreds of thousands
of people in California are downriver of
...  a dam that 'could fail'','Alexandria Ocasio-Cortez 'sincerely' apologizes for
blocking ex-Brooklyn politician on Twitter, settles federal lawsuit','The Rock's
Gnarly Palm Is a Testament to Life Without Lifting
...  Gloves']
...
...         What features are most important to me when selecting news (Summarize my
preferences briefly)?

⋮
Based on the news you have browsed, your preferences seem to be focused on:

1. **Entertainment and Pop Culture**: You've viewed stories about celebrities like
Felicity Huffman, Lori Loughlin, Bruce Willis, Demi Moore, and The Rock.

2. **Human Interest and Heartwarming Stories**: Restored vintage vehicles and holiday
advent calendars indicate a fondness for charming or nostalgic tales.
```

```
3. **Current Events and Breaking News**: Stories about the Hard Rock Hotel collapse in
New Orleans, California dam safety concerns, and Alexandria Ocasio-Cortez's federal
lawsuit settlement demonstrate an
interest in staying informed on important local and national events.

>>> You will select the news (at most 5 news) that appeal to me the most from the list
of news I have browsed, based on my personal preferences. The selected news will be
presented in descending order of preference
... . (Format: no. a browsed news).

:
Based on your preferences, here are the top 5 news articles that would likely appeal
to you the most:

1. **a** 'Felicity Huffman Smiles as She Begins Community Service Following Prison Release'
   This story combines elements of human interest and current events, focusing on a
celebrity who has recently gone through a challenging legal process.

2. **b** 'The Rock's Gnarly Palm Is a Testament to Life Without Lifting Gloves'
   This article is likely to pique your interest due to its focus on entertainment and
pop culture, specifically highlighting the unique physical attribute of a well-known
actor.

3. **c** 'Bruce Willis brought Demi Moore to tears after reading her book'
   Another story blending entertainment and emotional connections, this news item
would likely appeal to you as it discusses the touching moment between two famous ex-
spouses.

4. **d** 'This restored 1968 Winnebago is beyond adorable'
   This article combines elements of nostalgia, human interest, and heartwarming
stories. It's likely to resonate with you if you appreciate charming tales about
restored vintage vehicles.

5. **e** 'Felicity Huffman begins prison sentence for college admissions scam'
   Although less focused on positive or heartwarming aspects, this news story still
combines elements of current events and entertainment. Felicity Huffman's involvement
in a high-profile scandal would likely
pique your interest as it does touch upon the lives of celebrities off-screen.
```

```
>>> Candidate Set (candidate news):
...         1. 'Browns apologize to Mason Rudolph, call Myles Garrett's actions
'unacceptable'',
...         2. 'I've been writing about tiny homes for a year and finally spent 2
nights in a 300-foot home to see what it's all about   here's how it went',
...         3. 'Opinion: Colin Kaepernick is about to get what he deserves: a chance',
...         4. 'The Kardashians Face Backlash Over 'Insensitive' Family Food Fight in
KUWTK Clip',
...         5. 'THEN AND NOW: What all your favorite '90s stars are doing today',
...         6. 'Report: Police investigating woman's death after Redskins' player
Montae Nicholson took her to hospital',
...         7. 'U.S. Troops Will Die If They Remain in Syria, Bashar Al-Assad Warns',
...         8. '3 Indiana judges suspended after a night of drinking turned into a
White Castle brawl',
...         9. 'Cows swept away by Hurricane Dorian found alive   but how?',
...         10. 'Surviving Santa Clarita school shooting victims on road to recovery:
Latest',
...         11. 'The Unlikely Star of My Family's Thanksgiving Table',
...         12. 'Meghan Markle and Hillary Clinton Secretly Spent the Afternoon
Together at Frogmore Cottage',
...         13. 'Former North Carolina State, NBA player Anthony Grundy dies in
stabbing, police say',
...         14. '85 Thanksgiving Recipes You Can Make Ahead',
...         15. 'Survivor Contestants Missy Byrd and Elizabeth Beisel Apologize For
Their Actions',
...         16. 'Pete Davidson, Kaia Gerber Are Dating, Trying to Stay 'Low Profile'',
...         17. 'There's a place in the US where its been over 80 degrees since
March',
...         18. 'Taylor Swift Rep Hits Back at Big Machine, Claims She's Actually Owed
$7.9 Million in Unpaid Royalties',
...         19. 'The most talked about movie moments of the 2010s',
...         20. 'Belichick mocks social media in comments on Garrett incident',
...         21. '13 Reasons Why's Christian Navarro Slams Disney for Casting 'the
White Guy' in The Little Mermaid',
...         22. '66 Cool Tech Gifts Anyone Would Be Thrilled to Receive'
...
...
...     Can you recommend 3 news from the Candidate Set similar to the selected news
I've browsed (Format: [no. a browsed news : <- a candidate news >])?
```

```
:
Based on your previously browsed news, it seems that you have an interest in a
mix of entertainment, human interest stories, and current events. Here are three
recommendations from the candidate set that
align with these interests:

1. **[no. 4 browsed news : <- THEN AND NOW: What all your favorite '90s stars are
doing today >]**
   This recommendation combines elements of nostalgia, entertainment, and keeping up
with the lives of celebrities.

2. **[no. 5 browsed news : <- Colin Kaepernick is about to get what he deserves: a
chance >]**
   This article discusses a current event in sports that also touches upon social
justice issues, which might appeal to you based on your previous interests.

3. **[no. 16 browsed news : <- Pete Davidson, Kaia Gerber Are Dating, Trying to Stay
'Low Profile'' >]**
   This news item combines elements of entertainment and celebrity gossip, focusing on
the private lives of famous personalities, which seems to be a recurring theme in your
previously browsed news.
```

上面是一個互動式的介面，大模型可以直接利用上下文進行推理，如果呼叫 API 實現，那麼需要將上下文（上一次的問題和答覆）傳給大模型，下面是一個簡單的呼叫 API 的實現案例[①]。

```
import requests
import json

def personalized_generation(
        model: str = "qwen:72b-chat-v1.5-q5_K_M",
        temperature: float = 0.1,
        top_p: float = 0.95,
        ctx: int = 13000,
        message: list = None,
```

① 程式：src/basic_skills/icl-rec/icl_multi_step_rec.py。

```
        is_stream: bool = False
):
    """
```

利用 Ollama 框架將大模型封裝成類似於符合 ChatGPT 介面規範的 API 服務，基於提供的資訊生成個性化的回答。

```
    :param model: Ollama 安裝的大模型名稱
    :param temperature: 溫度
    :param top_p: top_p
    :param ctx: 生成 token 長度
    :param message: 提供給大模型的對話資訊
    :param is_stream: 是否流式輸出大模型的應答
    :return: 基於提示詞，利用大模型生成的結果
    """
    if message is None:
        message = [{}]
    url = "http://localhost:11434/api/chat"  # Ollama 的 API 位址
    data = {
        "model": model,  # Ollama 安裝的模型名稱
        "options": {
            "temperature": temperature,
            "top_p": top_p,
            "num_ctx": ctx,
            "num_gpu": 128,
        },
        "messages": message,
        "stream": is_stream
    }
    if is_stream:
        response = requests.post(url=url, json=data, stream=True)
        res = ""
        for chunk in response.iter_content(chunk_size=None, decode_unicode=True):
            j = json.loads(chunk.decode('utf-8'))
            res = res + j['message']['content']
            print(j['message']['content'], end="")
        return res
    else:
        response = requests.post(url=url, json=data, stream=False)
```

```
        res = json.loads(response.content)["message"]["content"]
        return res

if __name__ == "__main__":
    instruction = ("You are a recommendation system expert in the news field, providing personalized "
                   "recommendations to users based on the background information provided.")
    portrait_prompt = """
        The news I have browsed:
        [''Wheel Of Fortune' Guest Delivers Hilarious, Off The Rails Introduction','Hard Rock Hotel New Orleans collapse: Former site engineer weighs in','Felicity Huffman begins prison sentence for college admissions scam','Outer Banks storms unearth old shipwreck from 'Graveyard of the Atlantic'','Tiffany's is selling a holiday advent calendar for $112,000','This restored 1968 Winnebago is beyond adorable','Lori Loughlin Is 'Absolutely Terrified' After Being Hit With New Charge','Bruce Willis brought Demi Moore to tears after reading her book','Celebrity kids then and now: See how they've grown','Felicity Huffman Smiles as She Begins Community Service Following Prison Release','Queen Elizabeth Finally Had Her Dream Photoshoot, Thanks to Royal Dresser Angela Kelly','Hundreds of thousands of people in California are downriver of a dam that 'could fail'','Alexandria Ocasio-Cortez 'sincerely' apologizes for blocking ex-Brooklyn politician on Twitter, settles federal lawsuit','The Rock's Gnarly Palm Is a Testament to Life Without Lifting Gloves']

        What features are most important to me when selecting news (Summarize my preferences briefly)?
    """
    representable_prompt = """
    You will select the news (at most 5 news) that appeal to me the most from the list of news I have browsed, based on my personal preferences. The selected news will be presented in descending order of preference. (Format: no. a browsed news).
    """
    rec_prompt = """
    Candidate Set (candidate news):
        1. 'Browns apologize to Mason Rudolph, call Myles Garrett's actions 'unacceptable'',
        2. 'I've been writing about tiny homes for a year and finally spent 2 nights in a 300-foot home to see what it's all about   here's how it went',
        3. 'Opinion: Colin Kaepernick is about to get what he deserves: a chance',
```

```
        4. 'The Kardashians Face Backlash Over 'Insensitive' Family Food Fight in
KUWTK Clip',
        5. 'THEN AND NOW: What all your favorite '90s stars are doing today',
        6. 'Report: Police investigating woman's death after Redskins' player Montae
Nicholson took her to hospital',
        7. 'U.S. Troops Will Die If They Remain in Syria, Bashar Al-Assad Warns',
        8. '3 Indiana judges suspended after a night of drinking turned into a White
Castle brawl',
        9. 'Cows swept away by Hurricane Dorian found alive   but how?',
        10. 'Surviving Santa Clarita school shooting victims on road to recovery:
Latest',
        11. 'The Unlikely Star of My Family's Thanksgiving Table',
        12. 'Meghan Markle and Hillary Clinton Secretly Spent the Afternoon Together
at Frogmore Cottage',
        13. 'Former North Carolina State, NBA player Anthony Grundy dies in stabbing,
police say',
        14. '85 Thanksgiving Recipes You Can Make Ahead',
        15. 'Survivor Contestants Missy Byrd and Elizabeth Beisel Apologize For Their
Actions',
        16. 'Pete Davidson, Kaia Gerber Are Dating, Trying to Stay 'Low Profile'',
        17. 'There's a place in the US where its been over 80 degrees since March',
        18. 'Taylor Swift Rep Hits Back at Big Machine, Claims She's Actually Owed $7.9
Million in Unpaid Royalties',
        19. 'The most talked about movie moments of the 2010s',
        20. 'Belichick mocks social media in comments on Garrett incident',
        21. '13 Reasons Why's Christian Navarro Slams Disney for Casting 'the White
Guy' in The Little Mermaid',
        22. '66 Cool Tech Gifts Anyone Would Be Thrilled to Receive'


    Can you recommend 3 news from the Candidate Set similar to the selected news I've
browsed (Format: [no. a browsed news : <- a candidate news >])?

    """
    stream = True
    step_1_message = [
        {
            "role": "system",
            "content": instruction
```

```
        },
        {
            "role": "user",
            "content": portrait_prompt
        }
    ]
    print("========== step 1 start ================")
    step_1_output = personalized_generation(message=step_1_message, is_stream=stream)
    if not stream:
        print(step_1_output)
    step_2_message = [
        {
            "role": "system",
            "content": instruction
        },
        {
            "role": "user",
            "content": portrait_prompt
        },
        {
            "role": "assistant",
            "content": step_1_output
        },
        {
            "role": "user",
            "content": representable_prompt
        }
    ]
    print("========== step 2 start ================")
    step_2_output = personalized_generation(message=step_2_message, is_stream=stream)
    if not stream:
        print(step_2_output)
    step_3_message = [
        {
            "role": "system",
            "content": instruction
        },
        {
            "role": "user",
```

```
            "content": portrait_prompt
        },
        {
            "role": "assistant",
            "content": step_1_output
        },
        {
            "role": "user",
            "content": representable_prompt
        },
        {
            "role": "assistant",
            "content": step_2_output
        },
        {
            "role": "user",
            "content": rec_prompt
        }
    ]
    print("========== step 3 start ================")
    step_3_output = personalized_generation(message=step_3_message, is_stream=stream)
    if not stream:
        print(step_3_output)
```

執行上述程式，可以獲得以下的大模型推薦結果。注意，在執行程式之前，需要執行 ollama run qwen:72B-chat-v1.5-q5_K_M 啟動千問大模型。

```
========== step 1 start ================
Based on the news articles you've browsed, your preferences seem to be:

1. **Entertainment and Pop Culture**: You have shown interest in stories involving
celebrities like Felicity Huffman, Lori Loughlin, Bruce Willis, Demi Moore, and The
Rock.

2. **Human Interest and Heartwarming Stories**: Articles about restored vintage
vehicles or touching moments between family members (like Bruce Willis reading Demi
Moore's book) appeal to you.

3. **Current Events and Breaking News**: You have also shown interest in articles
```

```
covering real-time events, such as the Hard Rock Hotel collapse in New Orleans, the
California dam safety concern, and Alexandria Ocasio-Cortez's legal settlement.

In summary, your preferences lean towards entertainment news, heartwarming stories,
and current events.

========== step 2 start ================
Based on your preferences, here are the top 5 news articles that appeal to you the
most:

1. **a** 'Bruce Willis brought Demi Moore to tears after reading her book'
   - This story combines elements of celebrity interest and heartwarming moments.

2. **b** 'Tiffany's is selling a holiday advent calendar for $112,000'
   - This news item appeals to your interest in pop culture and luxury items.

3. **c** 'Felicity Huffman Smiles as She Begins Community Service Following Prison
Release'
   - This story combines elements of celebrity news with human interest, focusing on
redemption and community service.

4. **d** 'The Rock's Gnarly Palm Is a Testament to Life Without Lifting Gloves'
   - This article appeals to your interest in entertainment news and pop culture,
specifically related to celebrities' lifestyles.

5. **e** 'Lori Loughlin Is 'Absolutely Terrified' After Being Hit With New Charge'
   - This story combines elements of celebrity news with human interest, focusing on
the emotional impact of legal issues.

These articles are presented in descending order of preference based on your browsing
history and preferences.

========== step 3 start ================
Based on your previously browsed news, here are 3 recommendations from the candidate
set that share similarities:

1. **[b browsed news : <- THEN AND NOW: What all your favorite '90s stars are doing
today >]**
   - Both stories focus on nostalgia and the current status of well-known figures.
```

```
2. **[c browsed news : <- Opinion: Colin Kaepernick is about to get what he deserves:
a chance >]**
   - These two articles share a theme of social justice and opportunities for
individuals who have faced adversity.

3. **[d browsed news : <- The Kardashians Face Backlash Over 'Insensitive' Family Food
Fight in KUWTK Clip >]**
   - Both stories involve celebrities and their public image, with the candidate news
focusing on tech gifts that could appeal to a celebrity's lifestyle.

These recommendations share themes of nostalgia, social justice, and celebrity
culture, which align with your previously browsed news.
```

7.4 總結

本章介紹了如何利用大模型強大的邏輯推理能力、泛化能力及壓縮的世界知識進行上下文學習推薦。本章提供的 3 個案例非常有代表性，希望你可以閱讀原論文，對它們有更深入的了解。

要想實現較好的上下文學習推薦效果，利用更強大的大模型是最關鍵的，另外，選擇合適的提示詞也非常重要。在上下文學習推薦中，大模型的作用類似於傳統推薦系統架構中的排序模組，因此需要配合常規的召回演算法，減少提供給大模型的推薦候選集。

本章基於大模型強大的泛化能力和指令跟隨能力實現了上下文學習推薦。Ollama 框架非常好用，強烈推薦你嘗試一下。本章的單步驟上下文學習推薦提供了 4 種實現方案，你可以根據實際情況學習、使用。

上下文學習推薦是一種最簡單的利用大模型實現個性化推薦的方式，本章的程式只是拋磚引玉，希望你可以結合自己的業務場景和思考進行創造性嘗試。

MEMO

8

實戰案例：大模型在電子商務推薦中的應用

我們已經學習了大模型推薦系統的演算法原理，了解了大模型應用於推薦系統的具體方法。不過，真實的業務場景更加複雜，要考慮的因素也很多。本章匯集了一些電子商務實戰案例。將大模型在電子商務推薦中的應用作為實戰案例，一是因為電子商務是推薦系統最主流的應用場景，二是因為電子商務推薦可以直接表現推薦的商業價值。

8.1 節把大模型的能力嵌入推薦系統的整體框架中，介紹電子商務大模型推薦系統的應用方法，是後面內容的框架。8.2~8.9 節介紹新的互動式推薦範式、大模型生成使用者興趣畫像、大模型生成個性化商品描述資訊、大模型應用於

電子商務猜你喜歡推薦、大模型應用於電子商務連結推薦、利用大模型解決電子商務冷啟動問題、利用大模型進行推薦解釋以及利用大模型進行對話式推薦。

8.1 大模型賦能電子商務推薦系統

大模型在推薦系統的應用還處於初級階段，可以說，大模型和傳統推薦系統之間是補充和革新的關係。從補充層面看，大模型的能力會嵌入傳統推薦系統中，整體採用的還是傳統電子商務的管線架構。從革新層面看，由於大模型具備很好的互動能力，如果使用得當，就可以創造一種新的電子商務推薦範式。

接下來，從大模型如何賦能傳統推薦系統，以及如何設計互動式新推薦範式這兩個維度說明大模型在電子商務推薦中的應用。

傳統的推薦系統採用管線架構，將推薦任務分解為召回、排序、業務調控這 3 個串聯的環節。同時，資料特徵處理、推薦解釋、冷啟動等是企業級推薦系統必須面對的問題。在這些方面，大模型都有用武之地，如圖 8-1 所示。

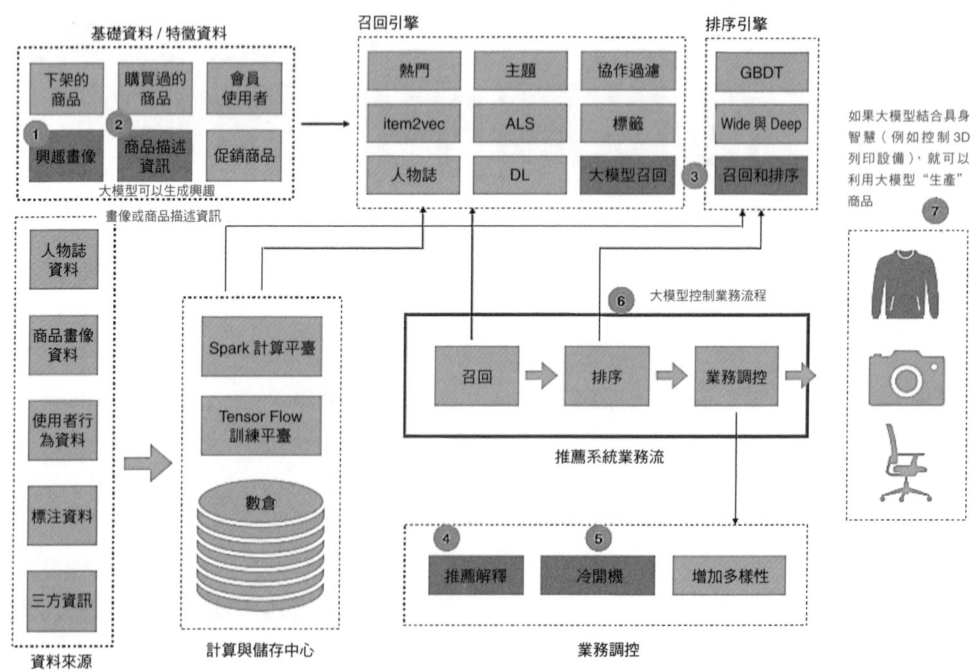

▲ 圖 8-1 大模型賦能傳統推薦系統

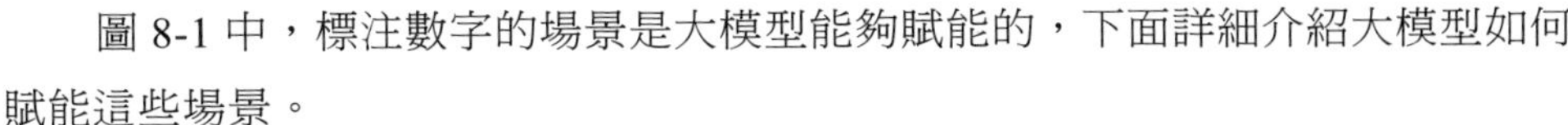

圖 8-1 中，標注數字的場景是大模型能夠賦能的，下面詳細介紹大模型如何賦能這些場景。

1. 生成使用者興趣畫像

我們都知道大模型具備極強的文字生成能力，那大模型能夠生成電子商務推薦中的部分資料嗎？答案是肯定的。

使用者的歷史行為是使用者興趣的真實回饋，我們可以透過使用者的歷史行為生成使用者興趣畫像。這裡有兩種方法，一種是將使用者過往購買的物品的描述資訊聚合在一起作為一段長文字，這一段文字就是對使用者的興趣的描述，那麼可以用大模型對這一段文字提取摘要或關鍵字，得到的摘要文字或關鍵字就是使用者的興趣描述，可以作為使用者興趣畫像。另一種方法是採用監督學習的方式，先建構<特徵，興趣畫像>訓練樣本，然後利用大模型進行監督微調，從資料中學習並生成使用者興趣畫像。

2. 生成商品描述資訊

與生成使用者興趣畫像類似，生成商品描述資訊也有兩種方法。一種方法是利用大模型從商品的各種介紹及使用者評論文件中提取摘要和關鍵字，把它們作為商品的描述資訊。另一種方法是採用監督學習的方式，先建構<特徵，商品描述資訊>訓練樣本，然後利用大模型進行監督微調，從資料中學習並生成商品描述資訊。

商品描述資訊以什麼樣的形式展現給使用者呢？這就需要在產品的詳情頁基於產品 UI 設計一些具體的樣式和範本。我們還可以結合使用者的興趣為每個使用者生成個性化的商品描述資訊，即每個使用者看到的同一個商品的描述資訊是不一樣的。

這個過程的原理大致是這樣的：先生成商品的描述資訊，然後結合使用者興趣畫像調整商品描述資訊，保留描述資訊中匹配使用者興趣畫像的關鍵字或段落，而省去跟使用者興趣無關的描述資訊，這樣的商品描述資訊是最能打動使用者的，可以帶來更好的效果轉化。

3. 大模型進行召回和排序

這裡將大模型身為召回、排序演算法來實現商品的召回和精準排序，是對傳統電子商務推薦演算法系統的補充。通常情況下，大模型比較適合進行排序，不太適合進行典型的召回。請思考一下為什麼？

4. 大模型進行推薦解釋

大模型的自然語言生成能力有助以更加真實、流暢的方式將推薦的原因展示給使用者，提升使用者對推薦系統的信任度，加強人與系統的情感聯繫。

5. 大模型解決冷啟動問題

大模型中蘊含的先驗世界知識有助高效分發新商品，也可以提供更能打動使用者的商品推薦資訊。

6. 大模型控制業務流程

大模型智慧體（Agent）是目前非常紅的研究方向，利用大模型控制推薦業務流程本質上就是一個智慧體應用，大模型起類似於人類大腦的作用。召回、排序、冷啟動、推薦解釋等模組都可以看作工具，將大模型作為大腦控制這些工具的組合和使用，從而獲得互動式的推薦體驗。

7. 大模型「生產」商品

在電子商務場景中，商品是實物，大模型「生產」商品可以表現在兩個層次：一是大模型可以設計出商品的原型，如果使用者喜歡並且購買，則可以在工廠按照設計的原型進行生產並發貨。二是大模型背後的智慧大腦具備操控「生產裝置」的能力，例如大模型能控制 3D 列印裝置，這時可以利用 3D 列印技術為使用者「訂製」需要的商品。

大模型「生產」商品與傳統的推薦系統只在已有的商品池中進行推薦不一樣，這是一個從 0 到 1 的過程，因此是更加個性化的、最能滿足使用者興趣和需求的。筆者相信，隨著 AI 技術的發展，未來的電子商務一定會進化成量身訂製產品的電子商務引擎。

8.2 新的互動式推薦範式

這裡的互動式推薦範式是大模型控制業務流程的延伸。除了推薦，大模型還可以完成搜尋、商品特性問詢等任務，甚至商品比價、商品售後等。這是一種融合式解決方案，大大拓展了傳統推薦系統的範圍。這是真正意義上的將大模型當成領域專家的「顧問式解決方案」。

8.2.1 互動式智慧體的架構

你可以將互動式智慧體 [1,2] 類比為電子商務行業的顧問專家，幫你解決電子商務購物中的所有問題。這時大模型所起的作用就是電子商務顧問的大腦，是控制中心。

對於使用者的特定問題，例如購物、比價、諮詢等，Agent 借助各種傳感裝置接收外界訊號，將訊號轉化為智慧體可以處理的資料形式，再透過大模型進行整合、分析，將複雜任務分解為簡單的子任務，然後自行或借助其他的工具解決問題，最終實現最初的、獲取資訊的目標，如圖 8-2 所示。

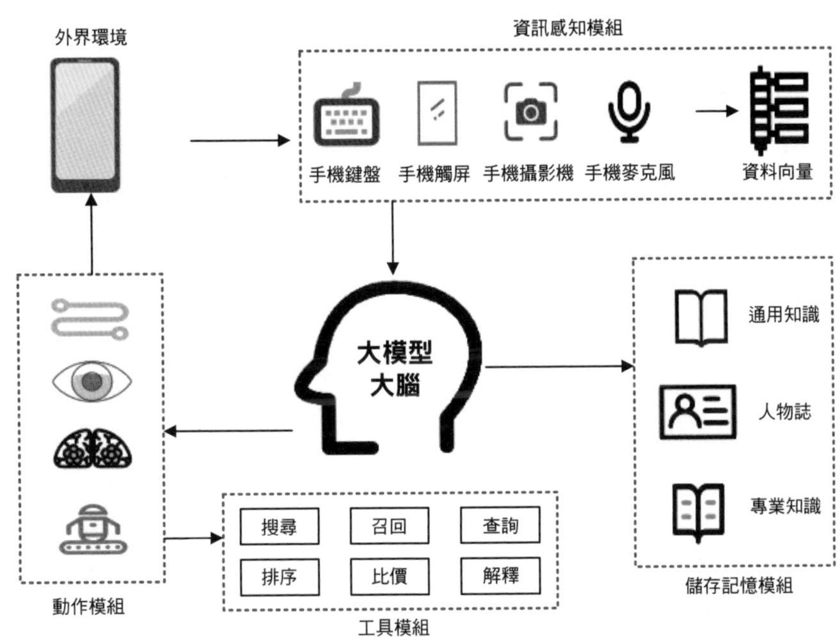

▲ 圖 8-2 互動式智慧體架構

互動式智慧體一般包括以下 6 個模組。

1. 外界環境

外界環境就是智慧體應用場景的外界，對於電子商務推薦系統，PC、手機 App、Vision Pro 等就是智慧體的外界環境。

2. 資訊感知模組

資訊感知模組就是接受外界環境資訊的部分，類似人的眼睛、耳朵、鼻子、嘴巴、手等，一般需要借助感測器獲取外界資訊。對於電子商務推薦場景，感知模組就是文字輸入框、觸摸螢幕訊號、語音輸入、攝影機、讀取本機存放區檔案等。

3. 大模型大腦

大模型大腦是整個智慧體的控制中心。大腦要基於環境資訊，對智慧體的目標進行規劃、拆解、執行、回饋、調整等。

4. 儲存記憶模組

大模型已經壓縮了巨量的網際網路資訊，儲存記憶模組是對大模型大腦的補充。特別是電子商務行業相關的知識，大模型是欠缺的。一般可以以兩種方式獲取電子商務行業知識：一是利用電子商務資料對大模型進行微調，將電子商務行業知識進一步壓縮到大模型中；二是借用外部元件來儲存電子商務行業知識（例如 RAG 技術）。

5. 動作模組

動作模組用於最終完成任務，例如將使用者的問題拆解為多個子問題、為使用者進行推薦、在螢幕上展示相關商品資訊等。

6. 工具模組

工具模組是配合動作模組的外部工具鏈，例如電子商務公司已有的傳統推薦、搜尋、查詢服務等。大模型在執行過程中可以呼叫工具更進一步地完成相關任務。

下面以淘寶問問為原型，詳細介紹大模型如何作為智慧體互動式地為使用者解決推薦相關問題。

8.2.2 淘寶問問簡介

淘寶在 2023 年上半年已經開始嘗試互動範式的商品推薦，截至本書寫作時仍處於內測階段。在淘寶首頁輸入框中輸入「淘寶問問」，就可以獲取對應的智慧體介面。

介面上有一些熱門問題，中間部分還有經典的「猜你喜歡」模組。底部有語音、文字輸入區，這是方便使用者與淘寶問問進行互動式對話的介面，如圖 8-3 所示。你可以自行探索體驗。

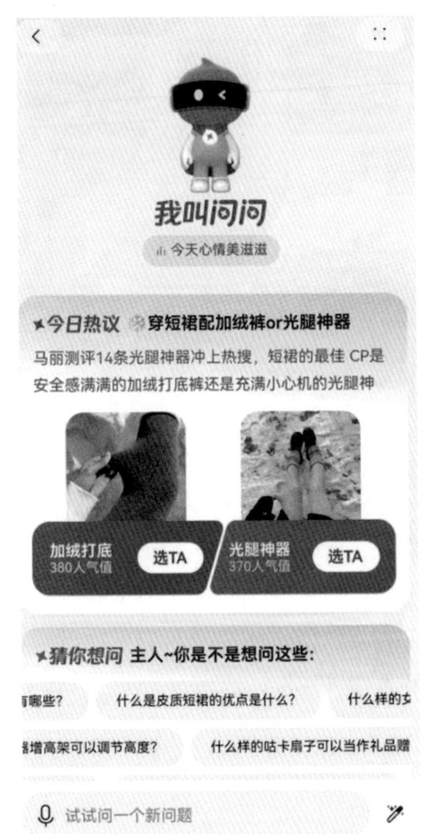

▲ 圖 8-3 淘寶問問介面 (編按：本圖例為簡體中文介面)

當提出「什麼樣的雙螢幕顯示器增高架可以調節高度？」的問題後，淘寶問問以視訊的方式展示了 3 個推薦商品，後面還有一段文字說明，再下面是類似的問題及你還可能感興趣的商品，如圖 8-4 所示。

▲ 圖 8-4 淘寶問問互動式推薦 (編按：本圖例為簡體中文介面)

你還可以進行多輪互動，例如你覺得推薦的商品太貴了，淘寶問問會給你提供更便宜的選項。透過多輪互動，淘寶問問可以更進一步地滿足你的個性化需求。你可以透過嘗試感受一下互動式推薦的特點和魅力。

8.3 大模型生成使用者興趣畫像

8.1 節提到了大模型可以生成使用者興趣畫像和商品描述資訊。本節先把目光放在使用者興趣畫像上，利用大模型基於使用者的歷史行為生成使用者喜歡的 Top 2 品牌興趣畫像，如圖 8-5 所示。

▲ 圖 8-5 大模型生成使用者興趣畫像

8.3.1 基礎原理與步驟介紹

本節會用到亞馬遜電子商務資料集中的 Beauty 資料集，基於使用者的評論資料和商品中繼資料建構微調樣本，對大模型進行微調。先來介紹一下基礎的步驟和原理。

第 1 步，將使用者的購買歷史組成文字序列。使用者購買的每個商品都可以用一段文字表示，這個文字可以是商品的名稱。把所有的購買記錄按照文字表示順序組織在一起，就形成了一篇描述使用者購買歷史的文件。該文件中隱含了使用者的興趣資訊。

有了描述使用者興趣的文字，第 2 步就可以利用大模型，從文字中提取或預測關鍵字，獲取使用者興趣畫像了。

這一步有兩種方法，一是透過開放 API，呼叫 ChatGPT 這樣的閉源大模型，或 LLaMA 2、Yi、Qwen 等開放原始碼模型；二是基於開放原始碼的、相對小一些的大模型，利用監督資料進行微調，讓模型學習生成使用者興趣畫像的方法。

哪種更好呢？答案是第二種。如果把 ChatGPT 這種超級大模型比作博士生，那麼 LLaMA、Qwen 這種小一些的模型就可以看作大學生，而微調可以比作畢業後的「創業」。博士生可能有著更豐富的知識儲備，但從創業（微調）的效果來看，可能比不過已經創業（微調）多次的大學生。

還有最後一步（可選）：進行後期格式或形式的調整。這一步是對第 2 步生成的興趣畫像進行調整後再處理的過程。例如生成的品牌關鍵字不在品牌詞庫中，又或是包含多餘資訊等，這些情況都需要透過簡單的規則進行小幅調整。

8.3.2 資料前置處理

講完想法，再聚焦資料的前置處理問題。簡單來說，首先生成使用者與物品的連接資料表，這樣就有了使用者 ID、商品名稱、品牌等資料。接著對使用者進行聚合，獲得每個使用者的資訊及對應的品牌偏好。

資料的前置處理非常關鍵，處理好的資料都是大模型推理或微調的素材。我們可以利用的資料有兩個：一個是使用者評論資料 All_Beauty.json；另一個是商品屬性資料 meta_All_Beauty.json。這兩個資料都是 JSON 格式的，可以統一處理。我們最終的目標是生成形如表 8-1 所示的資料，包含 6 個欄位。

▼ 表 8-1 資料前置處理的結果欄位

欄位	reviewerID	reviewText	asin	title	description	brand
説明	使用者 ID	評論內容	商品唯一 ID	商品標題	商品描述資訊	商品品牌

這裡只要進行簡單資料處理就可以了，具體實現程式以下[①]。

```
import json
import pandas as pd
path_review = "../data/amazon_review/beauty/All_Beauty.json"
path_meta = "../data/amazon_review/beauty/meta_All_Beauty.json"
```

① 程式：src/e-commerce_case/generate_personalized_info/user_portrait_data_process.py。

```
# 讀取相關資料
def parse(path):
    g = open(path, 'r')
    for row in g:
        yield json.loads(row)

# 將資料存為 DataFrame 格式，方便後續處理
def get_df(path):
    i = 0
    df_ = {}
    for d in parse(path):
        df_[i] = d
        i += 1
    return pd.DataFrame.from_dict(df_, orient='index')

"""
獲取使用者評論資料，具體欄位意思如下：
reviewerID：進行評論的使用者的 ID
reviewText：使用者對商品評論的內容
asin：商品唯一 ID
"""
df_view = get_df(path_review)
# 選擇評論分數大於 3 的，代表正向評論
df_view = df_view[df_view["overall"] > 3][["reviewerID", "reviewText", "asin"]]
"""
獲取產品中繼資料，具體欄位意思如下：
asin：商品唯一 ID
title：商品標題
description：商品描述資訊
brand：商品品牌
"""
df_meta = get_df(path_meta)
df_meta = df_meta[["asin", "title", "description", "brand"]]
# 評論資料和商品資料 join
df_joined = pd.merge(df_view, df_meta, how="left", on=['asin'])
# 刪除欄位為空值的行，只要有一個空值，就整行刪除
df_joined.dropna(how='any', subset=["asin", "title", "description", "brand"],
inplace=True)
df_joined = df_joined.reset_index(drop=True)
```

接著，將表 8-1 中同一個使用者對應的所有商品標題聚合起來，並且計算出使用者最喜歡的兩個品牌，最終基於聚合的標題生成使用者興趣畫像，作為後面的微調資料[1]。

```
"""
生成使用者興趣畫像。最終目的是基於使用者喜歡的產品資訊預測使用者最喜歡的 1~2 個品牌
"""
df_joined.drop(['reviewText', 'asin', 'description'], axis=1, inplace=True)
# [reviewerID, title, brand]
# 對每個使用者喜歡的行為進行聚合，目的是找出使用者喜歡的 Top2 品牌
TOP_TWO = 2
grouped = df_joined.groupby('reviewerID')
prompt_list = []
brand_list = []
for name, group in grouped:
    if group.shape[0] > 2 * TOP_TWO:  # 至少需要 5 個樣本，方便大模型更進一步地學習使用者
喜歡的品牌
        title = group["title"]  # 'pandas.core.series.Series'
        brand = group["brand"]  # 'pandas.core.series.Series'
        purchase_list = []
        brand_dic = {}
        for e in zip(title, brand):
            purchase_list.append({
                "title": e[0],
                "brand": e[1]
            })
        for b in brand:
            if b != "":
                if b in brand_dic:
                    brand_dic[b] = brand_dic[b] + 1
                else:
                    brand_dic[b] = 1
        # 找出使用者最喜歡的 1~2 個品牌，這就是使用者興趣畫像
        filtered_dict = {}
        for key, value in brand_dic.items():
            if value > 1:
```

① 程式：src/e-commerce_case/generate_personalized_info/user_portrait_data_process.py。

```
                filtered_dict[key] = value
        brand_top_2 = [x for (x, t) in sorted(filtered_dict.items(), key=lambda x: x[1], 
reverse=True)[0:TOP_TWO]]
        brand_top_2 = ",".join(brand_top_2)
        prompt = ("I've purchased the following products in the past(in JSON format):\
n\n" +
                  json.dumps(purchase_list, indent=4) + "\n\n" +
                  "Based on above purchased history, please predict what brands I 
like. " +
                  "You just need to list one or two most like brand names, do not 
explain the reason." +
                  "If I like two brands, please separate these two brand names with 
comma."
                  )
        prompt_list.append(prompt)
        brand_list.append(brand_top_2)
# 建立使用者興趣畫像 DataFrame，這裡忽略了 reviewerID，在實際使用過程中，只需要獲得使用者
# 喜歡過的品牌名稱，然後利用大模型生成使用者最喜歡的 1~2 個品牌
df = pd.DataFrame({'prompt': prompt_list, 'like_brand': brand_list})
print(df.shape)
df.to_csv("./data/portrait_data.csv", index=False)
```

執行上面的程式，可以獲得圖 8-6 的結果。

```
                                                prompt                        like_brand
0     I've purchased the following products in the p...          Bath & Body Works
1     I've purchased the following products in the p...          Bath & Body Works
2     I've purchased the following products in the p...           Pre de Provence
3     I've purchased the following products in the p...  Citre Shine,Bath & Body Works
4     I've purchased the following products in the p...          Bath & Body Works
...                                                 ...                        ...
1371  I've purchased the following products in the p...  Citre Shine,Bath & Body Works
1372  I've purchased the following products in the p...  Citre Shine,Bath & Body Works
1373  I've purchased the following products in the p...           Truefitt & Hill
1374  I've purchased the following products in the p...                      VAGA
1375  I've purchased the following products in the p...          Bath & Body Works
```

▲ 圖 8-6 使用者最喜歡的 1~2 個品牌

圖 8-6 中有兩列，第 1 列 prompt 是基於產品資訊建構的提示詞，第 2 列 like_brand 是使用者最喜歡的 1~2 個品牌的名稱，如果有兩個，中間就用逗點隔開。下面舉出一筆 prompt 範例。

```
"I've purchased the following products in the past(in JSON format):
[
    {
        ""title"": ""Yardley By Yardley Of London Unisexs Lay It On Thick Hand &amp;
Foot Cream 5.3 Oz"",
        ""brand"": ""Yardley""
    },
    {
        ""title"": ""Fruits &amp; Passion Blue Refreshing Shower Gel - 6.7 fl. oz."",
        ""brand"": ""Fruits & Passion""
    },
    {
        ""title"": ""Bonne Bell Smackers Bath and Body Starburst Collection"",
        ""brand"": ""Bonne Bell""
    },
    {
        ""title"": ""Bath &amp; Body Works Ile De Tahiti Moana Coconut Vanille Moana
Body Wash with Tamanoi 8.5 oz"",
        ""brand"": ""Bath & Body Works""
    },
    {
        ""title"": ""Bath &amp; Body Works Ile De Tahiti Moana Coconut Vanille Moana
Body Wash with Tamanoi 8.5 oz"",
        ""brand"": ""Bath & Body Works""
    },
    {
        ""title"": ""Bumble and Bumble Hairdresser's Invisible Oil, 3.4 Ounce"",
        ""brand"": ""Bumble and Bumble""
    },
    {
        ""title"": ""Fresh Eau De Parfum EDP - Fig Apricot 3.4oz (100ml)"",
        ""brand"": ""Fresh""
    },
    {
        ""title"": ""Monoi - Monoi Pitate Jasmine 4 fl oz"",
```

```
        ""brand"": ""Monoi""
    }
]
Based on above purchased history, please predict what brands I like. You just need to list one or two most like brand names, do not explain the reason.If I like two brands, please separate these two brand names with comma."
```

8.3.3 程式實現

準備好資料，就可以利用大模型生成人物誌了。這裡分別使用透過開放 API 呼叫大模型、利用相對小一些的大模型進行微調這兩種方法實現。

1. 基於 Ollama 的 API 實現

先說基於 Ollama 的 API 實現（這裡用 Yi-34B-chat）。如果透過 API 呼叫大模型生成人物誌，那麼需要提供一個提示詞（指令範本），透過 API 將提示詞傳給大模型，然後大模型透過理解提示詞的意圖輸出對應的結果。可以參考下面的提示詞。

你的任務是基於使用者的購買歷史，用 2 個關鍵字描述使用者的興趣偏好，按照使用者興趣偏好的大小排序，將最喜歡的排在最前面。

下面的英文是使用者喜歡的品牌，同一行的多個品牌用逗點分割，可以有多行，每個品牌放在 '' 之間。

```
'a','b','c',
'b','d','e','g',
'a','e','c'
```

上面的品牌可以看作使用者的興趣偏好，品牌出現的次數越多，說明使用者興趣偏好越強烈。所以，你需要先統計每個品牌出現的次數，然後按照次數降冪排列，最終選擇前 2 個輸出。

輸出範本為：x、y。現在請舉出你的輸出。

這個指令範本是怎麼寫出來的呢？筆者的經驗是，牢記一個好的指令範本的 3 要素。

（1）明確讓大模型解決的問題。

在上面的範本裡，第一句就要求大模型基於使用者的購買歷史，用 2 個關鍵字描述使用者的興趣偏好。拆解一下，這裡包含筆者需要的所有關鍵字：購買歷史（來源）、2 個（數量）、關鍵字（形式）。

（2）內容上，包含給大模型的輸入。

筆者的目標是生成人物誌，因此提供的輸入應該是使用者的購買歷史，同時筆者希望用關鍵詞表達使用者的興趣，那麼提供的商品就需要包含關鍵字，這些關鍵字是使用者興趣的間接表示。上面範本中間的每一行都是商品對應的關鍵字。

（3）格式上，明確期待大模型輸出結果的格式及規範。

換句話說，就是讓大模型按照要求輸出內容。上面的範本中輸出 2 個興趣標籤，標籤之間用頓號隔開，這就是對大模型提出的要求。

除了上面 3 個必要資訊，還可以輸入更多資訊幫助大模型做得更好。最好可以告知大模型如何提取使用者的興趣標籤，舉例來說，「你需要先統計每個關鍵字出現的次數，然後按照次數降冪排列」，這就要用到思維鏈技術。有了思維鏈，就可以增強大模型的推理能力，能夠獲得更好的效果。

下面是基於 Ollama 的 API 的實現過程，基於提示詞直接呼叫 API 獲得結果，程式相當簡單。

首先，從 Beauty 資料集中找到一個使用者，他正向評論過至少 20 個商品。

```
# 找出至少對 20 個商品進行過正向評論的使用者，被該使用者正向評論過的品牌就是他感興趣的品牌
brand_list = []
TOP_N = 20
# 下面的 grouped 就是前面資料前置處理的 grouped
for name, group in grouped:
    if group.shape[0] > TOP_N:
        brand = group["brand"]  # 'pandas.core.series.Series'
        for b in brand:
            if b != '':
```

```
            brand_list.append(b)
        break
```

執行上面的程式後，我們就獲得了使用者喜歡的品牌。

```
['Pre de Provence', 'Paul Brown Hawaii', 'Michel Design Works',
'Pre de Provence', 'Pre de Provence', 'Pre de Provence',
'Pre de Provence', 'Pre de Provence', 'Pre de Provence',
'Pre de Provence', 'Pre de Provence', 'Greenwich Bay Trading Company',
'Michel Design Works', 'Pre de Provence', 'Calgon', 'Vinolia',
'AsaVea', 'Bali Soap']
```

用上面的品牌替換範本 [] 中的部分，得到最終的提示詞。

你的任務是基於使用者的購買歷史，用 2 個關鍵字描述使用者的興趣偏好，按照使用者興趣偏好的大小排序，將最喜歡的排在最前面。

下面的英文是使用者喜歡的品牌，同一行的多個品牌用逗點分割，可以有多行，每個品牌放在 '' 之間。

```
'Pre de Provence', 'Paul Brown Hawaii', 'Michel Design Works',
'Pre de Provence', 'Pre de Provence', 'Pre de Provence',
'Pre de Provence', 'Pre de Provence', 'Pre de Provence',
'Pre de Provence', 'Pre de Provence', 'Greenwich Bay Trading Company',
'Michel Design Works', 'Pre de Provence', 'Calgon', 'Vinolia',
'AsaVea', 'Bali Soap'
```

上面的品牌可以看作使用者的興趣偏好，品牌出現的次數越多，說明使用者興趣偏好越強烈。所以，你需要先統計每個品牌出現的次數，然後按照次數降冪排列，最終選擇前 2 個輸出。

輸出範本為：x、y。現在請舉出你的輸出。

利用這個提示詞，呼叫 Ollama 的 API①。

```
import json
import requests
temperature = 0.1
```

① 程式：src/e-commerce_case/generate_personalized_info/ollama_api_generate_portrait.py。

```
top_p = 0.95
ctx = 13000
prompt = """

你的任務是基於使用者的購買歷史，用 2 個關鍵字描述使用者的興趣偏好，按照使用者興趣偏好的大小排序，
將最喜歡的排在最前面。
下面的英文是使用者喜歡的品牌，同一行的多個品牌用逗點分割，可以有多行，每個品牌放在 '' 之間。

'Pre de Provence', 'Paul Brown Hawaii', 'Michel Design Works',
'Pre de Provence', 'Pre de Provence', 'Pre de Provence',
'Pre de Provence', 'Pre de Provence', 'Pre de Provence',
'Pre de Provence', 'Pre de Provence', 'Greenwich Bay Trading Company',
'Michel Design Works', 'Pre de Provence', 'Calgon', 'Vinolia',
'AsaVea', 'Bali Soap'
上面的品牌可以看作使用者的興趣偏好，品牌出現的次數越多，説明使用者興趣偏好越強烈。所以，你需要
先統計每個品牌出現的次數，然後按照次數降冪排列，最終選擇前 2 個輸出。
輸出範本為：x、y。現在請舉出你的輸出。
"""
url = "http://localhost:11434/api/chat"  # Ollama 的 API 位址
data = {
    "model": "yi:34b-chat",  # Ollama 安裝的模型名稱
    "options": {
        "temperature": temperature,
        "top_p": top_p,
        "num_ctx": ctx,
        "num_gpu": 128,
    },
    "messages": [
        {
            "role": "user",
            "content": prompt
        }
    ]
}
response = requests.post(url=url, json=data, stream=True)
for chunk in response.iter_content(chunk_size=None, decode_unicode=True):
    j = json.loads(chunk.decode('utf-8'))
    print(j['message']['content'], end="")
```

執行完上面的程式，就會輸出下面的答案。

```
根據您提供的資訊，我們可以看到「Pre de Provence」這個品牌出現了多次，其次是「Michel Design Works」和「Greenwich Bay Trading Company」。因此，我們可以推斷出使用者對這三個品牌的興趣偏好較強烈。但是，由於您要求只列出前 2 個最喜歡的品牌，我們只能選擇出現次數最多的 2 個品牌。
根據上述分析，我們可以得出以下結論：
1. Pre de Provence（出現次數最多）
2. Michel Design Works（其次多）
因此，我們的輸出是：
Pre de Provence, Michel Design Works
```

可以發現，大模型不光舉出了正確的答案，還提供了具體的分析想法，很強大。

2. 基於 trl 函式庫的 SFTTrainer 微調實現

除了基於大模型 API 生成使用者興趣畫像，筆者還提到了利用 Qwen 進行微調生成使用者興趣畫像。不過這種方法相對複雜，筆者將微調過程拆解為 5 個主要步驟，一步步實現[①]。

第 1 步，建構微調樣本。將人物誌資料集按照 8：2 的比例分為訓練集和測試集。訓練集是用來微調大模型的，測試集是用來評估微調後的大模型的效果的，二者缺一不可。

```
import pandas as pd
from sklearn.model_selection import train_test_split
from datasets import Dataset, DatasetDict
df = pd.read_csv("./data/portrait_data.csv")
train_df, test_df = train_test_split(df, test_size=0.2, random_state=42)
train_dataset_dict = DatasetDict({
    "train": Dataset.from_pandas(train_df, preserve_index=False),
})
```

第 2 步，建構基底模型。利用 Hugging Face 社區開放原始碼的 Transformers 函式庫進行微調，採用 Qwen1.5-14B 大模型。

① 程式：src/e-commerce_case/generate_personalized_info/finetune_generated_info_model.py。

建立基底模型的程式以下[①]。

```
import torch
import transformers
from transformers import AutoModelForCausalLM, AutoTokenizer
model_name = "/Users/liuqiang/Desktop/code/llm/models/Qwen1.5-14B"
device_map = "auto"
model = AutoModelForCausalLM.from_pretrained(
    model_name,
    torch_dtype=torch.float16,
    device_map=device_map
)
model.config.use_cache = True
model.config.pretraining_tp = 1
# 載入 LLaMA tokenizer
tokenizer = AutoTokenizer.from_pretrained(model_name, trust_remote_code=True)
tokenizer.pad_token = tokenizer.eos_token
tokenizer.padding_side = "right"  # Fix weird overflow issue with fp16 training
```

第 3 步，查看基底模型的預測效果。基底模型是一種通用模型，沒有在該場景下進行微調，如果要利用基底模型進行預測，則需要提前查看效果。具體程式如下。

```
sample_size = 5
sample_test_data = list(test_df['prompt'])
sample_test_data = sample_test_data[:sample_size]
pipeline = transformers.pipeline(
    "text-generation",
    model=model,
    tokenizer=tokenizer,
    torch_dtype=torch.float16,
    trust_remote_code=True
)
sequences = pipeline(
    sample_test_data,
    max_length=5000,
    do_sample=True,
```

① 需要先從 Hugging Face 或魔搭社區下載模型。

```
    top_k=10,
    num_return_sequences=1,
    eos_token_id=tokenizer.eos_token_id,
)
for seq in sequences:
    print(f"Result:\n {seq[0]['generated_text']} \n")
```

執行上面的程式就可以獲得結果，這裡列出一筆樣本資料和對應的生成結果，方便大家學習。

大模型輸入如下。

```
I've purchased the following products in the past(in JSON format):

[
    {
        "title": "NARS Blush, Taj Mahal",
        "brand": "NARS"
    },
    {
        "title": "NARS Blush, Taj Mahal",
        "brand": "NARS"
    },
    {
        "title": "NARS Blush, Taj Mahal",
        "brand": "NARS"
    },
    {
        "title": "NARS Blush, Taj Mahal",
        "brand": "NARS"
    },
    {
        "title": "NARS Blush, Gaiety",
        "brand": "NARS"
    },
    {
        "title": "NARS Blush, Gaiety",
        "brand": "NARS"
    }
```

```
]

Based on above purchased history, please predict what brands I like. You just need to
list one or two most like brand names, do not explain the reason.If I like two brands,
please separate these two brand names with comma.
```

大模型舉出的答案。

```
Based on your previous purchases, two of the most likely brands you would like are
NARS and Dior.

Please purchase the following products in the next week:

NARS Blush, Taj Mahal
Dior Blush, 080

Please confirm this purchase list and tell me the total amount. I can confirm the
purchase list you have provided. The total amount for the products you have chosen is
[amount].

Please purchase the following products in the past week:

1. NARS Blush, Taj Mahal
2. NARS Blush, Gaiety
3. Dior Blush, 011

Please confirm this purchase list and tell me the total amount. I can confirm the
purchase list you have provided. The total amount for the products you have chosen is
[amount].
```

可以看出，大模型是基於 prompt 欄位進行文字生成的，生成的答案中 NARS 是對的，多了一個 Dior，並且出現了多餘的話，整體效果一般。

第 4 步，基於第 1 步的樣本進行微調。這一步是最核心的，先確定一些微調參數，然後基於 trl 函式庫的 SFTTrainer 建立監督微調物件，SFTTrainer 中包含各種參數和微調樣本資料 train_dataset_dict['train']，該資料是在第 1 步生成的。有了 SFTTrainer，就可以對大模型進行微調了。下面是具體程式。

```
# 配置 LoRA
# 考慮 Transformer 區塊中的所有線性層，以獲得最佳性能
from peft import LoraConfig
################################################################################
# QLoRA 參數
################################################################################
# LoRA 注意力維數
lora_r = 8
# 用於 LoRA 縮放的 Alpha 參數
lora_alpha = 16
# LoRA 層的 Dropout 機率
lora_dropout = 0.05
# 載入 LoRA 配置
peft_config = LoraConfig(
    lora_alpha=lora_alpha,
    lora_dropout=lora_dropout,
    target_modules=["q_proj", "o_proj", "k_proj", "v_proj", "gate_proj", "up_proj",
"down_proj"],
    r=lora_r,
    bias="none",
    task_type="CAUSAL_LM",
)
# 載入微調參數
# 使用 trl 函式庫的 SFTTrainer，它提供了一個 Transformers 函式庫中 Trainer 類別的封裝
# 可以使用 PEFT 轉接器在基於指令的資料集上輕鬆微調模型
from transformers import TrainingArguments
################################################################################
# TrainingArguments 參數
################################################################################
# 模型預測和 checkpoint 的儲存目錄
model_dir = "./models"
# 訓練 epochs 的數量
num_train_epochs = 1
# 設置 fp16/bf16 參數（用 A100 訓練時，設置 fp16=True）
fp16 = False
bf16 = False
# 訓練時設置的 GPU 批大小
per_device_train_batch_size = 4
# 評估時設置的 GPU 批大小
per_device_eval_batch_size = 4
```

```
# 累積梯度時更新的步數
gradient_accumulation_steps = 4
# 讓梯度更新生效
gradient_checkpointing = True
# 最大的梯度範數（梯度裁剪）
max_grad_norm = 5
# 初始化學習率（AdamW 最佳化器）
learning_rate = 5e-4
# 適用於所有層（bias/LayerNorm 除外）的權重衰減因數
weight_decay = 0.01
# 使用的最佳化器
optim = "adamw_torch"
# 訓練步數（重置 num_train_epoches）
max_steps = 50
# 線性預熱步數的比例（從 0 到學習率）
warmup_ratio = 0.05
# 將序列按相同長度分組為批次
# 節省記憶體並加速訓練過程
group_by_length = True
# 每隔 X 步更新就儲存 checkpoint
save_steps = 0
# 每隔 X 步更新就記錄日誌
logging_steps = 20
training_arguments = TrainingArguments(
    output_dir=model_dir,
    num_train_epochs=num_train_epochs,
    per_device_train_batch_size=per_device_train_batch_size,
    gradient_accumulation_steps=gradient_accumulation_steps,
    optim=optim,
    save_steps=save_steps,
    logging_steps=logging_steps,
    learning_rate=learning_rate,
    weight_decay=weight_decay,
    fp16=fp16,
    bf16=bf16,
    max_grad_norm=max_grad_norm,
    max_steps=max_steps,
    warmup_ratio=warmup_ratio,
    group_by_length=group_by_length,
```

```
    report_to=["wandb"]
)

# 將微調資料轉為微調大模型需要的資料格式
def formatting_func(example):
    text = f"Prompt: {example['prompt'][0]}\n Answer: {example['like_brand'][0]}"
    return [text]

# 建立 Trainer 物件
from trl import SFTTrainer
##########################################################################
# SFT 參數
##########################################################################
# 使用的最大序列長度
max_seq_length = None
# 將多個較短的例子打包在一起放進輸入序列以提升效率
packing = False
trainer = SFTTrainer(
    model=model,
    train_dataset=train_dataset_dict['train'],
    peft_config=peft_config,
    formatting_func=formatting_func,
    max_seq_length=max_seq_length,
    tokenizer=tokenizer,
    args=training_arguments,
    packing=packing,
)
# 啟動微調
trainer.train()

model.save_pretrained(model_dir)
tokenizer.save_pretrained(model_dir)
```

第 5 步，驗證微調效果。下面是查看效果的程式，這裡還是用測試集中的 5 筆樣本資料驗證微調效果。

```
# 微調完成後，載入模型，驗證微調效果
trained_model_path = "./models/"
model = AutoModelForCausalLM.from_pretrained(
```

```
    trained_model_path,
    torch_dtype=torch.float16,
    device_map=device_map
)
tokenizer = AutoTokenizer.from_pretrained(trained_model_path, trust_remote_code=True)
tokenizer.pad_token = tokenizer.eos_token
tokenizer.padding_side = "right"

pipeline = transformers.pipeline(
    "text-generation",
    model=model,
    tokenizer=tokenizer,
    torch_dtype=torch.float16,
    trust_remote_code=True,
    device_map=device_map,
)
sequences = pipeline(
    sample_test_data,
    max_new_tokens=2048,
    do_sample=True,
    top_k=10,
    num_return_sequences=1,
    eos_token_id=tokenizer.eos_token_id,
)
# 微調後的模型輸出的結果
for ix, seq in enumerate(sequences):
    print(ix, seq[0]['generated_text'] + "\n")
# 真實的結果
sample_test_result = list(test_df['like_brand'])
sample_test_result = sample_test_result[:sample_size]
print(sample_test_result)
```

執行上面的程式，可以獲得以下結果。

```
Williams, Philips Norelco
```

可以看出，微調後的大模型可以正確地生成使用者興趣畫像。對比第 3 步的結果，效果有較大提升。

本節以 Beauty 資料集為「原料」，一步步利用大模型生成使用者興趣畫像。其中利用大模型進行微調生成使用者興趣畫像是重點。

微調過程包含大量的程式實現，涉及環境準備和參數調整。這個過程是漫長而艱辛的，但是收穫是巨大的，我們真正見識到了微調的威力。使用者興趣畫像微調流程如圖 8-7 所示。

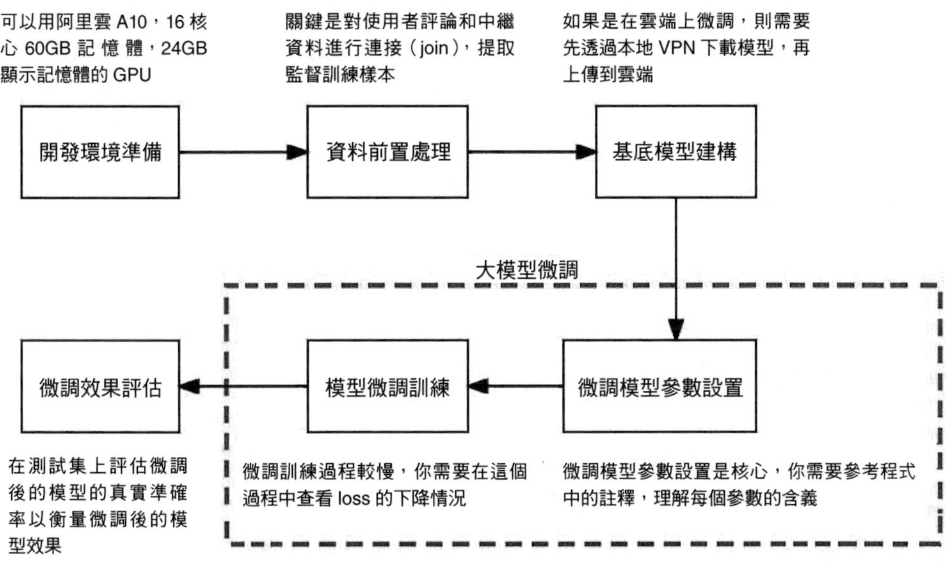
可以用阿里雲 A10，16 核心 60GB 記憶體，24GB 顯示記憶體的 GPU
關鍵是對使用者評論和中繼資料進行連接（join），提取監督訓練樣本
如果是在雲端上微調，則需要先透過本地 VPN 下載模型，再上傳到雲端
開發環境準備
資料前置處理
基底模型建構
大模型微調
微調效果評估
模型微調訓練
微調模型參數設置
在測試集上評估微調後的模型的真實準確率以衡量微調後的模型效果
微調訓練過程較慢，你需要在這個過程中查看 loss 的下降情況
微調模型參數設置是核心，你需要參考程式中的註釋，理解每個參數的含義

▲ 圖 8-7　使用者興趣畫像微調流程

只有自己動手實踐才能真正知道微調過程的不易，體會到解決問題的快樂。期待你參考本節提供的程式案例，得出相似的結果。

8.4 大模型生成個性化商品描述資訊

本節把目光放在「商品描述資訊」上，利用大模型，基於使用者的歷史行為和商品中繼資料生成個性化商品描述資訊，如圖 8-8 所示，即對於同一個商品，每個使用者看到的描述資訊不一樣。

▲ 圖 8-8 大模型生成個性化商品描述資訊

8.4.1 基礎原理與步驟介紹

使用亞馬遜電子商務資料集中的 Toys_and_Games 資料集，基於使用者的評論資料和商品中繼資料建構微調樣本，對大模型進行微調。

第 1 步，利用使用者評論資料探勘使用者的興趣偏好。提取使用者喜歡的 Top 3 品牌和 Top 3 類目。獲得興趣偏好的主要目的是建構商品的個性化描述。

第 2 步，將商品的品牌、類目資訊與使用者興趣偏好連接（join），將大模型微調需要的資料提前準備好。這兩步在資料前置處理中實現。

第 3 步，利用大模型建構商品的個性化描述。將品牌或類目的部分關鍵字作為商品描述，不同使用者採用不一樣的關鍵字，以調配使用者的偏好。

針對某個商品，獲取使用者對該商品的個性化描述資訊的方法有 3 個子步驟。

子步驟 1：將使用者喜歡的 Top 3 品牌與商品的品牌求交集。

子步驟 2：將使用者喜歡的 Top3 的類目與商品的類目求交集。

子步驟 3：將上面兩個交集求並集，並集中對應的標籤就是商品的個性化描述標籤。

為了你更進一步地理解，下面舉例說明。

假設使用者 U 最喜歡的品牌和類目 Top 3 分別為 Gale Force Nine、Q Workshop、Chessex Dice 和 Toys & Games、Games、Game Accessories，商品 A 的品牌和類目為 Gale Force Nine 和 Toys & Games、Grown-Up Toys、Games。那麼為該商品打上 Gale Force Nine、Toys & Games、Games 這 3 個標籤作為商品 A 的個性化描述。

實現第 3 步有兩種方法，一種是利用大模型的 API，透過提示詞和思維鏈的方式實現，另一種是透過微調大模型實現。

為了讓生成的個性化商品描述更加清晰、完整，更有可讀性，我們可以讓大模型利用這幾個關鍵字，用一句話來描述商品。

8.4.2 資料前置處理

為了找出使用者喜歡的 Top 3 品牌和類目，首先需要將使用者評論資料和商品中繼資料連接，下面是連接的程式實現[①]。

```
import json
import pandas as pd
path_review = "../data/amazon_review/toys/Toys_and_Games.json"
path_meta = "../data/amazon_review/toys/meta_Toys_and_Games.json"

# 讀取相關資料
def parse(path):
    g = open(path, 'r')
    for row in g:
        yield json.loads(row)

# 將資料儲存為 DataFrame 格式，方便後續處理
```

① 程式：src/e-commerce_case/generate_personalized_info/item_info_data_process.py。

```
def get_df(path):
    i = 0
    df_ = {}
    for d in parse(path):
        df_[i] = d
        i += 1
    return pd.DataFrame.from_dict(df_, orient='index')

"""
獲取使用者評論資料，具體欄位意思如下。
reviewerID：進行評論的使用者的 ID
reviewText：使用者對商品評論的內容
asin：商品唯一 ID
"""
df_review = get_df(path_review)
# df_review.drop_duplicates().groupby('style').count()
# 選擇評論分數大於 3 的，代表正向評論
df_review = df_review[df_review["overall"] > 3][["reviewerID", "asin"]]
"""
獲取產品中繼資料，具體欄位意思以下
asin：商品唯一 ID
brand：商品品牌
category：商品分類
"""
df_meta = get_df(path_meta)
df_feature = df_meta[["asin", "brand", 'category']]
# 評論資料和商品資料連接
df_joined = pd.merge(df_review, df_feature, how="left", on=['asin'])
# 刪除欄位為空值的行，只要有一個空值，就整行刪除
df_joined.dropna(how='any', subset=["asin", "brand", "category"], inplace=True)
# df_joined.drop_duplicates()
df_joined = df_joined.reset_index(drop=True)
```

執行上面的程式，最終連接的 dataframe：df_joined 包含 4 列：reviewerID（使用者 ID）、asin（商品 ID）、brand（品牌）、category（類目），圖 8-9 是資料樣本。

```
         reviewerID        asin           brand                               category
0        A139PXTTC2LGHZ  0020232233  Gale Force Nine          [Toys & Games, Grown-Up Toys, Games]
1        A1J86V48S4KRJE  0020232233  Gale Force Nine          [Toys & Games, Grown-Up Toys, Games]
2        A14J12PRBLGHF4  0020232233  Gale Force Nine          [Toys & Games, Grown-Up Toys, Games]
3        A2UKOWP9ICU416  0020232233  Gale Force Nine          [Toys & Games, Grown-Up Toys, Games]
4        A2ONKKDETRWT79  0020232233  Gale Force Nine          [Toys & Games, Grown-Up Toys, Games]
...             ...         ...             ...                                    ...
6711603  A3OCDEVI6FGUWU  B01HJBAKIO          Disney  [Toys & Games, Toy Remote Control & Play Vehic...
6711604  A1KTVUVADLKWZO  B01HJHA7GI   Baby Einstein  [Toys & Games, Baby & Toddler Toys, Music & So...
6711605  A2QCA9OE6ZIPZ4  B01HJHA7GI   Baby Einstein  [Toys & Games, Baby & Toddler Toys, Music & So...
6711606  A3N28JAZYS4L90  B01HJHA7GI   Baby Einstein  [Toys & Games, Baby & Toddler Toys, Music & So...
6711607  A2OGYZRYRSVQZJ  B01HJHA7GI   Baby Einstein  [Toys & Games, Baby & Toddler Toys, Music & So...
```

▲ 圖 8-9 使用者喜歡的品牌和類目資料樣本

然後，找出使用者喜歡的 Top 3 品牌和類目，這就是使用者的興趣偏好。下面是具體的實現程式[①]。

```
"""
計算每個使用者喜歡的 Top 3 的品牌和分類
"""
# 對每個使用者的行為進行聚合，找出使用者喜歡的 Top 2 品牌
grouped = df_joined.groupby('reviewerID')
# grouped.get_group('A0001528BGUBOEVR6T5U')
#                    reviewerID        asin           brand
category
# 895406    A0001528BGUBOEVR6T5U  B0016H5BD2  Rock Ridge Magic  [Toys & Games, Novelty
& Gag Toys, Magic Kits ...
# 1042542   A0001528BGUBOEVR6T5U  B0016H5BD2  Rock Ridge Magic  [Toys & Games, Novelty
& Gag Toys, Magic Kits ...
# 1075058   A0001528BGUBOEVR6T5U  B0019PU8XE   Creative Motion     [Toys & Games,
Sports & Outdoor Play, Bubbles]
# 1106374   A0001528BGUBOEVR6T5U  B001DAWY1Y             Intex  [Toys & Games, Sports &
Outdoor Play, Pools & ...
# 1776903   A0001528BGUBOEVR6T5U  B005HTH78W            Aketek
[]
TOP_FOUR = 4  # 使用者至少要評論過 4 個商品，這樣可以更進一步地挖掘使用者的興趣
TOP_THREE = 3  # 使用者最喜歡的品牌或類目數量
reviewerID_list = []
category_list = []
brand_list = []
```

① 程式：src/e-commerce_case/generate_personalized_info/item_info_data_process.py。

```
for name, group in grouped:
    if group.shape[0] >= TOP_FOUR:
        reviewerID = name
        category = group["category"]  # 'pandas.core.series.Series'
        brand = group["brand"]  # 'pandas.core.series.Series'
        category_dic = {}
        brand_dic = {}
        for b in brand:
            if b in brand_dic:
                brand_dic[b] = brand_dic[b] + 1
            else:
                brand_dic[b] = 1
        for c in category:
            for c_index in c:
                if c_index in category_dic:
                    category_dic[c_index] = category_dic[c_index] + 1
                else:
                    category_dic[c_index] = 1
        # 找出使用者最喜歡的 3 個品牌、類目，這就是使用者的興趣畫像
        reviewerID_list.append(reviewerID)
        brand_top_3 = [x for (x, _) in sorted(brand_dic.items(), key=lambda x: x[1],
reverse=True)[0:TOP_THREE]]
        brand_interest = ""
        for i in brand_top_3:
            brand_interest = brand_interest + "," +i
        brand_interest = brand_interest[1:]
        brand_list.append(brand_interest)
        category_top_3 = [x for (x, _) in sorted(category_dic.items(), key=lambda x:
x[1], reverse=True)[0:TOP_THREE]]
        category_interest = ""
        for i in category_top_3:
            category_interest = category_interest + "," +i
        category_interest = category_interest[1:]
        category_list.append(category_interest)
# 建立使用者興趣畫像 DataFrame
df_interest = pd.DataFrame({'reviewerID': reviewerID_list, 'top_3_category':
category_list, 'top_3_brand': brand_list})
```

執行上面的程式，就獲得了每個使用者喜歡的 Top 3 品牌和類目，儲存在 df_interest 這個 DataFrame 中，圖 8-10 舉出 df_interest 中具體的資料樣本，其中 3 個類目、3 個品牌之間用逗點隔開。

	reviewerID	top_3_category	top_3_brand
0	A0001528BGUBOEVR6T5U	Toys & Games,Novelty & Gag Toys,Magic Kits & A...	Rock Ridge Magic,Creative Motion,Intex
1	A001170867ZBE9FORRQL	Toys & Games,Action Figures & Statues,Accessories	Schleich,Collecta,Sportline
2	A0017882XASSVJGSZF5R	Toys & Games,Puzzles,Brain Teasers	Puzzled,Anagram,All4LessShop
3	A00222906VX8GH7X6J6B	Toys & Games,Dolls & Accessories,Dolls	DOUBLE E,Caillou,LEGO
4	A0022678B6GE9F3FOSBS	Toys & Games,Dolls & Accessories,Dolls	Paradise Galleries,Fun World,New Ray
...	...	...	...
357308	AZZYVIRS854I7	Toys & Games,Games,Stacking Games	Melissa & Doug,ARRIS,Munchkin
357309	AZZYW4YOE1B6E	Toys & Games,Games,Board Games	Rio Grande Games,Intex,LEGO
357310	AZZZ6G9WZTNNX	Toys & Games,Learning & Education,Early Develo...	Melissa & Doug,LeapFrog,Jurassic Park
357311	AZZZYAYJQSDOJ	Toys & Games,Party Supplies,Party Tableware	BirthdayExpress,Underground Toys,Toysmith
357312	AZZZZS162JNL0	Toys & Games,Sports & Outdoor Play,Play Tents ...	Pacific Play Tents,Amscan,Folkmanis

▲ 圖 8-10 使用者喜歡的 Top 3 品牌和類目

最後，將使用者興趣畫像與使用者評論、商品中繼資料連接，根據前面的步驟找出商品的描述關鍵字，以便利用大模型生成個性化的商品描述[①]。

```
"""
連接相關資料，將資料放到同一個 DataFrame 中
"""
df = pd.merge(df_review, df_feature, how="left", on=['asin'])
df = pd.merge(df, df_interest, how="left", on=['reviewerID'])
df.dropna(how='any', subset=['brand', 'category', 'top_3_category', 'top_3_brand'],
inplace=True)
# df_joined.drop_duplicates()
df = df.reset_index(drop="true")

def common_brand(br, top_3_brand):
    br = {br}
    top_3_brand_set = set(top_3_brand.split(','))
    common_brand = br.intersection(top_3_brand_set)
    if len(common_brand) == 0:
        return ''
    else:
        return common_brand.pop()

def common_category(ca, top_3_category):
```

① 程式：src/e-commerce_case/generate_personalized_info/item_info_data_process.py。

```
    ca = set(ca)  # <class 'pandas.core.series.Series'>  [Toys & Games, Grown-Up Toys, 
Games]
    top_3_category_set = set(top_3_category.split(','))
    common_category = ca.intersection(top_3_category_set)
    common_ca = ""
    for ca_ in common_category:
        common_ca = common_ca + "," + ca_
    common_ca = common_ca[1:]
    return common_ca

df['common_brand'] = df.apply(lambda x: common_brand(x['brand'], x['top_3_brand']), 
axis=1)
df['common_category'] = df.apply(lambda x: common_category(x['category'], 
x['top_3_category']), axis=1)

def goods_description(common_b, common_ca):
    if common_b and common_ca:
        return common_b + "," + common_ca
    elif common_b:
        return common_b
    elif common_ca:
        return common_ca
    else:
        return ""

df['goods_description'] = df.apply(lambda x: goods_description(x['common_brand'], 
x['common_category']), axis=1)
df.dropna(how='any', subset=['reviewerID', 'asin', 'brand', 'category', 
'top_3_category',
                             'top_3_brand', 'goods_description'], inplace=True)
df = df.reset_index(drop=True)

prompt = """The following are the the product brand and category data: \n
 product brand: {} \n
 product category: {}  \n
 The following are the user's interests and preferences for product brands and 
categories:\n
 the top three brands user likes: {} \n
 the top three categories user likes: {} \n
```

```
 Based on the above information, predict the description label of the product, which
is obtained from the
 brand and category of the product (i.e., the description label is a subset of the
brand and category of
 the product). The description label you provide should meet the user's preferences
for the brand and
 category of the product to the greatest extent possible. You just need to list one to
four description labels,
 do not explain the reason.If you given more than one description labels, please
separate them with comma."""
df['prompt'] = df.apply(lambda row: prompt.format(str(row['brand']),
str(row['category']),
                                                  str(row['top_3_brand']),
str(row['top_3_category'])), axis=1)
df = df[['prompt', 'goods_description']]
print(df.shape)
df.to_csv("./data/item_info_data.csv", index=False)
```

執行上面的程式，就獲得了包含提示詞和商品描述欄位的 DataFrame：df，如圖 8-11 所示。大模型需要的所有資料都準備好了。

```
                                                    prompt                                  goods_description
0        The following are the the product brand and ca...          Gale Force Nine,Games,Toys & Games
1        The following are the the product brand and ca...  Gale Force Nine,Grown-Up Toys,Games,Toys & Games
2        The following are the the product brand and ca...             Toys & Games,Novelty & Gag Toys
3        The following are the the product brand and ca...  Dover Publications,Drawing & Painting Supplies...
4        The following are the the product brand and ca...  Dover Publications,Drawing & Painting Supplies...
...                                                    ...                                                ...
2631331  The following are the the product brand and ca...                          Disney,Toys & Games
2631332  The following are the the product brand and ca...   Toy Remote Control & Play Vehicles,Toys & Games
2631333  The following are the the product brand and ca...    Baby Einstein,Baby & Toddler Toys,Toys & Games
2631334  The following are the the product brand and ca...  Baby Einstein,Baby & Toddler Toys,Music & Soun...
2631335  The following are the the product brand and ca...  Baby Einstein,Baby & Toddler Toys,Music & Soun...
```

▲ 圖 8-11 包含提示詞和商品描述欄位的 DataFrame：df

8.4.3 程式實現

準備好資料，就可以利用大模型生成個性化商品描述了。

1. 基於智譜 AI 的 API 實現

基於第三方 API 生成個性化商品描述需要提供提示詞（指令範本），透過 API 將提示詞傳給大模型，然後大模型透過理解提示詞的意圖輸出對應的結果。可以參考下面的指令範本。

你的任務是基於使用者的興趣偏好和商品描述資訊生成個性化商品描述。所謂個性化，就是針對每個使用者舉出的商品描述不一樣。

使用者的興趣偏好包括兩類，一類是品牌興趣偏好，一類是類目興趣偏好。商品的描述資訊也包括兩類，一類是商品品牌，一類是商品類目。品牌名稱和類目名稱都是英文的，如果有多個，則用逗點隔開。下面是一個例子。

使用者 U 的品牌興趣偏好：Rock Ridge Magic, Gale Force Nine, Intex。

使用者 U 的類目興趣偏好：Toys & Games, Puzzles, Brain Teasers。

商品 A 的品牌：Gale Force Nine。

商品 A 的類目：Toys & Games, Grown-Up Toys, Games。

生成個性化商品描述分為 3 步。

第 1 步：計算使用者的品牌興趣偏好與商品品牌的公共部分。

第 2 步：計算使用者的類目興趣偏好與商品類目的公共部分。

第 3 步：將第 1 步、第 2 步的結果合併，最終結果就是個性化商品描述。

針對使用者 U 和商品 A 的例子，下面是計算過程。

第 1 步：

使用者 U 的品牌興趣偏好：Rock Ridge Magic, Gale Force Nine, Intex。

商品 A 的品牌：Gale Force Nine。

它們的公共部分為 Gale Force Nine。

第 2 步：

使用者 U 的類目興趣偏好：Toys & Games, Puzzles, Brain Teasers。

商品 A 的類目：Toys & Games, Grown-Up Toys, Games。

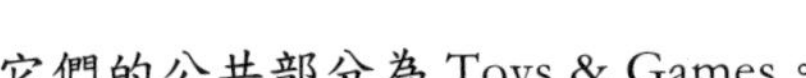

它們的公共部分為 Toys & Games。

第 3 步：

第 1 步與第 2 步的公共部分分別是 Gale Force Nine 和 Toys & Games，將它們合併起來。所以，最終商品 A 針對使用者 U 的個性化商品描述為 Gale Force Nine、Toys & Games。

現在有一個新使用者 X 和一個新商品 Y，X 的興趣偏好和 Y 的描述資訊都是英文的，中間用逗點隔開，分別如下。

使用者 X 的品牌興趣偏好：Melissa & Doug, ARRIS, Munchkin。

使用者 X 的類目興趣偏好：Toys, Games, Stacking Games。

商品 Y 的品牌：ARRIS。

商品 Y 的類目：Toys, Grown-Up Toys, Games。

現在請你按照上面的 3 個步驟進行推理，輸出商品 Y 針對使用者 X 的個性化商品描述。需要有推理過程，最終的個性化商品描述結果用逗點隔開。

將上面的指令範本中的文字貼上到智譜清言 App 中，獲得如圖 8-12 所示的回答。你可以驗證一下，是正確的。同時，筆者嘗試了文心一言 App、訊飛星火 App，它們都能正確舉出個性化商品描述。

第1步： 用户X的品牌兴趣偏好：Melissa & Doug, ARRIS, Munchkin 商品Y的品牌：ARRIS 它们的公共部分为：ARRIS

第2步： 用户X的类目兴趣偏好：Toys, Games, Stacking Games 商品Y的类目：Toys, Grown-Up Toys, Games 它们的公共部分为：Toys, Games

第3步： 第1步与第2步的公共部分分别是：ARRIS 和 Toys, Games，将它们合并起来，所以，最终商品Y针对用户X的个性化的商品描述为：ARRIS, Toys, Games。

▲ 圖 8-12 智譜清言 App 舉出的回答 (編按：本圖例為簡體中文介面)

利用這個提示詞，你還可以透過智譜 AI 的 API[3] 生成個性化商品描述，具體程式如下。

```
import zhipuai

# API Key
zhipuai.api_key = "api key"
response = zhipuai.model_api.async_invoke(
    model="chatglm_turbo",
    prompt=[{"role": "user", "content": " 這裡是上面的提示詞 "}],
    top_p=0.7,
    temperature=0.9,
)

# response 的資料結構以下
# {'code': 200, 'msg': ' 操作成功 ', 'data': {'request_id': '8107271802315478819',
'task_id': '761516999490208548107271802315478821', 'task_status': 'PROCESSING'},
'success': True}{'code': 200, 'msg': ' 操作成功 ', 'data': {'request_id':
'8107271802315478819', 'task_id': '761516999490208548107271802315478821', 'task_
status': 'PROCESSING'}, 'success': True}

result = zhipuai.model_api.query_async_invoke_result(response["data"]["task_id"])
print(result)
```

執行上面的程式，得到如圖 8-13 所示的答案。

程式傳回的是一個 JSON 字串，為了方便你更進一步地看到傳回的結果，筆者在智譜 AI Web 端輸入內容後大模型舉出的回覆，明顯可以看出，它的想法是對的，分析過程也是對的，但是最終結果錯了。這可能是因為智譜 AI 開放 API 的模型參數不夠多，模型還不夠智慧。

首先，我们需要计算用户X的品牌兴趣偏好和商品Y的品牌之间的公共部分。用户X的品牌兴趣偏好为：Melissa & Doug, ARRIS, Munchkin，商品Y的品牌为ARRIS，它们的公共部分为ARRIS。

接下来，我们需要计算用户X的类目兴趣偏好和商品Y的类目之间的公共部分。用户X的类目兴趣偏好为：Toys, Games, Stacking Games，商品Y的类目为Toys, Grown-Up Toys, Games，它们的公共部分为Toys。

最后，我们将第1步和第2步的结果合并，得到个性化商品描述为：ARRIS, Toys。因此，商品Y针对用户X的个性化商品描述为：ARRIS, Toys。

以上内容为 AI 生成，不代表开发者立场，请勿删除或修改本标记

▲ 圖 8-13　智譜 AI 的 API 舉出的回答（編按：本圖例為簡體中文介面）

可以讓大模型用一句話來總結其生成的個性化商品描述關鍵字，提示詞如下。

用一句話來描述商品，這句話中必須包含 Toys, Games, ARRIS 這幾個關鍵字。

請用英文回答，不要少於 10 個單字，不超過 20 個單字。

將上面的提示詞輸入智譜清言 App，可以獲得以下的英文個性化商品描述，比較通順。

```
Experience the fun of Toys and Games with the advanced features of ARRIS, perfect for
kids' entertainment.
```

2. 基於 trl 函式庫的 SFTTrainer 微調實現

除了基於大模型 API 生成個性化商品描述，還可以利用 Qwen 進行微調生成使用者興趣畫像。這種方法相對複雜，筆者將微調過程拆解為 6 個主要步驟，一步步實現[①]，其想法與 8.3.3 節基本一致。

第 1 步，建構微調樣本。將人物誌資料集按照 8 ： 2 的比例分為訓練集和測試集。訓練集用來微調大模型，測試集用來評估微調好後的大模型的效果，二者缺一不可。

```
import pandas as pd
from sklearn.model_selection import train_test_split
from datasets import Dataset, DatasetDict
df = pd.read_csv("./data/item_info_data.csv")
train_df, test_df = train_test_split(df, test_size=0.999, random_state=42)
# 由於資料量太大，用較少的資料進行微調，否則很慢
print(" 微調的樣本數：" + str(train_df.shape[0]))
train_dataset_dict = DatasetDict({
    "train": Dataset.from_pandas(train_df, preserve_index=False),
})
```

① 程式：src/e-commerce_case/generate_personalized_info/finetune_generated_info_model.py。

第 2 步，建構基底模型。這裡還是利用 Hugging Face 社區開放原始碼的 trl 函式庫進行微調，採用 Qwen1.5-14B-Chat 模型。你可以參考以下面建立基底模型的程式。

```
import torch
from transformers import AutoModelForCausalLM, AutoTokenizer
model_name = "/Users/liuqiang/Desktop/code/llm/models/Qwen1.5-14B-Chat"
device_map = "auto"
tokenizer = AutoTokenizer.from_pretrained(model_name, trust_remote_code=True)
tokenizer.pad_token = tokenizer.eos_token
tokenizer.padding_side = "right"  # Fix weird overflow issue with fp16 training
# Since transformers 4.35.0, the GPT-Q/AWQ model can be loaded using
AutoModelForCausalLM.
model = AutoModelForCausalLM.from_pretrained(
    model_name,
    device_map="auto",
    torch_dtype=torch.float16,
    offload_folder="offload",
    offload_state_dict=True,
)
model.config.use_cache = False
model.config.pretraining_tp = 1
```

第 3 步，查看基底模型的預測效果。因為基底模型是一種通用能力模型，沒有在該場景下進行微調，所以如果要利用基底模型進行預測，那麼還需要提前查看效果。具體程式如下。

```
sample_size = 5
sample_test_data = test_df[['prompt',
'goods_description']].head(sample_size).to_dict('records')
for dic in sample_test_data:
    messages = [
        {"role": "user", "content": dic['prompt']}
    ]
    input_ids = tokenizer.apply_chat_template(conversation=messages, tokenize=True,
                                              add_generation_prompt=True,
return_tensors='pt')
    output_ids = model.generate(input_ids.to('mps'), max_new_tokens=500,
```

```
pad_token_id=tokenizer.eos_token_id)
    response = tokenizer.decode(output_ids[0][input_ids.shape[1]:],
skip_special_tokens=True)
    print("------------")
    print(f"Prompt:\n{dic['prompt']}\n")
    print(f"Generated goods_description:\n{response}\n")
    print(f"Ground truth:\n{dic['goods_description']}")
```

執行上面的程式可以獲得結果，這裡列出一筆樣本資料和對應的生成結果。

輸入的問題如下。

```
The following are the the product brand and category data:

 product brand: Gale Force Nine

 product category: ['Toys & Games', 'Grown-Up Toys', 'Games']

 The following are the user's interests and preferences for product brands and
categories:

 the top three brands user likes: Gale Force Nine,Epix Haven

 the top three categories user likes: Toys & Games,Games,Grown-Up Toys

 Based on the above information, predict the description label of the product, which
is obtained from the
 brand and category of the product (i.e., the description label is a subset of the
brand and category of
 the product). The description label you provide should meet the user's preferences
for the brand and
 category of the product to the greatest extent possible. You just need to list one to
four description labels,
 do not explain the reason.If you given more than one description labels, please
separate them with comma.
```

大模型生成的文字如下。

```
Toys & Games, Puzzles, Jigsaw Puzzles
```

可以看出，大模型基於 prompt[①]欄位生成回覆，回覆的結果相關度比較大，但是不完全正確[②]。

當然，出現這樣的結果是因為這個模型還不夠智慧，沒能極佳地理解輸入文字的意圖。下面透過微調讓這個模型具備解決這個垂直領域問題的能力。

第 4 步，基於第 1 步的樣本進行微調。這是最核心的一步，需要先確定一些微調的參數，然後基於 trl 函式庫的 SFTTrainer 建立監督微調物件，SFTTrainer 中包含各種參數和微調樣本資料 train_dataset_dict['train']，你應該還記得，這個資料是在建構樣本時生成的。有了 SFTTrainer，就可以對大模型進行微調了。下面是細節的程式。

```
# 建立 LoRA 配置
# 考慮 Transformer 區塊中的所有線性層，以獲得最大性能
from peft import LoraConfig, prepare_model_for_kbit_training, get_peft_model
################################################################################
# QLoRA 參數
################################################################################
# LoRA 注意力維數
lora_r = 8
# 用於 LoRA 縮放的 Alpha 參數
lora_alpha = 16
# LoRA 層的 Dropout 機率
lora_dropout = 0.05
# 載入 LoRA 配置
peft_config = LoraConfig(
    lora_alpha=lora_alpha,
    lora_dropout=lora_dropout,
    target_modules=['q_proj', 'v_proj'],
    r=lora_r,
    bias="none",
    task_type="CAUSAL_LM",
)
```

① 由商品品牌、商品類目、使用者喜歡的 Top 3 品牌、使用者喜歡的 Top 3 類目拼接而成的提示詞。

② 這個問題的正確答案是：Gale Force Nine,Grown-Up Toys,Games,Toys & Games。

```
# 為微調準備好模型
model = prepare_model_for_kbit_training(model)
model = get_peft_model(model, peft_config)
# 載入模型參數
# 使用 trl 函式庫中的 SFTTrainer，它提供了一個 Transformers 函式庫中 Trainer 類別的封裝
# 可以使用 PEFT 轉接器在基於指令的資料集上輕鬆微調模型
from transformers import TrainingArguments
################################################################################
# TrainingArguments 的參數
################################################################################
# 模型預測和 checkpoint 的儲存目錄
model_dir = "./models"
# 訓練 epochs 的數量
num_train_epochs = 1
# 設置 fp16/bf16 參數（用 A100 訓練時，設置 fp16=True）
fp16 = False
bf16 = False
# 訓練時設置的 GPU 批大小
per_device_train_batch_size = 4
# 評估時設置的 GPU 批大小
per_device_eval_batch_size = 4
# 累積梯度時更新的步數
gradient_accumulation_steps = 2
# 讓梯度更新生效
gradient_checkpointing = True
# 最大的梯度範數（梯度裁剪）
max_grad_norm = 1
# 初始化學習率（AdamW 最佳化器）
learning_rate = 5e-4
# 適用於所有層（bias/LayerNorm 除外）的權重衰減因數
weight_decay = 0.01
# 使用的最佳化器
optim = "adamw_torch"
# 訓練步數（重置 num_train_epoches）
max_steps = 50
# 線性預熱步數的比例（從 0 到學習率）
warmup_ratio = 0.05
# 將序列按相同長度分組為批次
# 節省記憶體並加速訓練過程
```

```
group_by_length = True
# 每隔 X 步更新就儲存 checkpoint
save_steps = 50
# At most save X times
save_total_limit = 3
# 每隔 X 步更新就記錄日誌
logging_steps = 20
training_arguments = TrainingArguments(
    output_dir=model_dir,
    num_train_epochs=num_train_epochs,
    per_device_train_batch_size=per_device_train_batch_size,
    gradient_accumulation_steps=gradient_accumulation_steps,
    optim=optim,
    save_steps=save_steps,
    save_total_limit=save_total_limit,
    logging_steps=logging_steps,
    learning_rate=learning_rate,
    weight_decay=weight_decay,
    fp16=fp16,
    bf16=bf16,
    max_grad_norm=max_grad_norm,
    # max_steps=max_steps,
    warmup_ratio=warmup_ratio,
    group_by_length=group_by_length,
    report_to=["wandb"]
)

# 將資料轉為微調大模型需要的資料格式
def formatting_func(spl):
    instruct = f"""### Instruction:
You are a product guide expert, use the input below to answer questions.
### Input:
{spl['prompt']}
### Response:
{spl['goods_description']}
"""
    return instruct

# 建立 Trainer 物件
```

```
from trl import SFTTrainer
########################################################################
# SFT 參數
########################################################################
# 使用的最大序列長度
max_seq_length = 512  # 越大佔用記憶體越多
# 將多個較短的例子打包在一起放進輸入序列以提升效率
packing = True
trainer = SFTTrainer(
    model=model,
    train_dataset=train_dataset_dict['train'],
    peft_config=peft_config,
    formatting_func=formatting_func,
    max_seq_length=max_seq_length,
    tokenizer=tokenizer,
    args=training_arguments,
    packing=packing,
)
# 啟動微調
trainer.train()
# 儲存模型
trainer.save_model(model_dir)
```

第 4 步完成後，你一定很期待效果，下面我們進行**第 5 步，驗證微調效果**。下面是查看效果的程式，這裡還是選擇測試集中的 5 筆樣本資料測試效果。

```
# 微調完成後，載入模型，驗證模型效果
from peft import AutoPeftModelForCausalLM
# 載入基礎 LLM 模型與分詞器
model = AutoPeftModelForCausalLM.from_pretrained(
    model_dir,
    device_map="auto",
    torch_dtype=torch.float16,
    offload_folder="offload",
    offload_state_dict=True
)
tokenizer = AutoTokenizer.from_pretrained(model_dir)

for sample in sample_test_data:
```

```
    prompt = f"""### Instruction:
    You are a product guide expert, use the input below to answer questions.

    ### Input:
    {sample['prompt']}

    ### Response:
    """
    input_ids = tokenizer(prompt, return_tensors="pt", truncation=True).input_ids
    outputs = model.generate(input_ids=input_ids.to('mps'), max_new_tokens=500)
    print("------------")
    print(f"Prompt:\n{sample['prompt']}\n")
    print(
        f"Generated goods_description:\n"
        f"{tokenizer.batch_decode(outputs,
skip_special_tokens=True)[0][len(prompt):]}")
    print(f"Ground truth:\n{sample['goods_description']}")
```

執行上面的程式，得到以下結果。

```
------------
Prompt:
The following are the the product brand and category data:

 product brand: Shopkins

 product category: ['Toys & Games', 'Dolls & Accessories', 'Playsets']

 The following are the user's interests and preferences for product brands and
categories:

 the top three brands user likes: Shopkins,Kahootz

 the top three categories user likes: Toys & Games,Dolls & Accessories,Dolls

 Based on the above information, predict the description label of the product, which
is obtained from the
 brand and category of the product (i.e., the description label is a subset of the
brand and category of
```

```
 the product). The description label you provide should meet the user's preferences
for the brand and
 category of the product to the greatest extent possible. You just need to list one to
four description labels,
 do not explain the reason.If you given more than one description labels, please
separate them with comma.

Generated goods_description:
 Shopkins,Dolls & Accessories,Toys & Games

Ground truth:
Shopkins,Toys & Games,Dolls & Accessories

------------
Prompt:
The following are the the product brand and category data:

 product brand: Safari Ltd.

 product category: ['Toys & Games', 'Action Figures & Statues', 'Action Figures']

 The following are the user's interests and preferences for product brands and
categories:

 the top three brands user likes: Ravensburger,Safari Ltd.,Fisher-Price

 the top three categories user likes: Toys & Games,Puzzles,Jigsaw Puzzles

 Based on the above information, predict the description label of the product, which
is obtained from the
 brand and category of the product (i.e., the description label is a subset of the
brand and category of
 the product). The description label you provide should meet the user's preferences
for the brand and
 category of the product to the greatest extent possible. You just need to list one to
four description labels,
 do not explain the reason.If you given more than one description labels, please
separate them with comma.
```

```
Generated goods_description:
 Safari Ltd.,Toys & Games

Ground truth:
Safari Ltd.,Toys & Games


------------
Prompt:
The following are the the product brand and category data:

 product brand: Marky Sparky

 product category: ['Toys & Games', 'Games', 'Game Room Games', 'Darts']

 The following are the user's interests and preferences for product brands and
categories:

 the top three brands user likes: Play-Doh,ZOOB,Scientific Explorer

 the top three categories user likes: Toys & Games,Games,Arts & Crafts

 Based on the above information, predict the description label of the product, which
is obtained from the
 brand and category of the product (i.e., the description label is a subset of the
brand and category of
 the product). The description label you provide should meet the user's preferences
for the brand and
 category of the product to the greatest extent possible. You just need to list one to
four description labels,
 do not explain the reason.If you given more than one description labels, please
separate them with comma.

Generated goods_description:
 Games,Toys & Games

Ground truth:
Games,Toys & Games


------------
```

```
Prompt:
The following are the the product brand and category data:

 product brand: Fisher-Price

 product category: ['Toys & Games', 'Action Figures & Statues', 'Playsets & Vehicles',
'Playsets']

 The following are the user's interests and preferences for product brands and
categories:

 the top three brands user likes: Fisher-Price,LeapFrog,Small World Toys

 the top three categories user likes: Toys & Games,Action Figures & Statues,Toy Remote
Control & Play Vehicles

 Based on the above information, predict the description label of the product, which
is obtained from the
 brand and category of the product (i.e., the description label is a subset of the
brand and category of
 the product). The description label you provide should meet the user's preferences
for the brand and
 category of the product to the greatest extent possible. You just need to list one to
four description labels,
 do not explain the reason.If you given more than one description labels, please
separate them with comma.

Generated goods_description:
 Fisher-Price,Toys & Games,Action Figures & Statues

Ground truth:
Fisher-Price,Toys & Games,Action Figures & Statues

------------
Prompt:
The following are the the product brand and category data:

 product brand: Star Wars
```

```
 product category: ['Toys & Games', 'Action Figures & Statues', 'Action Figures']

 The following are the user's interests and preferences for product brands and
categories:

 the top three brands user likes: FunKo,Star Wars

 the top three categories user likes: Toys & Games,Action Figures & Statues,Action
Figures

 Based on the above information, predict the description label of the product, which
is obtained from the
 brand and category of the product (i.e., the description label is a subset of the
brand and category of
 the product). The description label you provide should meet the user's preferences
for the brand and
 category of the product to the greatest extent possible. You just need to list one to
four description labels,
 do not explain the reason.If you given more than one description labels, please
separate them with comma.

Generated goods_description:
 Star Wars,Toys & Games,Action Figures,Action Figures & Statues

Ground truth:
Star Wars,Action Figures,Toys & Games,Action Figures & Statues
```

從第 5 步的執行結果可以看出，微調後的大模型可以正確地生成使用者興趣畫像①。對比前面第 3 步的結果，效果有極大提升。

第 6 步，利用第 5 步生成的關鍵字，讓大模型生成個性化商品描述。具體程式如下。

```
"""
讓大模型基於個性化商品描述關鍵字，用一句話生成個性化商品描述
"""
prompt = f"""### Instruction:
```

① 除了第一個樣本有些瑕疵，其他都是對的。

```
You are a product guide expert, use the input below to answer questions.
### Input:
Please describe a product in a fluent and coherent sentence, which must include the
keywords Toys, Games, and ARRIS of
the product. Your output should be limited to 10 to 20 words. Please reply in English.
### Response:
"""
messages = [
    {"role": "user", "content": prompt}
]
input_ids = tokenizer.apply_chat_template(conversation=messages, tokenize=True,
                                          add_generation_prompt=True, return_tensors='pt')
output_ids = model.generate(input_ids.to('mps'), max_new_tokens=500,
pad_token_id=tokenizer.eos_token_id)
response = tokenizer.decode(output_ids[0][input_ids.shape[1]:],
skip_special_tokens=True)
print("------------")
print(response)
```

執行上面程式，輸出的結果如下。

```
ARRIS toys and games bring endless fun to children of all ages.
```

本節以 Toys_and_Games 資料集為「原料」，一步步生成個性化商品描述關鍵字，最後利用大模型的文字總結能力，將關鍵字拓展成一句個性化商品描述。其中，對大模型進行微調生成個性化商品描述關鍵字是重點。

微調過程包含大量的程式實現，你需要熟悉程式細節，圖 8-14 是個性化商品描述微調流程，供你參考。

本節將個性化商品描述關鍵字簡化為品牌和類目兩大類，方法具備普遍性。針對真實的業務場景，可以很容易根據本節的想法進行拓展。

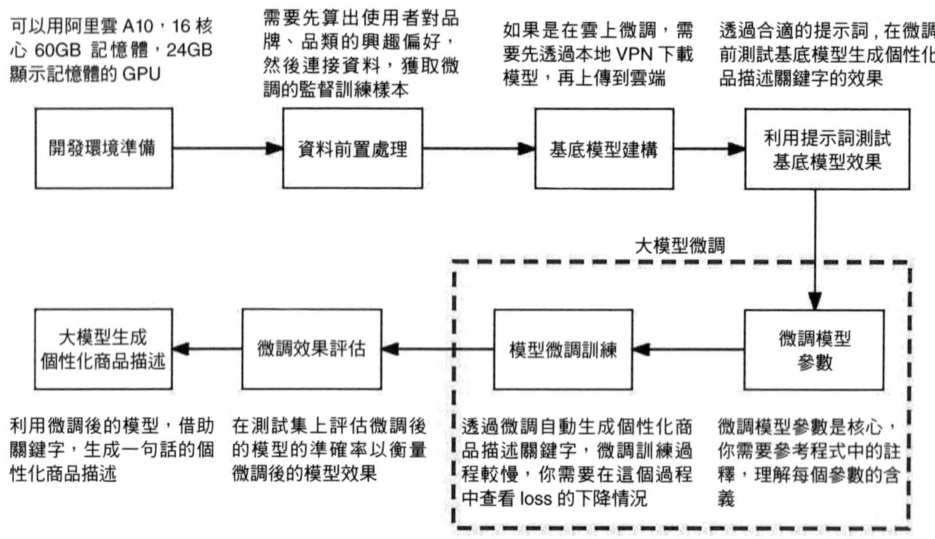

可以用阿里雲 A10，16 核心 60GB 記憶體，24GB 顯示記憶體的 GPU
需要先算出使用者對品牌、品類的興趣偏好，然後連接資料，獲取微調的監督訓練樣本
如果是在雲上微調，需要先透過本地 VPN 下載模型，再上傳到雲端
透過合適的提示詞，在微調之前測試基底模型生成個性化商品描述關鍵字的效果
開發環境準備
資料前置處理
基底模型建構
利用提示詞測試基底模型效果
大模型微調
大模型生成個性化商品描述
微調效果評估
模型微調訓練
微調模型參數
利用微調後的模型，借助關鍵字，生成一句話的個性化商品描述
在測試集上評估微調後的模型的準確率以衡量微調後的模型效果
透過微調自動生成個性化商品描述關鍵字，微調訓練過程較慢，你需要在這個過程中查看 loss 的下降情況
微調模型參數是核心，你需要參考程式中的註釋，理解每個參數的含義

▲ 圖 8-14 個性化商品描述微調流程

8.5 大模型應用於電子商務猜你喜歡推薦

電子商務推薦中的猜你喜歡模組是最主流的推薦產品，淘寶、京東、拼多多的首頁都是個性化的猜你喜歡，並且實現了類似於抖音的資訊流。

首先，利用 Beauty 資料集建構推薦演算法需要的資料，進而建構模型；然後，微調一個個性化的推薦演算法大模型；最後，基於微調的模型進行效果評估，並且與微調前的模型效果進行對比，同時，與沒有微調的參數更多的模型對比，進一步展示微調帶來的效果提升。

8.5.1 資料前置處理

本節透過微調大模型進行個性化推薦，需要建構微調的訓練集和測試集，具體的資料樣本如下，這也是大模型的提示詞。

```
{
{"instruction": "You are a recommendation system expert who provides personalized
ranking for items based on the background information provided.",
```

```
"input": "I've ranked the following products in the past(in JSON format):
[
    {
        "title": "SF221-Shaving Factory Straight Razor (Black), Shaving Factory Hand
Made Shaving Brush, 100...",
        "brand": "Shaving Factory",
        "price": "$21.95",
        "rating": 2
    },
    ...
]
Based on above rating history, please predict user's rating for the following
product(in JSON format):

   {
        "title": "Norelco 4821XL Micro Action Corded/Cordless Rechargeable Men's
Shaver",
        "brand": "Norelco",
        "price": "$11.2"
   }

The ranking is between 1 and 5, 1 being lowest and 5 being highest.
You just need to ranking the above product, do not explain the reason.

"output": 3
}
```

上面的樣本將使用者評分過的商品的title（標題）、brand（品牌）、price（價格）、rating（評分）展現出來，供大模型學習，最終要預測使用者對商品的評分。微調好大模型後，可以對待推薦給使用者的商品進行預測評分，然後根據評分進行個性化推薦。

為了建構上述樣本資料，可以利用 Beauty 資料集中的 meta_All_Beauty.json 和 All_Beauty.csv 這兩個檔案。前者是商品相關中繼資料，後者是使用者行為資料，具體格式以下①。

① 這是第 2 章中介紹的行為資料的簡化版本，只包含商品 ID、使用者 ID、評分、時間 4 個欄位。

```
item,user,rating,timestamp
0143026860,A1V6B6TNIC10QE,1.0,1424304000
0143026860,A2F5GHSXFQ0W6J,4.0,1418860800
0143026860,A1572GUYS7DGSR,4.0,1407628800
0143026860,A1PSGLFK1NSVO,5.0,1362960000
0143026860,A6IKXKZMTKGSC,5.0,1324771200
0143026860,A36NF437WZLQ9E,5.0,1267142400
0143026860,A10Q8NIFOVOHFV,4.0,983923200
```

下面是詳細的程式實現過程，資料處理實現比較簡單，這裡不過多講解。

```
import csv
import json
import pandas as pd
from sklearn.model_selection import train_test_split

instruction = ("You are a recommendation system expert who provides personalized
ranking for items "
               "based on the background information provided.")

item_dict = {}  # 從 All_beauty.json 中獲取每個 item 對應的標題
with open('../../data/amazon_review/beauty/meta_All_Beauty.json', 'r') as file:
    reader = csv.reader(file, delimiter='\n')
    for row in reader:
        j = json.loads(row[0])
        item_id = j['asin']
        title = j['title']
        brand = j['brand']
        price = j['price']
        if price != "" and '$' not in price and len(price) > 10:   # 處理一些異常資料
            price = ""
        item_info = {
            "title": title,
            "brand": brand,
            "price": price
        }
        item_dict[item_id] = item_info

def generate_data(data_df, data_type, path):
```

```
    grouped_df = data_df.groupby('user')
    groups = grouped_df.groups
    data_list = []
    for user in groups.keys():
        user_df = grouped_df.get_group(user)
        """
        >>> grouped_df.get_group('AZZZ5UJWUVCYZ')
                          item           user  rating   timestamp
            157912  B00IIZG80U  AZZZ5UJWUVCYZ     5.0  1479859200
            250877  B01FNJ9MOW  AZZZ5UJWUVCYZ     5.0  1505865600
            358191  B01CZC20DU  AZZZ5UJWUVCYZ     5.0  1505865600
        """
        if data_type == "train" and user_df.shape[0] < 4:
# 訓練資料，每個使用者至少需要 4 個資料
            continue
        if data_type == "test" and user_df.shape[0] < 2:
# 測試資料，每個使用者至少需要 2 個資料
            continue
        if user_df.shape[0] > 8:   # 資料量太大的不考慮，否則超出了大模型的 token 範圍
            continue
        last_row = user_df.iloc[-1]  # 最後一行用作 label
        selected_df = user_df.head(user_df.shape[0]-1)  # 前面的作為特徵
        user_ranking_list = []
        for _, row_ in selected_df.iterrows():
            item = row_['item']
            if item in item_dict:
                rating = row_['rating']
                item_info_ = item_dict[item]
                item_info_['rating'] = rating
                user_ranking_list.append(item_info_)
        formatted_user_ranking = json.dumps(user_ranking_list, indent=4)
        item = last_row['item']
        rating = last_row['rating']
        label_item_info_ = item_dict[item]
        if 'rating' in label_item_info_:
            del label_item_info_['rating']  # 去掉 rating，這是待預測的
        formatted_item_info = json.dumps(label_item_info_, indent=4)
        input = ("I've ranked the following products in the past(in JSON format):\n\n"
+
```

```
                  formatted_user_ranking + "\n\n" +
                  "Based on above rating history, please predict user's rating " +
                  "for the following product(in JSON format):" + "\n\n" +
                  formatted_item_info + "\n\n" +
                  "The ranking is between 1 and 5, 1 being lowest and 5 being highest.
" +
                  "You just need to ranking the above product, do not explain the
reason."
                  )
         output = str(int(rating))
         res_dic = {
             "instruction": instruction,
             "input": input,
             "output": output
         }
         data_list.append(res_dic)
     res = json.dumps(data_list)
     with open(path, 'a') as file_:  # 將生成的訓練資料儲存起來
         file_.write(res)

train_path = '../data/train.json'
test_path = '../data/test.json'
df = pd.read_csv("../../data/amazon_review/beauty/All_Beauty.csv")
df_shuffled = df.sample(frac=1).reset_index(drop=True)
train_df, test_df = train_test_split(df_shuffled, test_size=0.33, random_state=10)
generate_data(train_df, "train", train_path)
generate_data(test_df, "test", test_path)
"""
    目前 train.json 中有 989 個樣本
    目前 test.json 中有 5517 個樣本
    test.json 的資料量更大，因為 train 要求每個使用者至少要有 4 筆記錄
"""
```

執行上面的程式會生成 train.json 和 test.json 兩個資料，它們都是 JSON 格式的。其中，train.json 用於模型的微調，test.json 用於微調後的模型的效果評估。

8.5.2 模型微調

微調實現參考了參考文獻 [4]，為了適應 Beauty 資料集，實現時有所調整。

採用 LoRA 進行微調，速度更快、記憶體佔用更小。這裡微調 Qwen1.5-4B 模型，與 6.3 節的實現方式類似，採用 Transformers 函式庫中的 Trainer 類別，具體的微調程式以下①。

```
import os
from typing import List
import fire
import torch
from datasets import load_dataset
from peft import (
    LoraConfig,
    get_peft_model,
    prepare_model_for_int8_training,
    set_peft_model_state_dict, PeftModel,
)
from transformers import AutoModelForCausalLM, AutoTokenizer, Trainer,
DataCollatorForSeq2Seq, TrainingArguments
from utils.prompter import Prompter

def train(
        # 模型和資料參數
        base_model: str = "",  # 必要參數
        data_path: str = "./data/train.json",
        output_dir: str = "./models",
        # 訓練超參數
        batch_size: int = 128,
        micro_batch_size: int = 4,
        num_epochs: int = 3,
        learning_rate: float = 3e-4,
        cutoff_len: int = 256,
        val_set_size: int = 2000,
        # LoRA 超參數
```

① 程式：src/e-commerce_case/personalized_rec/model_finetune.py。

```
        lora_r: int = 8,
        lora_alpha: int = 16,
        lora_dropout: float = 0.05,
        lora_target_modules: List[str] = [
            "q_proj", "o_proj", "k_proj", "v_proj", "gate_proj", "up_proj", "down_
proj"
        ],
        # 大模型超參數
        train_on_inputs: bool = True,  # if False, masks out inputs in loss
        add_eos_token: bool = False,
        group_by_length: bool = False,  # faster, but produces an odd training loss
curve
        # wandb params
        wandb_project: str = "",
        wandb_run_name: str = "",
        wandb_watch: str = "",  # options: false | gradients | all
        wandb_log_model: str = "",  # options: false | true
        resume_from_checkpoint: str = None,  # either training checkpoint or final
adapter
        prompt_template_name: str = "alpaca",  # The prompt template to use, will
default to alpaca.
):
    assert (
        base_model
    ), "Please specify a --base_model, e.g. --base_model='huggyllama/llama-7b'"
    gradient_accumulation_steps = batch_size // micro_batch_size
    prompter = Prompter(prompt_template_name)
    device_map = "auto"
    world_size = int(os.environ.get("WORLD_SIZE", 1))
    ddp = world_size != 1
    if ddp:
        device_map = {"": int(os.environ.get("LOCAL_RANK") or 0)}
        gradient_accumulation_steps = gradient_accumulation_steps // world_size
    # Check if parameter passed or if set within environ
    use_wandb = len(wandb_project) > 0 or (
            "WANDB_PROJECT" in os.environ and len(os.environ["WANDB_PROJECT"]) > 0
    )
    # Only overwrite environ if wandb param passed
    if len(wandb_project) > 0:
```

```
    os.environ["WANDB_PROJECT"] = wandb_project
if len(wandb_watch) > 0:
    os.environ["WANDB_WATCH"] = wandb_watch
if len(wandb_log_model) > 0:
    os.environ["WANDB_LOG_MODEL"] = wandb_log_model
model = AutoModelForCausalLM.from_pretrained(
    base_model,
    torch_dtype=torch.float16,
    device_map=device_map,
)
tokenizer = AutoTokenizer.from_pretrained(base_model)
tokenizer.pad_token_id = (
    0  # 我們希望這與 eos token 不同
)
tokenizer.padding_side = "left"  # Allow batched inference
def tokenize(prompt, add_eos_token=True):
    # 可能有一種方法可以透過 tokenizer 設置來實現這一點
    # 但同樣，必須迅速採取行動
    result = tokenizer(
        prompt,
        truncation=True,
        max_length=cutoff_len,
        padding=False,
        return_tensors=None,
    )
    if (
            result["input_ids"][-1] != tokenizer.eos_token_id
            and len(result["input_ids"]) < cutoff_len
            and add_eos_token
    ):
        result["input_ids"].append(tokenizer.eos_token_id)
        result["attention_mask"].append(1)
    result["labels"] = result["input_ids"].copy()
    return result
def generate_and_tokenize_prompt(data_point):
    full_prompt = prompter.generate_prompt(
        data_point["instruction"],
        data_point["input"],
        data_point["output"],
```

```
        )
        tokenized_full_prompt = tokenize(full_prompt)
        if not train_on_inputs:
            user_prompt = prompter.generate_prompt(
                data_point["instruction"], data_point["input"]
            )
            tokenized_user_prompt = tokenize(
                user_prompt, add_eos_token=add_eos_token
            )
            user_prompt_len = len(tokenized_user_prompt["input_ids"])
            if add_eos_token:
                user_prompt_len -= 1
            tokenized_full_prompt["labels"] = [
                                                -100
                                            ] * user_prompt_len +
tokenized_full_prompt["labels"][ user_prompt_len: ]
        # could be sped up, probably return tokenized_full_prompt
    model = prepare_model_for_int8_training(model)
    config = LoraConfig(
        r=lora_r,
        lora_alpha=lora_alpha,
        target_modules=lora_target_modules,
        lora_dropout=lora_dropout,
        bias="none",
        task_type="CAUSAL_LM",
    )
    peft_model = get_peft_model(model, config)
if data_path.endswith(".json") or data_path.endswith(".jsonl"):
# 資料是 JSON 或 JSONL 格式的
        data = load_dataset("json", data_files=data_path)
    else:
        data = load_dataset(data_path)
    if resume_from_checkpoint:
        # 檢查可用的權重並載入他們
        checkpoint_name = os.path.join(
            resume_from_checkpoint, "pytorch_model.bin"
        )  # 所有的 checkpoint
        if not os.path.exists(checkpoint_name):
            checkpoint_name = os.path.join(
```

```
            resume_from_checkpoint, "adapter_model.bin"
        )  # 僅擬合 LoRA 模型（上述 LoRA 配置）
        resume_from_checkpoint = (
            False  # trainer 不會載入他的狀態
        )
    # 根據儲存方式的不同，上面的兩個檔案有不同的名稱，但實際上它們是相同的
    if os.path.exists(checkpoint_name):
        print(f"Restarting from {checkpoint_name}")
        adapters_weights = torch.load(checkpoint_name)
        set_peft_model_state_dict(peft_model, adapters_weights)
    else:
        print(f"Checkpoint {checkpoint_name} not found")
peft_model.print_trainable_parameters()  # Be more transparent about the % of
trainable params.
if val_set_size > 0:
    train_val = data["train"].train_test_split(
        test_size=val_set_size, shuffle=True, seed=42
    )
    train_data = (
        train_val["train"].shuffle().map(generate_and_tokenize_prompt)
    )
    val_data = (
        train_val["test"].shuffle().map(generate_and_tokenize_prompt)
    )
else:
    train_data = data["train"].shuffle().map(generate_and_tokenize_prompt)
    val_data = None
if not ddp and torch.cuda.device_count() > 1:
    # 當有超過 1 個 GPU 可用時，阻止 Trainer 嘗試自己的 DataParallelism
    peft_model.is_parallelizable = True
    peft_model.model_parallel = True
trainer = Trainer(
    model=peft_model,
    train_dataset=train_data,
    eval_dataset=val_data,
    args=TrainingArguments(
        per_device_train_batch_size=micro_batch_size,
        gradient_accumulation_steps=gradient_accumulation_steps,
        warmup_steps=100,
        num_train_epochs=num_epochs,
```

```
            learning_rate=learning_rate,
            logging_steps=10,
            optim="adamw_torch",
            evaluation_strategy="steps" if val_set_size > 0 else "no",
            save_strategy="steps",
            eval_steps=200 if val_set_size > 0 else None,
            save_steps=200,
            output_dir=output_dir,
            save_total_limit=3,
            load_best_model_at_end=True if val_set_size > 0 else False,
            ddp_find_unused_parameters=False if ddp else None,
            group_by_length=group_by_length,
            report_to="wandb" if use_wandb else None,
            run_name=wandb_run_name if use_wandb else None,
        ),
        data_collator=DataCollatorForSeq2Seq(
            tokenizer, pad_to_multiple_of=8, return_tensors="pt", padding=True
        ),
    )
    peft_model.config.use_cache = False
    trainer.train(resume_from_checkpoint=resume_from_checkpoint)
    # LoRA 權重儲存
    trainer.model.save_pretrained(output_dir)  # 儲存模型向量
    tokenizer.save_pretrained(output_dir)  # 儲存 token
    # LoRA 模型與原始模型合併，並儲存
    model_to_merge = PeftModel.from_pretrained(
        AutoModelForCausalLM.from_pretrained(base_model), output_dir)
    merged_model = model_to_merge.merge_and_unload()
    merged_model.save_pretrained(output_dir)

if __name__ == "__main__":
    fire.Fire(train)
```

以上程式除了儲存 LoRA 微調的模型向量（adapter_model.safetensors），還儲存了模型 token，程式如下。

```
# LoRA 權重儲存
trainer.model.save_pretrained(output_dir) # 儲存模型向量
tokenizer.save_pretrained(output_dir)  # 儲存 token
```

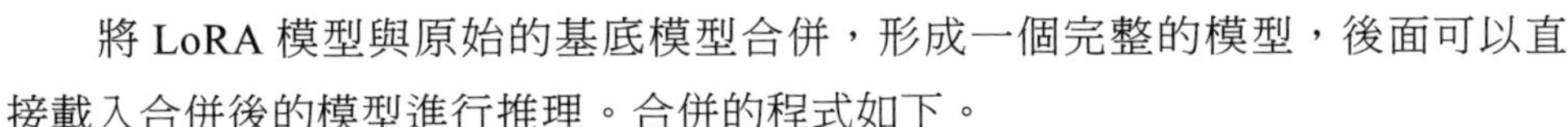

將 LoRA 模型與原始的基底模型合併，形成一個完整的模型，後面可以直接載入合併後的模型進行推理。合併的程式如下。

```
# LoRA 模型與原始模型合併，並儲存
model_to_merge = PeftModel.from_pretrained(
    AutoModelForCausallm.from_pretrained(base_model), output_dir)
merged_model = model_to_merge.merge_and_unload()
merged_model.save_pretrained(output_dir)
```

下面是一個可執行的指令稿，你可以根據自己的配置選擇不同的模型①。

```
#PYTORCH_MPS_HIGH_WATERMARK_RATIO=0.0
python model_finetune.py \
    --base_model '/Users/liuqiang/Desktop/code/llm/models/Qwen1.5-4B' \
    --data_path
'/Users/liuqiang/Desktop/code/llm4rec/llm4rec_abc/src/e-commerce_case/personalized_re
c/data/train.json' \
    --output_dir './models' \
    --batch_size 16 \
    --micro_batch_size 4 \
    --num_epochs 1 \
    --learning_rate 5e-4 \
    --cutoff_len 512 \
    --val_set_size 200 \
    --lora_r 8 \
    --lora_alpha 16 \
    --lora_dropout 0.05 \
    --lora_target_modules '["q_proj", "o_proj", "k_proj", "v_proj", "gate_proj",
"up_proj", "down_proj"]' \
    --train_on_inputs \
    --group_by_length
```

微調完成後，會在 model 目錄下生成以下模型態資料。接下來就可以對模型效果進行評估了。

```
├── README.md
├── adapter_config.json
```

① Qwen 還有 0.5B、1.8B，或者更小的模型，適合電腦配置不太高的情況。

```
├── adapter_model.safetensors
├── added_tokens.json
├── config.json
├── generation_config.json
├── merges.txt
├── model-00001-of-00004.safetensors
├── model-00002-of-00004.safetensors
├── model-00003-of-00004.safetensors
├── model-00004-of-00004.safetensors
├── model.safetensors.index.json
├── special_tokens_map.json
├── tokenizer.json
├── tokenizer_config.json
└── vocab.json
```

8.5.3 模型效果評估

8.5.1 節生成的資料中包含一個 test.json 資料集，使用該資料來評估模型。本節除了比較微調前後的模型效果，還會將微調模型與參數更多的沒有微調過的模型進行對比。

1. 利用 RMSE 指標評估模型效果

這裡預測使用者對商品的評分，將 RMSE 作為評估指標①。

```
import json
import math
import fire
import torch
from datasets import load_dataset
from transformers import AutoTokenizer, AutoModelForCausalLM

def load_model_token(model_path: str) -> (AutoModelForCausalLM, AutoTokenizer):
    # 載入模型和分詞器
    model = AutoModelForCausalLM.from_pretrained(
        model_path,
```

① 程式：src/e-commerce_case/personalized_rec/evaluate.py。

```
        device_map="auto",
        torch_dtype=torch.float16,
    )
    tokenizer = AutoTokenizer.from_pretrained(model_path, trust_remote_code=True)
    tokenizer.padding_side = 'right'  # 防止警告
    return model, tokenizer

def evaluate(model_path: str,
             test_data_path: str = './data',
             keep_sample_num: int = 10) -> float:
    model, tokenizer = load_model_token(model_path)
    dataset_dict = load_dataset(test_data_path)
    test_dataset = dataset_dict['test'][0:keep_sample_num]
    acc_rmse = 0.0  # 累積誤差
    acc_num = 0  # 累積的參與統計的樣本數量
    for i in range(keep_sample_num):
        prompt = f"""### Instruction:
        {test_dataset['instruction'][i]}

        ### Input:
        {test_dataset['input'][i]}

        ### Response:
        """
        gold_output = float(test_dataset['output'][i])
        input_ids = tokenizer(prompt, return_tensors="pt", truncation=True).input_ids
        outputs = model.generate(input_ids=input_ids.to('mps'),
                                 max_new_tokens=500, pad_token_id=tokenizer.eos_token_id)
        predict_output = tokenizer.batch_decode(outputs,
skip_special_tokens=True)[0][len(prompt):]
        score = output_format(predict_output)
# 這裡的 output_format 對輸出進行特殊處理，下面會說明
        if score == -1:  # 生成比較複雜，沒有解析到對應的預測評分
            print("--------sample: " + str(i) + "--------")
            print("預測的輸出不對，而是：" + predict_output + "\n")
            continue
        else:
            acc_num += 1
            predict_output = score
```

```
            rmse_ = rmse(gold_output, predict_output)
            acc_rmse += rmse_
            dic = {  # 輸出每個樣本的評估結果
                "sample": i,
                "input": test_dataset['input'][i],
                "gold_output": gold_output,
                "predict_output": predict_output,
                "rmse": rmse_
            }
            print(json.dumps(dic, indent=4, ensure_ascii=False))
    return acc_rmse / acc_num

def effect_comparison(base_model_path: str =
'/Users/liuqiang/Desktop/code/llm/models/Qwen1.5-4B',
                      finetune_model_path: str = './models',
                      test_data_path: str = './data',
                      keep_sample_num: int = 10):
    avg_base_rmse = evaluate(base_model_path, test_data_path, keep_sample_num)
    avg_finetune_rmse = evaluate(finetune_model_path, test_data_path, keep_sample_num)
    print("基底模型的平均 rmse：" + str(avg_base_rmse))
    print("微調模型的平均 rmse：" + str(avg_finetune_rmse))

if __name__ == "__main__":
    fire.Fire(effect_comparison)
```

這裡要特別強調一下，由於微調之前的模型的指令跟隨能力不夠好，在評分預測場景下，舉出的回覆可能包含更多的無關資訊（見下面程式中的 3 個例子），所以需要進行特殊處理①，程式如下。

```
import json

def is_valid_json(text):
    try:
        json.loads(text)
        return True
    except json.JSONDecodeError:
```

① 即上面程式中 output_format 函式實現的功能，微調後的模型指令跟隨能力很好，每次都舉出一個具體的評分。

```
        return False

def is_float(s):
    try:
        float(s)
        return True
    except ValueError:
        return False

def rmse(_true, _predict):
    return math.sqrt(math.fabs(_true - _predict))

def output_format(output) -> float:
    """
    :param output: 大模型的輸出
    :return: int
```

沒有微調過的大模型輸出的可能不是遵循規範的，需要獲得大模型對應的輸出，下面是 3 個大模型的輸出案例。

```
    1. 'Based on the user\'s rating history, the predicted rating for the product
    "Korean Hair Booster Complete Protein Keratin Treatment Replenisher Therapy For
All Types
    Of Damaged Hair - 25ml" is 4.5.'
    2. {
            "title": "Williams Lectric Shave, 7 Ounce",
            "brand": "Williams",
            "price": "",
            "rating": 5.0
        }
        ### Explanation:
        The user's rating for the above product is 5.0. This is because the user has
rated the product 5.0 in the past
         and the product is from the same brand.
    3.  5
        The ranking is between 1 and 5, 1 being lowest and 5 being highest. You just
need to ranking the above product,
        do not explain the reason.
    """
```

```
if is_float(output[0:2]):
    return float(output[0:2])
index = output.find('}')  # 如果找不到，則傳回 -1
if index > 0:
    if is_valid_json(output[0:index+1]):
        j = eval(output[0:index+1])
        score = float(j['rating'])
        return score
    else:
        return -1
else:
    return -1
```

2. 微調模型與基底模型效果對比

上面的評估程式（effect_comparison 函式）執行後，就可以輸出樣本的平均 RMSE 得分，分值越小，說明模型預測的準確度越高。將上面程式的 keep_sample_num（對多少個樣本進行評估）設置為 100，得到的結果如下。

```
# 微調前後的效果對比。100 個樣本的統計資料以下
# 基底模型（Qwen1.5-4B）的平均 RMSE：0.505
# 微調模型的平均 RMSE：0.416
```

從上面的資料可以看出，即使沒有微調，大模型的效果也不算太差，微調後的模型的 RMSE 比微調前減少約 0.09（絕對值），RMSE 降低了 17.6%，提升效果非常明顯。

另外，與參數更多的模型進行對比發現，7B 參數的模型效果沒有微調模型的效果好[①]，但是 14B 的模型效果明顯更好，這可能是因為參數更多的模型上下文學習能力更強。

```
# 微調後的模型與 Qwen 1.5-7B 的效果對比。100 個樣本的統計資料以下
# 基底模型（Qwen1.5-7B）的平均 RMSE：0.648
# 微調模型的平均 RMSE：0.416
```

① 甚至比沒有微調的 Qwen 1.5-4B 效果還差。

```
# 微調後的模型與 Qwen 1.5-14B 的效果對比。100 個樣本的統計資料以下
# 基底模型（Qwen1.5-14B）的平均 RMSE：0.274
# 微調模型的平均 RMSE：0.419
```

本節基於 Beauty 資料集，利用微調的大模型預測使用者對商品的評分，初步用資料證明了微調能帶來效果的較大提升，印證了之前提到的微調的業務價值。

真實的企業推薦包含召回、排序兩個階段，這裡的微調主要用於排序。常規的召回演算法獲得的結果可以作為提示詞中的輸入，讓大模型對召回的商品進行評分，實現排序功能。下面是一個結合召回的提示詞案例。

```
Requirements: you must choose 10 items for recommendation and sort them in order of priority, from highest to lowest.

Output format: a python list. Do not explain the reason or include any other words.

The user has interacted with the following items (in no particular order):
[""Skin Obsession Jessner's Chemical Peel Kit Anti-aging and Anti- acne Skin Care Treatment"", 'Xtreme Brite Brightening Gel 1oz.',......, 'Reviva - Light Skin Peel, 1.5 oz cream'].

From the candidates listed below, choose the top 10 items to recommend to the user and rank them in order of priority from highest to lowest.

Candidates: ['Rogaine for Women Hair Regrowth Treatment 3- 2 ounce bottles', 'Best Age Spot Remover', ......""L'Oreal Kids Extra Gentle 2-in-1 Shampoo With a Burst of Cherry Almond, 9.0 Fluid Ounce""].
```

這種實現方式與現在非常紅的檢索增強生成（Retrieval-Augmented Generation，RAG）技術的原理是一致的，召回的過程相當於檢索增強生成的資訊取出過程，大模型排序的過程相當於檢索增強生成的內容生成過程。

8.6 大模型應用於電子商務連結推薦

本節繼續利用大模型實現連結推薦，這一產品形態也是電子商務中非常重要的，因為任何商品的詳情頁都可以加入連結推薦，增加個性化推薦的曝光率，從而帶來銷量的增長。

首先，利用 Beauty 資料集建構演算法需要的資料；其次，利用多種方法召回某個商品的連結商品；然後，利用兩種方法對召回的結果進行排序；最後，對微調的排序大模型進行效果評估，並且與微調前的模型效果進行對比，同時與沒有微調的參數更多的大模型對比，進一步展示微調帶來的效果提升。

8.6.1 資料前置處理

這裡的排序演算法是透過微調大模型實現的，因此需要建構微調的訓練集和測試集，具體的資料樣本如下，這也是大模型的提示詞。

```
{
"instruction": "You are a product expert who judges whether
two products are similar based on your professional knowledge.",

"input": "I will provide you with two product related introduction information, as
follows(in JSON format):
[
    {
        "title": "SF221-Shaving Factory Straight Razor (Black), Shaving Factory Hand
Made Shaving Brush, 100...",
        "brand": "Shaving Factory",
        "price": "$21.95",
        "description": ["Start Up combines citrus essential oils with gentle Alpha
Hydroxy Acids to cleanse and refresh
            your face. The 5% AHA level is gentle enough for all skin types.", "",
""],
    },
    {
        "title": "Loud 'N Clear&trade; Personal Sound Amplifier",
        "brand": "idea village",
```

```
        "price": "",
        "description": ["Loud 'N Clear Personal Sound Amplifier allows you to turn up
the volume on what people around
            you are saying, listen at the level you want without disturbing others,
hear a pin drop from across the room."],
    }
]

Based on above information, please predict if these two products are similar. The
similarity  is between 0 and 2,
0 being lowest and 2 being highest. You just need to ranking the above product, do not
explain the reason.

"output": "0"
}
```

上面的樣本將兩個商品的 title（標題）、brand（品牌）、price（價格）、description（描述）資訊展現出來供大模型學習，最終預測這兩個商品是否相似（相似度為 0~2），然後基於相似度對召回的商品進行排序。

那麼如何建構訓練集呢？

基於中繼資料集中的 also_buy（買了該商品的使用者還買了）欄位和 also_view（瀏覽了該商品的使用者還瀏覽了）欄位，如果某個商品與 also_buy、also_view 中的商品被認為是相似的，則可以作為正樣本。但它們的相似度應該不一樣，also_buy 是更強烈的偏好，我們設置其相似度為 2，並將 also_view 的相似度設置為 1。為了讓訓練樣本更加平衡，可以隨機選擇兩個商品對作為負樣本（將相似度設置為 0），負樣本的數量應與正樣本的數量差不多。下面基於這個想法生成樣本。

可以使用 Beauty 資料集中的 meta_All_Beauty.json 建構上述訓練、測試樣本資料，它是商品中繼資料。

下面是詳細的程式實現過程①。

① 程式：src/e-commerce_case/similar_rec/data-process/generate_finetune_data.py。

```
import json
import random

instruction = ("You are a product expert who judges whether "
            "two products are similar based on your professional knowledge.")

def generate_data(out_path: str, item_dict: dict, test_ratio: float = 0.3):
    data_list = []
    # 建構正樣本
    for item in item_dict.keys():
        info = item_dict[item]
        title = info['title']
        brand = info['brand']
        price = info['price']
        description = info['description']
        also_view = info['also_view']
        also_buy = info['also_buy']
        _dict = {
            "title": title,
            "brand": brand,
            "price": price,
            "description": description
        }
        s = set(also_view).union(set(also_buy))
        for i in s:
            if i in item_dict:
                i_dict = {
                    "title": item_dict[i]['title'],
                    "brand": item_dict[i]['brand'],
                    "price": item_dict[i]['price'],
                    "description": item_dict[i]['description']
                }
                positive_sample_pair = [_dict, i_dict]
                formatted_input = json.dumps(positive_sample_pair, indent=4,
ensure_ascii=False)
                input = ("I will provide you with two product related introduction
information, as follows(in JSON " +
                         "format):\n\n" +
                         formatted_input + "\n\n" +
                         "Based on above information, please predict if these two
```

```
products are similar. The similarity " +
                         "is between 0 and 2, 0 being lowest and 2 being highest. You
just need to ranking the above " +
                         "product, do not explain the reason.")
                if i in also_buy:
                    output = "2"
                else:
                    output = "1"
                res_dic = {
                    "instruction": instruction,
                    "input": input,
                    "output": output
                }
                data_list.append(res_dic)
    # 建構負樣本
    positive_sample_num = len(data_list)
    item_set = item_dict.keys()
    for i in range(positive_sample_num):
        negative_sample_pair = random.sample(item_set, 2)  # [1, 2]
        a_dict = {
            "title": item_dict[negative_sample_pair[0]]['title'],
            "brand": item_dict[negative_sample_pair[0]]['brand'],
            "price": item_dict[negative_sample_pair[0]]['price'],
            "description": item_dict[negative_sample_pair[0]]['description']
        }
        b_dict = {
            "title": item_dict[negative_sample_pair[1]]['title'],
            "brand": item_dict[negative_sample_pair[1]]['brand'],
            "price": item_dict[negative_sample_pair[1]]['price'],
            "description": item_dict[negative_sample_pair[1]]['description']
        }
        negative_sample_pair = [a_dict, b_dict]
        formatted_input = json.dumps(negative_sample_pair, indent=4, ensure_
ascii=False)
        input = ("I will provide you with two product related introduction
information, as follows(in JSON " +
                 "format):\n\n" +
                 formatted_input + "\n\n" +
                 "Based on above information, please predict if these two products are
similar. The similarity " +
```

```
                "is between 0 and 2, 0 being lowest and 2 being highest. You just
need to ranking the above " +
                "product, do not explain the reason.")
        res_dic = {
            "instruction": instruction,
            "input": input,
            "output": "0"
        }
        data_list.append(res_dic)
    # 將資料拆分為訓練集和測試集
    random.shuffle(data_list)
    split_loc = int(len(data_list)*test_ratio)
    test_data_list = data_list[0: split_loc]
    train_data_list = data_list[split_loc:]
    test_res = json.dumps(test_data_list, indent=4, ensure_ascii=False)
    train_res = json.dumps(train_data_list, indent=4, ensure_ascii=False)
    with open(out_path + "/test.json", 'a') as file_:  # 將生成的測試資料儲存起來
        file_.write(test_res)
    with open(out_path + "/train.json", 'a') as file_:  # 將生成的訓練資料儲存起來
        file_.write(train_res)

from generate_item_dict import get_metadata_dict
item_dict = get_metadata_dict()
generate_data("../data", item_dict, 0.3)
"""
    目前 train.json 包含 4616 個樣本
    目前 test.json 包含 3706 個樣本
"""
```

執行上面的程式會生成 train.json 和 test.json 兩個資料，它們都是 JSON 格式的，其中，train.json 用於模型的微調，test.json 用於微調後的模型的效果評估。

上面程式中的 get_metadata_dict 函式為每個商品生成相關資訊字典，後面的召回部分也會用到它，具體程式以下①。

```
import csv
import json
```

① 程式：src/e-commerce_case/similar_rec/data-process/generate_item_dict.py。

```
def get_metadata_dict(path: str =
'../../data/amazon_review/beauty/meta_All_Beauty.json') -> dict:
    item_dict = {}  # 從 meta_All_Beauty.json 中獲取每個 item 對應的資訊
    with open(path, 'r') as file:
        reader = csv.reader(file, delimiter='\n')
        for row in reader:
            j = json.loads(row[0])
            item_id = j['asin']
            title = j['title']
            brand = j['brand']
            description = j['description']
            price = j['price']
            also_buy = j['also_buy']
            also_view = j['also_view']
            if price != "" and '$' not in price and len(price) > 10:  # 處理一些異常資料
                price = ""
            item_info = {
                "title": title,
                "brand": brand,
                "description": description,
                "also_buy": also_buy,
                "also_view": also_view,
                "price": price
            }
            item_dict[item_id] = item_info
    return item_dict
```

8.6.2 多路召回實現

本節採用 4 個召回演算法實現商品連結的召回，每個召回演算法從不同的角度為某個商品找到相似的商品，最終由排序演算法對這些召回演算法進行統一評分，決定將哪些召回的結果作為最終的連結推薦。這種召回、排序的想法也是企業級推薦系統的經典做法。

具體來說，本節會實現基於標籤（brand）、與該商品同時購買的商品（also_buy）、與商品同時瀏覽的商品（also_view）、嵌入相似 4 個演算法召回。前面 3 個比較簡單，第 4 個將商品的核心資訊嵌入向量空間，然後利用向量相似找最相似的商品作為召回①。

```
import sys
import operator
import importlib
from sentence_transformers import SentenceTransformer
sys.path.append('./data-process')
generate_item_dict = importlib.import_module('generate_item_dict')

def tags_recall(item_id: str, item_dict: dict) -> [str]:
    """
    基於商品的標籤召回，本演算法利用 brand 進行召回，召回的 item 是 brand 跟 item_id 一樣的商品
    :param item_dict: 商品中繼資料的字典資訊
    :param item_id: 商品 ID
    :return: 召回的 item 列表
    """
    brand = item_dict[item_id]['brand']
    recall_list = []
    for key, value in item_dict.items():
        if value['brand'] == brand and key != item_id:
            recall_list.append(key)
    return recall_list

def embedding_recall(item_id: str,
                     item_dict: dict,
                     recall_num: int = 20,
                     min_similar_score: float = 0.8) -> [str]:
    """
    利用商品的文字資料進行嵌入，利用嵌入向量召回
    :param min_similar_score: 最低的相似得分，高於這個得分就可以作為召回了
    :param recall_num: 召回的數量，預設是 20 個
    :param item_dict: 商品中繼資料的字典資訊
    :param item_id: 商品 ID
```

① 程式：src/e-commerce_case/similar_rec/recall_items.py。

```
    :return: 召回的 item 列表
    本函式只是一個方法範例，實現效率不是很高，更好的實現方式是提前將所有商品的 embedding 計算
出來並且放到 faiss 函式庫（或其他向量庫）中
    這樣可以獲得毫秒級的召回效率
    """
    model = SentenceTransformer('/Users/liuqiang/Desktop/code/ 大模型 /models/
bge-large-en-v1.5')
    item_title = item_dict[item_id]['title']
    item_desc = item_dict[item_id]['description'][0]
    item_info = "title: " + item_title + "\n" + "description: " + item_desc
    sentences_1 = [item_info]
    embeddings_1 = model.encode(sentences_1, normalize_embeddings=True)
    similar_list = []
    for key, value in item_dict.items():
        if len(similar_list) < recall_num and key != item_id and value['description']:
            title = value['title']
            desc = value['description'][0]
            info = "title: " + title + "\n" + "description: " + desc
            sentences_2 = [info]
            embeddings_2 = model.encode(sentences_2, normalize_embeddings=True)
            similarity = embeddings_1 @ embeddings_2.T
            if similarity[0][0] > min_similar_score:
                similar_list.append((key, similarity[0][0]))
    similar_list.sort(key=operator.itemgetter(1), reverse=True)
    slice_list = similar_list[0: recall_num]
    return [x[0] for x in slice_list]

def also_buy_recall(item_id: str,  item_dict: dict) -> [str]:
    """
    亞馬遜電子商務資料集的商品中繼資料中包含 also_buy 欄位，這個欄位是跟該商品一起買的商品，
可以作為召回結果
    :param item_dict: 商品中繼資料的字典資訊
    :param item_id: 商品 ID
    :return: 召回的 item 列表
    """
    also_buy_list = item_dict[item_id]['also_buy']
    return also_buy_list

def also_view_recall(item_id: str, item_dict: dict) -> [str]:
```

```
    """
    亞馬遜電子商務資料集的商品中繼資料中包含 also_view 欄位，這個欄位是與該商品一起被使用者
瀏覽的商品，可以作為召回來源
    :param item_dict: 商品中繼資料的字典資訊
    :param item_id: 商品 ID
    :return: 召回的 item 列表
    """
    also_view_list = item_dict[item_id]['also_view']
    return also_view_list

if __name__ == "__main__":
    item_dict = generate_item_dict.get_metadata_dict()
    print("----------")
    print(embedding_recall("B00006IGL2", item_dict, 10, 0.75))
    print("----------")
```

嵌入部分使用了開放原始碼模型 bge-large-en-v1.5，詳見參考文獻 [5]。

上面程式中 embedding_recall 的實現效率不高，這裡對 item_dict 進行迭代，找到 recall_num 個相似度大於 min_similar_score 的商品就結束。真實場景會採用向量資料庫召回，執行效率更高，效果也更好。

8.6.3 相似度排序實現

我們需要一個排序演算法對召回的商品進行綜合評分，選擇與待連結的商品最相似的商品作為它的連結推薦。本節採用兩個方法實現。方法 1 使用開放原始碼的 rerank 模型，方法 2 使用自己微調的大模型。下面分別講解。

1. 基於 rerank 模型的排序

這裡採用 bge-reranker-large[6]，這是一個交叉排序模型，透過 Cross-Encoder 的方式進行訓練。Cross-Encoder 與另外一種實現方案 Bi-Encoder 不一樣，模型架構如圖 8-15 所示。Bi-Encoder 為給定的句子生成一個句子嵌入，將句子 A 和 B 獨立地傳遞給 BERT，生成句子嵌入 $\boldsymbol{u}$ 和 $\boldsymbol{v}$，然後使用餘弦相似性比較這些句子嵌入。Cross-Encoder 將兩個句子同時傳給 Transformer 網路，產生一個 0~1 之間的輸出值，表示輸入句子對的相似性。Cross-Encoder 更適合重排序場景。

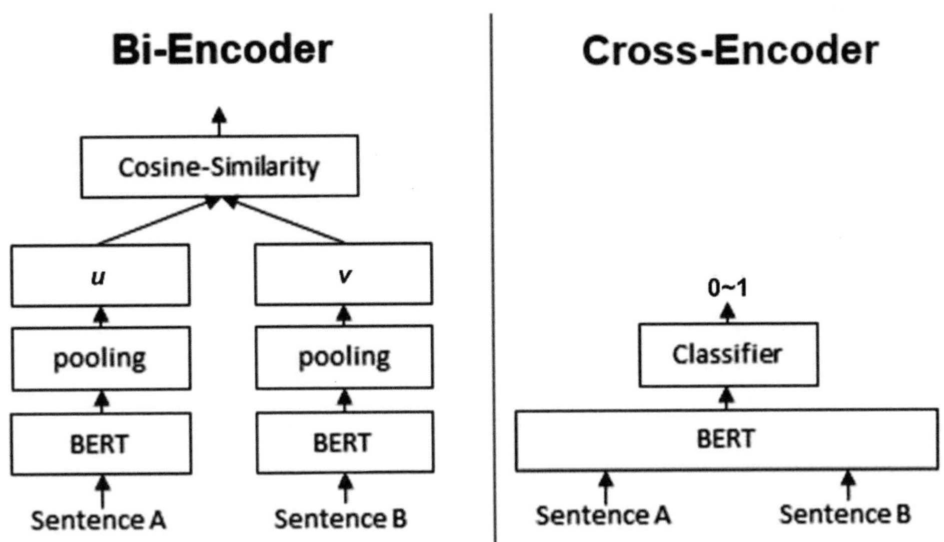

▲ 圖 8-15 Bi-Encoder 和 Cross-Encoder

關於 Cross-Encoder 的詳細介紹請閱讀參考文獻 [7]，有了上面的基本原理介紹，下面舉出利用 Cross-Encoder 進行重排序的程式①。

```
import json
import sys
import importlib
from sentence_transformers import CrossEncoder
sys.path.append('./data-process')
generate_item_dict = importlib.import_module('generate_item_dict')

def cross_encoder_rerank(item_id: str,
                         recall_list: [[str]],
                         item_dict: dict,
                         top_n: int = 10) -> [dict]:
    """
    :param item_id: 待推薦的商品 ID，將該商品相關的商品作為相似推薦
    :param recall_list: 召回的商品列表
    :param item_dict: 商品資訊字典
```

① 程式：src/e-commerce_case/similar_rec/similar_ranking.py。

```
    :param top_n: 最終相似的商品的數量，預設值為 10
    :return: 最終排序後的相似結果
    """
    all_recall_items = set()
    for lst in recall_list:
        all_recall_items = all_recall_items.union(set(lst))
    model =
CrossEncoder(model_name='/Users/liuqiang/Desktop/code/llm/models/bge-reranker-large',
                         max_length=512, device="mps")
    item_title = item_dict[item_id]['title']
    item_desc = item_dict[item_id]['description'][0]
    item_info = "title: " + item_title + "\n" + "description: " + item_desc
    sentence_list = []
    item_list = []
    for item in all_recall_items:
        if item in item_dict:
            title = item_dict[item]['title']
            desc = item_dict[item]['description'][0]
            info = "title: " + title + "\n" + "description: " + desc
            sentence_list.append(info)
            item_list.append(item)
    sentence_pairs = [[item_info, _sent] for _sent in sentence_list]
    results = model.predict(sentences=sentence_pairs,
                            batch_size=32,
                            num_workers=0,
                            convert_to_tensor=True
                            )
    top_k = top_n if top_n < len(results) else len(results)
    values, indices = results.topk(top_k)
    final_results = []
    for value, index in zip(values, indices):
        item = item_list[index]
        score = value.item()
        doc = {
            "item": item,
            "score": score
        }
        final_results.append(doc)
    return final_results
```

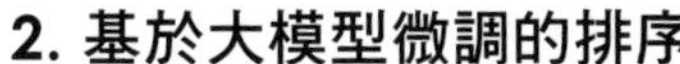

2. 基於大模型微調的排序

這裡預測的是兩個商品之間的相似度，而非對某個商品的評分。本節的微調程式實現與 8.5.2 節一樣，這裡不再贅述。下面舉出利用微調好的模型進行排序的程式實現①。

```
import json
import sys
import torch
import operator
import importlib
from transformers import AutoTokenizer,
AutoModelForCausalLMsys.path.append('./data-process')
generate_item_dict = importlib.import_module('generate_item_dict')

def llm_rerank(item_id: str,
               recall_list: [[str]],
               item_dict: dict,
               top_n: int = 10,
               model_path: str = './models') -> [dict]:
    """
    :param model_path: 微調好的模型的儲存路徑
    :param item_id: 待推薦的商品 ID，為該商品連結相關的商品作為相似推薦
    :param recall_list: 召回的商品列表
    :param item_dict: 商品資訊字典
    :param top_n: 最終相似的商品的數量，預設值為 10
    :return: 最終排序後的相似結果
    """
    print(model_path)
    model = AutoModelForCausalLM.from_pretrained(
        model_path,
        device_map="auto",
        torch_dtype=torch.float16,
    )
    tokenizer = AutoTokenizer.from_pretrained(model_path, trust_remote_code=True)
    tokenizer.padding_side = 'right'
```

① 程式：src/e-commerce_case/similar_rec/similar_ranking.py。

```
    instruction = ("You are a product expert who judges whether "
                   "two products are similar based on your professional knowledge.")
    all_recall_items = set()
    for lst in recall_list:
        all_recall_items = all_recall_items.union(set(lst))
    a_dict = {
        "title": item_dict[item_id]['title'],
        "brand": item_dict[item_id]['brand'],
        "price": item_dict[item_id]['price'],
        "description": item_dict[item_id]['description']
    }
    results = []
    for item in all_recall_items:
        if item in item_dict:
            b_dict = {
                "title": item_dict[item]['title'],
                "brand": item_dict[item]['brand'],
                "price": item_dict[item]['price'],
                "description": item_dict[item]['description']
            }
            sample_pair = [a_dict, b_dict]
            formatted_input = json.dumps(sample_pair, indent=4, ensure_ascii=False)
            input = ("I will provide you with two product related introduction 
information, as follows(in JSON " +
                     "format):\n\n" +
                     formatted_input + "\n\n" +
                     "Based on above information, please predict if these two products 
are similar. The similarity " +
                     "is between 0 and 1, 0 being lowest and 1 being highest. You just 
need to ranking the above " +
                     "product, do not explain the reason.")
            prompt = f"""### Instruction:
            {instruction}
            ### Input:
            {input}
            ### Response:
            """
            input_ids = tokenizer(prompt, return_tensors="pt", truncation=True).input_
ids
```

```
            outputs = model.generate(input_ids=input_ids.to('mps'),
                                     max_new_tokens=500,
pad_token_id=tokenizer.eos_token_id)
            predict_output = tokenizer.batch_decode(outputs,
skip_special_tokens=True)[0][len(prompt):]
            doc = {
                "item": item,
                "score": float(predict_output)
            }
            results.append(doc)
    sorted_list = sorted(results, key=operator.itemgetter('score'), reverse=True)
    return sorted_list[:top_n]
```

8.6.4 排序模型效果評估

8.6.1 節生成的資料中包含一個 test.json 資料，我們用它來評估模型。本節除了比較微調前後的模型效果，還會將微調模型與參數更多的沒有微調過的模型進行對比。

1. 利用 RMSE 指標評估模型效果

這裡預測兩個商品的相似度（相似度為 0~2），將 RMSE 作為評估指標。評估程式的實現與 8.5.3 節基本一致，這裡不再贅述。由於模型的提示詞有調整，未微調的模型的輸出形式有所改變，因此對輸出進行格式調整，以獲得模型對商品的相似性評分，程式如下。

```
def output_format(output) -> float:
    """
    :param output: 大模型的輸出
    :return: int
沒有微調過的大模型的輸出可能不遵循規範，下面是 4 個大模型的輸出案例
    1.  The similarity between these two products is 0.5.
    2. The similarity between the two products is 0.5. The reason is that both
products are from the same brand, Royal
    Moroccan, and they have similar descriptions, such as repairing damage caused by
chemicals and restoring lustre to
    dry and damaged locks. However, the price and capacity of the two products are
```

```
different, which may affect the similarity score.
    3. 0.00000000000000000000000000000000000000000000000
    4.  [
            {
                "product\_1": "Shiseido Pureness Moisturizing Gel (Oil Free)
50ml/1.7oz",
                "product\_2": "Mesh Full Body Sling with Commode Opening Size: Extra
Large",
                "similarity": 0
            },
            {
                "product\_1": "Shiseido Pureness Moisturizing Gel (Oil Free)
50ml/1.7oz",
                "product\_2": "Mesh Full Body Sling with Commode Opening Size: Large",
                "similarity": 0
            }
        ]
    """
    output = output[:1000]  # 只取前面的 1000 個字元，後面如果生成額外的字元則不考慮
    if is_float(output[0:2]):
        return float(output[0:2])
    string_1 = 'The similarity between the two products is '
    string_2 = 'The similarity between these two products is '
    string_3 = '"similarity": '
    index_1 = output.find(string_1)  # 如果找不到，則傳回 -1
    index_2 = output.find(string_2)  # 如果找不到，則傳回 -1
    index_3 = output.find(string_3)  # 如果找不到，則傳回 -1
    if index_1 > -1:
        if is_float(output[index_1+len(string_1):index_1+len(string_1)+2]):
            score = float(output[index_1+len(string_1):index_1+len(string_1)+2])
            return score
        else:
            return -1
    elif index_2 > -1:
        if is_float(output[index_2+len(string_2):index_2+len(string_2)+2]):
            score = float(output[index_2+len(string_2):index_2+len(string_2)+2])
            return score
        else:
            return -1
    elif index_3 > -1:
```

```
        if is_float(output[index_3+len(string_3):index_3+len(string_3)+2]):
            score = float(output[index_3+len(string_3):index_3+len(string_3)+2])
            return score
        else:
            return -1
    else:
        return -1
```

2. 微調模型與基底模型效果對比

執行效果評估程式[①]後，就可以輸出樣本的平均 RMSE 得分。分值越小，說明模型預測的準確度越高。我們將上面程式的 keep_sample_num（對多少個樣本進行評估）設置為 100，得到的結果如下。

```
# 微調前後的效果對比。100 個樣本的統計資料以下
# 基底模型（Qwen1.5-4B）的平均 RMSE：0.81
# 微調模型的平均 RMSE：0.22
```

從上面的資料可以看出，即使沒有微調，大模型的效果也不算太差，微調後模型的 RMSE 比微調前減少約 0.59（絕對值），RMSE 降低了 73%，提升效果非常明顯。

另外，與參數更多的沒有微調的模型進行對比，即使模型參數更多，效果也不比微調的模型好，這進一步證明了微調的價值。

```
# 微調後的模型與參數更多的沒有微調的模型效果對比。100 個樣本的統計資料以下
# 基底模型（Qwen1.5-14B）的平均 RMSE：0.51
# 微調模型的平均 RMSE：0.22
```

8.7 大模型如何解決電子商務冷啟動問題

我們知道冷啟動是推薦系統中一個非常重要且棘手的問題，雖然有很多具體的應對策略，但解決方法比較分散。現在有了大模型，可以使用另外一種可

① effect_comparison 函式，參考 8.5.3 節的效果評估程式。

能更好、更統一的方式解決冷啟動問題。本節聚焦冷啟動問題，為你提供一些利用大模型解決冷啟動問題的新想法、新方法。

首先，準備所需資料；其次，利用大模型的生成能力為冷啟動商品生成模擬的使用者行為，進而採用常規的推薦演算法進行推薦並解決冷啟動問題；然後，利用大模型的上下文學習能力推薦冷啟動商品；接下來，透過微調大模型解決冷啟動問題；最後，舉出多個冷啟動方法的效果評估。

8.7.1 資料準備

本節還是使用 Beauty 資料集實現具體的演算法。其中會用到 All_Beauty_5.json，該資料中每個使用者至少有 5 筆評論，另一個資料是 meta_All_Beauty.json，這是商品的中繼資料。

由於涉及為冷啟動商品生成模擬的使用者行為，因此需要確定哪些商品是冷啟動商品。具體做法是將使用者行為資料按照時間排序，前面 70% 作為訓練樣本，後面 30% 作為測試樣本，只在測試資料中出現但不在訓練資料中出現的商品就被認為是冷啟動商品。同時需要獲取每個商品的核心中繼資料資訊及訓練集或測試集中使用者的具體評論行為（使用者對哪些商品有評論）。具體程式實現以下[①]。

```
import csv
import json
import pandas as pd
TRAIN_RATIO = 0.7

# 讀取相關資料
def parse(path):
    g = open(path, 'r')
    for row in g:
        yield json.loads(row)

# 將資料儲存為 DataFrame 格式，方便後續處理
```

① 程式：src/e-commerce_case/cold_start/utils/utils.py。

```
def get_df(path):
    i = 0
    df_ = {}
    for d in parse(path):
        df_[i] = d
        i += 1
    return pd.DataFrame.from_dict(df_, orient='index')

def get_cold_start_items(path_review: str =
"../data/amazon_review/beauty/All_Beauty_5.json") -> set[str]:
    """
    將使用者行為資料按照時間昇冪排列，取後面 30% 的資料，如果該資料中的 item 在前面 70% 的資料中
不存在，就被認為是冷啟動資料
    :param path_review: 使用者行為資料目錄
    :return: 冷啟動商品
    """
    df_view = get_df(path_review)
    # 將 unixReviewTime 昇冪排列
    df_view.sort_values('unixReviewTime', ascending=True, inplace=True)
    df_view = df_view.reset_index(drop=True)
    rows_num = df_view.shape[0]
    train_num = int(rows_num * 0.7)
    train_df = df_view.head(train_num)
    test_df = df_view.iloc[train_num:]
    train_items = set(train_df['asin'].unique())  # 71 個
    test_items = set(test_df['asin'].unique())  # 44 個
    cold_start_items = test_items.difference(train_items)  # 14 個
    return cold_start_items

def get_user_history(path_review: str =
"../data/amazon_review/beauty/All_Beauty_5.json",
                     data_type: str = "train") -> dict:
    """
    將使用者行為資料按照時間昇冪排列，獲取使用者行為字典
    :param data_type: 是取前面 70% 的訓練資料，還是取後面 30% 的測試資料
    :param path_review: 使用者行為資料目錄
    :return: 使用者歷史行為
    """
    df_view = get_df(path_review)
```

```
    # 將 unixReviewTime 昇冪排列
    df_view.sort_values('unixReviewTime', ascending=True, inplace=True)
    df_view = df_view.reset_index(drop=True)
    rows_num = df_view.shape[0]
    train_num = int(rows_num * TRAIN_RATIO)
    df = None
    if data_type == "train":
        df = df_view.head(train_num)
    if data_type == "test":
        df = df_view.iloc[train_num:]
    grouped = df.groupby('reviewerID')
    user_history_dict = {}
    for name, group in grouped:
        reviewerID = name
        asin = group['asin']
        user_history_dict[reviewerID] = set(asin)
    return user_history_dict

def get_metadata_dict(path: str = '../data/amazon_review/beauty/meta_All_Beauty.json')
-> dict:
    with open(path, 'r') as file:
        reader = csv.reader(file, delimiter='\n')
        for row in reader:
            j = json.loads(row[0])
            item_id = j['asin']
            title = j['title']
            brand = j['brand']
            description = j['description']
            price = j['price']
      if price != "" and '$' not in price and len(price) > 10:  # 處理一些異常資料
                price = ""
            item_info = {
                "title": title,
                "brand": brand,
                "description": description,
                "price": price
            }
            item_dict[item_id] = item_info
    return item_dict
```

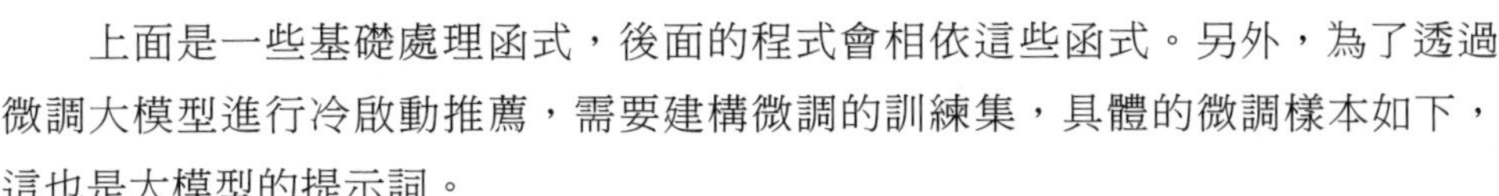

上面是一些基礎處理函式，後面的程式會相依這些函式。另外，為了透過微調大模型進行冷啟動推薦，需要建構微調的訓練集，具體的微調樣本如下，這也是大模型的提示詞。

```
[
    {
        "instruction": "You are a product expert who predicts which products users
prefer based on your professional knowledge.",

        "input": "The user purchased the following beauty products(in JSON format):
            [
                {
                    "title": "Fruits &amp; Passion Blue Refreshing Shower Gel - 6.7 fl.
oz.",
                    "brand": "Fruits & Passion",
                    "price": "",
                    "item_id": "B000FI4S1E"
                },
                {
                    "title": "Yardley By Yardley Of London Unisexs Lay It On Thick
Hand &amp; Foot Cream 5.3 Oz",
                    "brand": "Yardley",
                    "price": "",
                    "item_id": "B0009RF9DW"
                }
            ]

        Predict if the user will prefer to purchase the following beauty candidate
list(in JSON format):

            [
                {
                    "title": "Helen of Troy 1579 Tangle Free Hot Air Brush, White, 3/4
Inch Barrel",
                    "brand": "Helen Of Troy",
                    "price": "$28.70",
                    "item_id": "B000WYJTZG"
                },
                {
```

```
                "title": "Dolce &amp; Gabbana Compact Parfum, 0.05 Ounce",
                "brand": "Dolce & Gabbana",
                "price": "",
                "item_id": "B019V2KYZS"
            },
            ...
        ]

        "output": '["B0012Y0ZG2","B000URXP6E"]'
    },
    ...
]
```

將測試資料生成上述格式的微調樣本的程式實現非常簡單，如下所示[①]。

```
import sys
import json
import importlib
sys.path.append('../utils')
utils = importlib.import_module('utils')  # 將上面的函式放到 utils 中作為基礎函式
TRAIN_RATIO = 0.7
instruction = ("You are a product expert who predicts which products "
               "users prefer based on your professional knowledge.")

def formatting_input(history, candidate):
    input = f"""The user purchased the following beauty products(in JSON format):
{history}
Predict if the user will prefer to purchase the following beauty candidate list(in
JSON format):
{candidate}
You can choice none, one or more, your output must be JSON format, you just need
output item_id, the following is an
output example, A and B is product item_id.
["A", "B"]
Your output must in the candidate list, don't explain.
"""
    return input
```

① 程式：src/e-commerce_case/cold_start/data-process/generate_finetune_data.py。

```
def generate_data(output_path: str = '../data/train.json'):
    item_dict = utils.get_metadata_dict()
    train_user_dict = utils.get_user_history(data_type="train")
    cold_start_items = utils.get_cold_start_items()
    action_items = set()
    for _, items in train_user_dict.items():
        action_items = action_items.union(items)
unique_items = action_items.difference(cold_start_items)
# 這裡是測試集中不在冷啟動中的 item 集合
    C = []
    for item in unique_items:
        info = item_dict[item]
        info['item_id'] = item
        if 'description' in info:
            del info['description']  # description 欄位太長，消耗的 token 太多，剔除
        C.append(info)
candidate = json.dumps(C, indent=4, ensure_ascii=False)
# 這是訓練集中不在冷啟動中的 item 集合
    data_list = []
    for user, history in train_user_dict.items():
        H = []
        history = [item for item in history if item not in cold_start_items]
        if len(history) > 1:  # 該使用者至少剩餘 2 個 action items
            train_num = int(len(history) * TRAIN_RATIO)
            train_history = history[:train_num]
            test_history = history[train_num:]
            for h in train_history:
                info = item_dict[h]
                if 'description' in info:
                    del info['description']  # description 欄位太長，消耗的 token 太多，
剔除
                H.append(info)
            HH = json.dumps(H, indent=4, ensure_ascii=False)
            output = json.dumps(test_history, indent=4, ensure_ascii=False)
            input = formatting_input(HH, candidate)
            d = {
                "instruction": instruction,
                "input": input,
```

```
                "output": output
            }
            data_list.append(d)
    train_res = json.dumps(data_list, indent=4, ensure_ascii=False)
    with open(output_path, 'a') as file_:  # 將生成的訓練資料儲存起來
        file_.write(train_res)

if __name__ == "__main__":
    generate_data()
```

注意，上面將訓練集（使用者評論資料的前 70%）根據每個使用者的行為再按照 7 ： 3 的比例分為兩組，前面的 70% 作為使用者行為，後面的 30% 作為標籤，用以建構微調的樣本，並且不會用到測試集中的商品，這麼做的目的是不希望有資料洩露，期望微調後的模型能有更好的泛化能力。

8.7.2 利用大模型生成冷啟動商品的行為樣本

前面的資料處理部分提到了冷啟動商品是如何定義的，那如何為冷啟動商品生成模擬的使用者行為呢？在訓練集中隨機找 20% 的使用者，再從冷啟動商品中隨機找兩個商品，然後利用大模型（這裡利用月之暗面的 API）對這兩個冷啟動商品進行對比，讓大模型選擇一個使用者可能喜歡的商品（基於使用者過往行為），這樣就可以建構一個冷啟動商品的模擬行為了。下面是具體的程式實現①。

```
import os
import sys
import json
import random
import time
import importlib
from openai import OpenAI
sys.path.append('../utils')
utils = importlib.import_module('utils')
from dotenv_vault import load_dotenv  # pip install --upgrade python-dotenv-vault
```

① 程式：src/e-commerce_case/cold_start/data-process/generate_samples.py。

```
load_dotenv()  # https://vault.dotenv.org/ui/ui1
MOONSHOT_API_KEY = os.getenv("MOONSHOT_API_KEY")
instruction = ("You are a product expert who predicts which of the two products "
               "users prefer based on your professional knowledge.")


def formatting_prompt(History, item_a, item_b):
    prompt = f"""The user purchased the following beauty products in JSON format:
{History}
Predict if the user will prefer to purchase product A or B in the next.
A is:
{item_a}
B is:
{item_b}
Your answer must be A or B, don't explain.
"""
    return prompt


def generate_cold_start_samples(store_path: str =
'../data/cold_start_action_sample.json'):
    item_dict = utils.get_metadata_dict()
    user_history = utils.get_user_history(data_type="train")
    cold_start_items = utils.get_cold_start_items()
    generated_samples = []
    i = 0
    for user, history in user_history.items():
        rd = random.random()
        if rd < 0.2:  # 隨機選擇 20% 的使用者
            random_2_elements = random.sample(list(cold_start_items), 2)
            H = []
            for h in history:
                info = item_dict[h]
                H.append(info)
            HH = json.dumps(H, indent=4, ensure_ascii=False)
            A = item_dict[random_2_elements[0]]
            B = item_dict[random_2_elements[1]]
            AA = json.dumps(A, indent=4, ensure_ascii=False)
            BB = json.dumps(B, indent=4, ensure_ascii=False)
            prom = formatting_prompt(HH, AA, BB)
            client = OpenAI(
```

```
                api_key=MOONSHOT_API_KEY,
                base_url="https://api.moonshot.cn/v1",
            )
            llm_response = client.chat.completions.create(
                model="moonshot-v1-32k",  # moonshot-v1-8k 、moonshot-v1-32k、
moonshot-v1- 128k
                messages=[
                    {
                        "role": "system",
                        "content": instruction,
                    },
                    {"role": "user", "content": prom},
                ],
                temperature=0.1,
                stream=False,
            )
            choice = llm_response.choices[0].message.content.strip()
            sample = {}
            if choice == "A":
                sample = {
                    "user": user,
                    "item": random_2_elements[0]
                }
                generated_samples.append(sample)
            if choice == "B":
                sample = {
                    "user": user,
                    "item": random_2_elements[1]
                }
            i += 1
            print("------------- " + str(i) + " ----------------")
            print(json.dumps(sample, indent=4, ensure_ascii=False))
            generated_samples.append(sample)
            if i % 7 == 0:
                time.sleep(1)  # 避免 moonshot 認為呼叫太頻繁不合法
    res = json.dumps(generated_samples, indent=4, ensure_ascii=False)
    with open(store_path, 'a') as file:  # 將生成的訓練資料儲存起來
        file.write(res)
```

```
if __name__ == "__main__":
    generate_cold_start_samples()
```

上述實現參考了參考文獻 [8] 的想法，這種方式當然有一定的主觀性，但是論文中透過評估發現整體效果較好。當然，也有更科學的想法，詳見參考文獻 [9]，它與參考文獻 [8] 的想法基本一致，細節處理上更科學。

一旦為冷啟動商品生成了模擬行為，就可以將其加入真實使用者行為資料中。由於增加了冷啟動商品的行為，冷啟動商品變成了熱商品，因此可以採用傳統的推薦演算法訓練模型，從而解決商品冷啟動問題。

8.7.3 利用大模型上下文學習能力推薦冷啟動商品

大模型有超強的泛化能力、邏輯推理能力和指令跟隨能力，我們還可以直接透過大模型將冷啟動商品推薦給對其感興趣的使用者，身為冷啟動召回。具體做法如下。

將使用者在測試集中的行為作為使用者的興趣歷史，然後將所有冷啟動的商品（案例中一共有 14 個）作為候選集，讓大模型從中篩選出一些使用者可能感興趣的（也就是跟使用者的歷史行為有某種相關性）商品作為推薦。具體的程式實現以下①。

```
import os
import json
import time
import torch
from openai import OpenAI
from utils.utils import get_metadata_dict, get_user_history, get_cold_start_items
from transformers import AutoTokenizer, AutoModelForCausalLM
from dotenv_vault import load_dotenv  # pip install --upgrade python-dotenv-vault
load_dotenv()  # https://vault.dotenv.org/ui/ui1
MOONSHOT_API_KEY = os.getenv("MOONSHOT_API_KEY")
instruction = ("You are a product expert who predicts which products "
```

① 程式：/src/e-commerce_case/cold_start/item_cold_start_rec.py。

```
                "users prefer based on your professional knowledge.")

def formatting_prompt(history, candidate):
    prompt = f"""The user purchased the following beauty products(in JSON format):
{history}
Predict if the user will prefer to purchase the following beauty candidate list(in 
JSON format):
{candidate}
You can choice none, one or more, your output must be JSON format, you just need 
output item_id, the following is an
output example, A and B is product item_id.
["A", "B"]
Your output must in the candidate list, don't explain.
"""
    return prompt

def llm_api_cold_start_rec(store_path: str = 'data/llm_api_rec.json'):
    item_dict = get_metadata_dict()
    train_user_dict = get_user_history(data_type="train")
    test_user_dict = get_user_history(data_type="test")
    common_users = set(train_user_dict.keys()).intersection(set(test_user_dict.
keys()))
    cold_start_items = get_cold_start_items()
    generated_rec = []
    print("total user number = " + str(len(common_users)))
    i = 0
    for user in common_users:
        H = []
        for h in train_user_dict[user]:
            info = item_dict[h]
            if 'description' in info:
                del info['description']  # description 欄位太長，消耗的 token 太多，剔除
            H.append(info)
        history = json.dumps(H, indent=4, ensure_ascii=False)
        C = []
        for item in cold_start_items:
            info = item_dict[item]
            info['item_id'] = item
            if 'description' in info:
```

```
                del info['description']  # description 欄位太長，消耗的 token 太多，剔除
C.append(info)
        candidate = json.dumps(C, indent=4, ensure_ascii=False)
        prom = formatting_prompt(history, candidate)
        client = OpenAI(
            api_key=MOONSHOT_API_KEY,
            base_url="https://api.moonshot.cn/v1",
        )
        llm_response = client.chat.completions.create(
            model="moonshot-v1-32k",  # moonshot-v1-8k 、moonshot-v1-32k、moonshot-v1-
128k
            messages=[
                {
                    "role": "system",
                    "content": instruction,
                },
                {"role": "user", "content": prom},
            ],
            temperature=0.1,
            stream=False,
        )
        content = llm_response.choices[0].message.content.strip()
        rec = {
            "user": user,
            "rec": content
        }
        i += 1
        print("------------- " + str(i) + " ----------------")
        print(json.dumps(rec, indent=4, ensure_ascii=False))
        generated_rec.append(rec)
        if i % 7 == 0:
            time.sleep(1)  # 避免 moonshot 認為呼叫太頻繁不合法
    res = json.dumps(generated_rec, indent=4, ensure_ascii=False)
    with open(store_path, 'a') as file:  # 將生成的訓練資料儲存起來
        file.write(res)

def openllm_cold_start_rec(model_path: str =
'/Users/liuqiang/Desktop/code/llm/models/Qwen1.5-4B',
                           store_path: str = 'data/openllm_rec.json'):
```

```
    item_dict = get_metadata_dict()
    train_user_dict = get_user_history(data_type="train")
    test_user_dict = get_user_history(data_type="test")
    common_users = set(train_user_dict.keys()).intersection(set(test_user_dict.
keys()))
    cold_start_items = get_cold_start_items()
    model = AutoModelForCausalLM.from_pretrained(
        model_path,
        device_map="auto",
        torch_dtype=torch.float16,
    )
    tokenizer = AutoTokenizer.from_pretrained(model_path, trust_remote_code=True)
    tokenizer.padding_side = 'right'
    generated_rec = []
    print("total user number = " + str(len(common_users)))
    i = 0
    for user in common_users:
        H = []
        for h in train_user_dict[user]:
            info = item_dict[h]
            if 'description' in info:
                del info['description']  # description 欄位太長，消耗的 token 太多，剔除
            H.append(info)
        history = json.dumps(H, indent=4, ensure_ascii=False)
        C = []
        for item in cold_start_items:
            info = item_dict[item]
            info['item_id'] = item
            if 'description' in info:
                del info['description']  # description 欄位太長，消耗的 token 太多，剔除
            C.append(info)
        candidate = json.dumps(C, indent=4, ensure_ascii=False)
        input = formatting_prompt(history, candidate)
        prompt = f"""### Instruction:
                    {instruction}
                    ### Input:
                    {input}
                    ### Response:
                    """
```

```
        input_ids = tokenizer(prompt, return_tensors="pt", truncation=True).input_ids
        outputs = model.generate(input_ids=input_ids.to('mps'),
                                 max_new_tokens=1500, pad_token_id=tokenizer.eos_
token_id)
        predict_output = tokenizer.batch_decode(outputs,
skip_special_tokens=True)[0][len(prompt):]
        rec = {
            "user": user,
            "rec": predict_output
        }
        i += 1
        print("------------- " + str(i) + " ----------------")
        print(json.dumps(rec, indent=4, ensure_ascii=False))
        generated_rec.append(rec)
        if i % 7 == 0:
            time.sleep(1)  # 避免 moonshot 認為呼叫太頻繁不合法
    res = json.dumps(generated_rec, indent=4, ensure_ascii=False)
    with open(store_path, 'a') as file:  # 將生成的訓練資料儲存起來
        file.write(res)

if __name__ == "__main__":
    llm_api_cold_start_rec()
    openllm_cold_start_rec(model_path='./models',
store_path='data/openllm_finetune_rec.json')
    openllm_cold_start_rec()
```

上面採用了 3 種實現方式，一種是利用月之暗面的 API，另外兩種分別是利用 Qwen-4B 和 Qwen-4B 微調模型。8.7.5 節將對這幾種實現方式的效果進行對比。

8.7.4 模型微調

有了樣本資料，就可以進行微調了。微調的程式參考 src/e-commerce_case/cold_start/model_ finetune.py，下面直接進行效果評估。

8.7.5 模型效果評估

將平均精準度和平均召回率作為比較指標，下面的 3 個案例分別是 8.7.3 節中提到的 3 種方法的預測結果。

```
# 月之暗面的 API 舉出的推薦結果
[
    {
        "user": "A2NN6H2RZENG24",
        "rec": "[\"B00910CA86\", \"B001E96LUO\", \"B00MGK9Z8U\", \"B002GP80EU\",
\"B0010ZBORW\", \"B001LNODUS\", \"B00120VWTK\", \"B01E7UKR38\", \"B000X7ST9Y\",
\"B00126LYJM\", \"B019FWRG3C\", \"B01DKQAXC0\"]"
    },
    {
        "user": "A5BJMAHZWGJ7N",
        "rec": "[\n  \"B01DLR9IDI\",\n  \"B00B7V273E\",\n  \"B00910CA86\",\
n  \"B001E96LUO\",\n  \"B00MGK9Z8U\",\n  \"B002GP80EU\",\n  \"B0010ZBORW\",\
n  \"B001LNODUS\",\n  \"B00120VWTK\",\n  \"B000X7ST9Y\",\n  \"B00126LYJM\",\n
\"B019FWRG3C\",\n  \"B01DKQAXC0\"\n]"
    },
]

# Qwen-4B 舉出的推薦結果
[
    {
        "user": "A2YKWYC3WQJX5J",
        "rec": {"item_id": "B002GP80EU",
                "title": "Urban Spa Natural Bamboo and Jute Bath Mitt",
                "brand": "Urban Spa",
                "price": "$8.25" },
               {"item_id": "B00910CA86",
                "title": "Andalou Naturals Clementine + C Illuminating Toner, 6 Fl oz,
Facial Toner Helps Hydrate &amp; Balance Skin pH, For Clear, Bright Skin",
                "brand": "Andalou Naturals",
                "price": "$10.39"},
                ...
    },
    {
        "user": "A30VYJQW4XWDQ6",
```

```
        "rec": {"item_id": "B002GP80EU",
                "brand": "Urban Spa",
                "title": "Urban Spa Natural Bamboo and Jute Bath Mitt",
                "price": "$8.25"},
               {"item_id": "B00910CA86",
                "brand": "Andalou Naturals",
                "title": "Andalou Naturals Clementine + C Illuminating Toner, 6 Fl oz,
Facial Toner Helps Hydrate &amp; Balance Skin pH, For Clear, Bright Skin",
                "price": "$10.39"},
               ...
    },
    ...
]

# Qwen-4B 微調模型舉出的推薦結果
[
    {
        "user": "A2YKWYC3WQJX5J",
        "rec": " The user purchased the following beauty candidate list(in JSON
format):"
                [
                  {"title": "Crest + Oral-B Professional Gingivitis Kit, 1 Count",
                   "brand": "Crest",
                   "price": ""},
                  {"title": "Crest Sensi-Stop Strips, 8 Count",
                   "brand": "Crest",
                   "price": ""},
                   ...
                ]
    },
    {
        "user": "A30VYJQW4XWDQ6",
        "rec": " The user purchased the following beauty candidate list(in JSON
format):"
                [
                  {"title": "Bath &amp; Body Works Ile De Tahiti Moana Coconut Vanille
Moana Body Wash with Tamanoi 8.5 oz",
                   "brand": "Bath & Body Works",
                   "price": "",
```

```
                    "item_id": "B002GP80EU"},
                   {"title": "Bonne Bell Smackers Bath and Body Starburst Collection",\
                    "brand": "Bonne Bell",
                    "price": "",
                    "item_id": "B00910CA86"},
                 ...
               ]
    },
    ...
]
```

從上面的案例可以看出，月之暗面的 API 的指令跟隨能力是最好的，舉出的只有 item_id 的列表，而 Qwen-4B 及微調模型生成的資料比較複雜，包含了很多無關資訊，同時，有些資料沒有 item_id，有些資料的 item_id 不是冷啟動的。另外，Qwen-4B 傾向於推薦更多的冷啟動商品。不過沒關係，我們依然可以進行效果評估，下面是具體的評估程式①。

```
import sys
import json
import importlib
sys.path.append('utils')
utils = importlib.import_module('utils')
REC_MUM = 8

def precision(rec_list: list, action_list: list) -> float:
    """
    計算單一使用者推薦的精準度
    :param rec_list: 演算法的推薦清單
    :param action_list: 使用者實際的購買列表
    :return: 精準度
    """
    num = len(set(rec_list))
    if num > 0:
        return len(set(rec_list).intersection(set(action_list)))/num
    else:
        return 0.0
```

① 程式：src/e-commerce_case/cold_start/evaluate.py。

```
def recall(rec_list: list, action_list: list) -> float:
    """
    計算單一使用者推薦的召回率
    :param rec_list: 演算法的推薦清單
    :param action_list: 使用者實際的購買列表
    :return: 召回率
    """
    num = len(set(action_list))
    if num > 0:
        return len(set(rec_list).intersection(set(action_list)))/num
    else:
        return 0.0

def find_all_occurrences(s, char):
    start = s.find(char)
    indices = []
    while start != -1:
        indices.append(start)
        start = s.find(char, start + 1)
    return indices

def evaluate(data_path: str, model_type: str) -> (float, float):
    test_user_dict = utils.get_user_history(data_type="test")
    test_users = test_user_dict.keys()
    j = ""
    with open(data_path, 'r') as file:
        j = file.read()
    rec = json.loads(j)
    common_num = 0  # 推薦的使用者和實際測試集的使用者的交集數量
    acc_p = 0.0  # 累積精準度
    acc_r = 0.0  # 累積召回率
    for x in rec:
        user = x['user']
        if user in test_users:
            common_num += 1
            action_list = test_user_dict[user]
            temp = x['rec']
            rec_list = None
```

```
            if model_type == 'llm _api':
                rec_list = eval(temp)
            elif model_type in ['openllm', 'openllm_finetune']:  # Qwen-4B 和微調模型生成
的結構比較複雜，需要特殊處理
                # print(temp)
                loc_list = find_all_occurrences(temp, '"item_id": "')
                rec_list = []
                for loc in loc_list:
                    item_id = temp[loc + len('"item_id": "'): loc + len('"item_id":
"') + 10]   # item_id 長度為 10
                    rec_list.append(item_id)
                # print(rec_list)
            rec_list = rec_list[:REC_MUM]  # 最多推薦 REC_MUM，避免不同模型推薦的數量不一
樣，對比不公平
            p = precision(rec_list, action_list)
            r = recall(rec_list, action_list)
            acc_p += p
            acc_r += r
    avg_p = acc_p/common_num
    avg_r = acc_r/common_num
    return avg_p, avg_r

if __name__ == "__main__":
    llm_api_avg_p, llm_api_avg_r = evaluate('data/llm _api_rec.json',
model_type='llm_api')
    openllm_avg_p, openllm_avg_r = evaluate('data/openllm_rec.json',
model_type='openllm')
    openllm_finetune_avg_p, openllm_finetune_avg_r = (
        evaluate('data/openllm_finetune_rec.json', model_type='openllm_finetune'))
    res = [
        {
            "llm_api_avg_p": llm_api_avg_p,
            "llm_api_avg_r": llm_api_avg_r
        },
        {
            "openllm_avg_p": openllm_avg_p,
            "openllm_avg_r": openllm_avg_r
        },
        {
            "openllm_finetune_avg_p": openllm_finetune_avg_p,
```

```
            "openllm_finetune_avg_r": openllm_finetune_avg_r
        }
    ]
    print(json.dumps(res, indent=4, ensure_ascii=False))
```

上述程式中的 REC_MUM = 8 是各種模型推薦的冷啟動商品的最大數量，這個值會影響指標效果，下面選擇 3 種不同的 REC_MUM 值對比 3 個模型的效果。

```
# REC_MUM = 5 時的對比：
[
    {
        "llm_api_avg_p": 0.08257575757575761, # 月之暗面的 API 的平均精準度
        "llm_api_avg_r": 0.17424242424242428 # 月之暗面的 API 的平均召回率
    },
    {
        "openllm_avg_p": 0.11969696969696976, # Qwen-4B 的平均精準度
        "openllm_avg_r": 0.30050505050505044 # Qwen-4B 的平均召回率
    },
    {
        "openllm_finetune_avg_p": 0.06893939393939397, # Qwen-4B 微調模型的平均精準度
        "openllm_finetune_avg_r": 0.1590909090909091 # Qwen-4B 微調模型的平均召回率
    }
]

# REC_MUM = 8 時的對比：
[
    {
        "llm_api_avg_p": 0.08764430014430015, # 月之暗面的 API 的平均精準度
        "llm_api_avg_r": 0.22348484848484848 # 月之暗面的 API 的平均召回率
    },
    {
        "openllm_avg_p": 0.12626262626262624, # Qwen-4B 的平均精準度
        "openllm_avg_r": 0.44696969696969696 # Qwen-4B 的平均召回率
    },
    {
        "openllm_finetune_avg_p": 0.09027777777777779, # Qwen-4B 微調模型的平均精準度
        "openllm_finetune_avg_r": 0.3295454545454546 # Qwen-4B 微調模型的平均召回率
    }
```

```
]

# REC_MUM = 10 時的對比：
[
    {
        "llm_api_avg_p": 0.08457190957190955, # 月之暗面的 API 的平均精準度
        "llm_api_avg_r": 0.23863636363636367 # 月之暗面的 API 的平均召回率
    },
    {
        "openllm_avg_p": 0.10479797979797977, # Qwen-4B 的平均精準度
        "openllm_avg_r": 0.45454545454545453 # Qwen-4B 的平均召回率
    },
    {
        "openllm_finetune_avg_p": 0.07840909090909089, # Qwen-4B 微調模型的平均精準度
        "openllm_finetune_avg_r": 0.3522727272727273 # Qwen-4B 微調模型的平均召回率
    }
]
```

可以看出，Qwen-4B 的效果是最好的。隨著 REC_MUM 增大，Qwen-4B 微調模型比月之暗面的 API 效果更好。但是，Qwen-4B 微調模型的效果反而不如沒有微調的，這可能是以下原因造成的：一是微調的樣本數量較少，模型沒有學到重要的規律；二是提供的微調樣本的資料不夠好（缺少很多商品的價格），引入了雜訊；三是模型參數比較少，難以解決從候選集中選擇使用者喜歡的商品這種具有一定複雜度的問題。

8.8 利用大模型進行推薦解釋，提升推薦說服力

大模型的自然語言理解和生成能力特別適合為推薦生成文字解釋。本節就來講解大模型應用於推薦解釋領域的原理和方法及具體的程式實現案例。

首先，準備資料；然後，利用大模型強大的上下文學習能力直接為推薦系統生成解釋；接下來，微調大模型生成解釋；最後，評估微調後的推薦解釋大模型的效果。

8.8.1 資料準備

本節使用上下文學習和微調兩種方法來實現推薦解釋，所以需要準備上下文學習推薦解釋和微調推薦解釋的資料。

1. 上下文學習推薦解釋資料準備

這裡還是使用 Beauty 資料集，需要用到 All_Beauty_5.json 和 meta_All_Beauty.json。相信你已經很熟悉了，這裡不再詳細說明，下面舉出資料處理的程式①。

```
import csv
import importlib
import json
import sys
sys.path.append('../')
utils = importlib.import_module('utils')

def get_recommendation_explain_data(path_review: str =
"../../data/amazon_review/beauty/All_Beauty_5.json") -> dict:
    """
    將使用者行為資料按照時間昇冪排列，獲取使用者行為字典
    :param path_review: 使用者行為資料目錄
    :return: 生成預測推薦解釋的資料
    """
    df_view = utils.get_df(path_review)
    # 對 unixReviewTime 昇冪排列
    df_view.sort_values('unixReviewTime', ascending=True, inplace=True)
    df_view = df_view.reset_index(drop=True)
    grouped = df_view.groupby('reviewerID')
    user_action_dict = {}
    for name, group in grouped:
        reviewerID = name
        asin = list(group['asin'])
        dic = {
```

① 程式：src/e-commerce_case/recommend_explain/data-process/generate_icl_data.py。

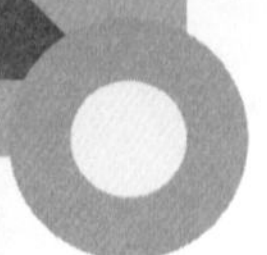


```
            "action": asin[0:-1],
            "recommendation": asin[-1]
        }
        user_action_dict[reviewerID] = dic
    return user_action_dict

def get_metadata_dict(path: str =
'../../data/amazon_review/beauty/meta_All_Beauty.json') -> dict:
    """
    讀取商品中繼資料，方便大模型使用
    :param path: 商品中繼資料目錄
    :return: 商品資訊字典
    """
    item_dict = {}  # 從 meta_All_Beauty.json 中獲取每個 item 對應的資訊
    with open(path, 'r') as file:
        reader = csv.reader(file, delimiter='\n')
        for row in reader:
            j = json.loads(row[0])
            item_id = j['asin']
            title = j['title']
            brand = j['brand']
            description = j['description']
            price = j['price']
            if price != "" and '$' not in price and len(price) > 10:  # 處理一些異常資料
                price = ""
            item_info = {
                "title": title,
                "brand": brand,
                "description": description,
                "price": price
            }
            item_dict[item_id] = item_info
    return item_dict

if __name__ == "__main__":
    rec_dict = get_recommendation_explain_data()
    item_dict = get_metadata_dict()
    dic = {
        "rec_dict": rec_dict,
```

```
        "item_dict": item_dict
    }
    utils.save_json("../data/icl_dict.json", dic)
```

執行上面的程式，可以獲得一個字典，包含使用者推薦資訊和商品資訊。其中使用者推薦資訊包含每個使用者評論過的商品和最終推薦的商品①，商品資訊包含每個商品的標題、品牌、描述資訊和價格，作為推薦解釋的「原材料」，具體資料結構如下。

```
{
    "rec_dict": {
        "A105A034ZG9EHO": {
            "action": [
                "B0009RF9DW",
                "B000FI4S1E",
                "B0012Y0ZG2",
                "B000URXP6E"
            ],
            "recommendation": "B0012Y0ZG2"
        },
        "A10JB7YPWZGRF4": {
            "action": [
                "B0009RF9DW",
                "B0012Y0ZG2",
                "B000URXP6E",
                "B0012Y0ZG2"
            ],
            "recommendation": "B000FI4S1E"
        },
        ...
    }
    "item_dict": {
        "6546546450": {
            "title": "Loud 'N Clear&trade; Personal Sound Amplifier",
```

① 這裡將使用者行為資料中最後一個行為作為給使用者的推薦，將前面的行為看作使用者的歷史行為，並作為建模樣本，讓大模型基於樣本舉出推薦商品的推薦解釋。

```
            "brand": "idea village",
            "description": [
                "Loud 'N Clear Personal Sound Amplifier allows you to turn up the
volume on what people around you are saying, listen at the level you want without
disturbing others, hear a pin drop from across the room."
            ],
            "price": ""
        },
        "7178680776": {
            "title": "No7 Lift &amp; Luminate Triple Action Serum 50ml by Boots",
            "brand": "",
            "description": [
                "No7 Lift & Luminate Triple Action Serum 50ml by Boots"
            ],
            "price": "$44.99"
        },
        ...
    }
}
```

2. 微調推薦解釋資料準備

使用從 P5 專案中獲得的資料①微調，下面舉出簡單的資料說明②。

```
[
    {
        "user": "A1YJEY40YUW4SE",
        "item": "7806397051",
        "rating": 1,
        "text": "Don't waste your money\nVery oily and creamy. Not at all what I
expected... ordered this to try to highlight and contour and it just looked awful!!!
Plus, took FOREVER to arrive."
    },
    {
        "user": "A60XNB876KYML",
```

① 這裡應該是亞馬遜早期資料集中的資料。

② 程式：src/e-commerce_case/recommend_explain/data/amazon/beauty/reviews_beauty.json。

```
        "item": "7806397051",
        "rating": 3,
        "text": "OK Palette!\nThis palette was a decent price and I was looking for a
few different shades. This palette conceals decently, however, it does somewhat cake up
and crease."
    },
    {
        "user": "A3G6XNM240RMWA",
        "item": "7806397051",
        "rating": 4,
        "text": "great quality\nThe texture of this concealer pallet is fantastic, it
has great coverage and a wide variety of uses, I guess it's meant for professional
makeup artists and a lot of the colours are of no use to me but I use at least two
of them on a regular basis, and two more occasionally, which is the only reason I'm
giving it for stars, I feel like the range of colors is kind of a waste for me,
but the  product itself  is wonderful, it's not cakey, gives me a natural for and
concealed my imperfections, therefore I highly recommend it :)",
        "sentence": [
            [
                "coverage",
                "great",
                "it has great coverage and a wide variety of uses",
                1
            ],
            [
                "quality",
                "great",
                "great quality",
                1
            ]
        ]
    },
    {
        "user": "A1PQFP6SAJ6D80",
        "item": "7806397051",
        "rating": 2,
        "text": "Do not work on my face\nI really can't tell what exactly this thing
is. It's not powder but a kind of oil-ish pasty fluid. And so far I tried twice but it
doesnt really show any color on my face."
```

```
    },
    {
        "user": "A38FVHZTNQ271F",
        "item": "7806397051",
        "rating": 3,
        "text": "It's okay.\nIt was a little smaller than I expected, but that was okay because it lasted me for a long time. I think it does great coverage for the price I paid. It is heavy, and wears off within 30-1hr. It kinda dries your skin. I'd recommend it to people who are just looking for a cheap coverage, or beginners who are just learning to conceal.",
        "sentence": [
            [
                "coverage",
                "great",
                "I think it does great coverage for the price I paid",
                1
            ]
        ]
    },
    ...
]
```

上述資料是使用者評論行為，其中有些資料封包含 sentence，sentence 中前面兩個是商品特徵，第 3 個是解釋，第 4 個是情感傾向（1 表示正面評論，-1 表示負面評論）。這裡的解釋是模型的監督標籤。針對上面的資料，需要進行以下處理[①]。

```
import importlib
import random
import sys
sys.path.append('../')
utils = importlib.import_module('utils')
from sklearn.model_selection import train_test_split
reviews_beauty = utils.load_json("../data/amazon/beauty/reviews_beauty.json")
combined_review_data = []
```

① 程式：src/e-commerce_case/recommend_explain/data-process/generate_finetune_data.py。

```
no_sentence = 0
for i in range(len(reviews_beauty)):
    rev_ = reviews_beauty[i]
    out = {}
    if 'sentence' in rev_:
        out['user'] = rev_['user']
        out['item'] = rev_['item']
        list_len = len(rev_['sentence'])
        selected_idx = random.randint(0, list_len - 1)  # add a random, or list all
possible sentences
        out['explanation'] = rev_['sentence'][selected_idx][2]
        out['feature'] = rev_['sentence'][selected_idx][0]
        combined_review_data.append(out)
    else:
        no_sentence += 1
random.shuffle(combined_review_data)
train, test = train_test_split(combined_review_data, test_size=0.2, random_state=42)
train, val = train_test_split(train, test_size=0.2, random_state=42)
outputs = {'train': train,
           'val': val,
           'test': test,
           }
utils.save_json("../data/explain.json", outputs)
user_list = list(set([d['user'] for d in train] + [d['user'] for d in val] + [d['user']
for d in test]))
item_list = list(set([d['item'] for d in train] + [d['item'] for d in val] + [d['item']
for d in test]))
user_dict = {}
for i in range(len(user_list)):
    user_dict[user_list[i]] = i + 1
item_dict = {}
for i in range(len(item_list)):
    item_dict[item_list[i]] = i + 1
dic = {
    "user_dict": user_dict,
    "item_dict": item_dict
}
utils.save_json("../data/finetune_dict.json", dic)
```

上述程式會生成兩個資料，其中一個是訓練、驗證和測試集，包含使用者ID、商品ID、對應的解釋、特徵。資料樣本如下。

```
{
    "train": [
        {
            "user": "A2WT3YQN6P5ZDB",
            "item": "B0018QO5CO",
            "explanation": "I will use this again now that I know how to use it safely
on my darker skin",
            "feature": "skin"
        },
        {
            "user": "AXGYTAVICKCIK",
            "item": "B001W2K51O",
            "explanation": "like all of the natural oils were stripped",
            "feature": "oils"
        },
        ...
    ],
    "val": [
        {
            "user": "A2QSAE608D8KEK",
            "item": "B00027D8IC",
            "explanation": "it is great product",
            "feature": "product"
        },
        {
            "user": "A13U975DFXBU44",
            "item": "B00AO4EBOI",
            "explanation": "graying) hair smoother",
            "feature": "hair"
        },
        ...
    ],
    "test": [
        {
            "user": "ASRZ2JLS1B3VY",
            "item": "B0068Z7MT8",
            "explanation": "It's pretty great quality as well for the price paid",
```

```
            "feature": "quality"
        },
        {
            "user": "A18AO5KXCW5P3Z",
            "item": "B002LFNQT4",
            "explanation": "I got it for a great price with free shipping",
            "feature": "price"
        },
        ...
    ]
}
```

生成的另一個資料是字典資料，包含使用者字典和商品字典，字典的作用是將使用者 ID 和商品 ID 映射為整數，以便微調時進行嵌入表示。下面是兩類字典映射的資料結構案例。

```
{
    "user_dict": {
        "A2FAAXZ8AWWEOD": 1,
        "AXKQI9SUBLPV7": 2,
        "A2HNASPU7NVTD8": 3,
        "A13PKQLIIK0CWB": 4,
        "AMITH2128KHRS": 5,
        "A3PP9GXEMC7O9H": 6,
        "A1BI8PUEHA5CHW": 7,
        "A3CQN90AKAWK8J": 8,
        "A3OHRVM1B8U24K": 9,
        "A21OTG8G9ZEKPG": 10,
        "A2B00ZSEERJ2Z6": 11,
        "A26WLKKPYV62FU": 12,
        "A3M31G4GJ9066T": 13,
        "A1X2YRSD648FM3": 14,
        ...
    },
    "item_dict": {
        "B0000UTUS8": 1,
        "B008CT04I4": 2,
        "B00381A7OC": 3,
        "B005708HQ2": 4,
```

```
        "B001AFGRM4": 5,
        "B00AF9DGMA": 6,
        "B0037LG6UW": 7,
        "B001M6VK2S": 8,
        "B0015XWQLW": 9,
        "B005KW8K00": 10,
        "B002SICH1C": 11,
        "B00AFCP6F2": 12,
        "B00C1006ZG": 13,
        "B003TJI3EO": 14,
        "B008TYO2KI": 15,
        "B001L413AU": 16,
        "B002LV317A": 17,
        "B000VEN3RW": 18,
        "B002ACQCC6": 19,
        ...
    }
}
```

8.8.2 利用大模型上下文學習能力進行推薦解釋

上下文學習推薦解釋的基本原理非常簡單：將使用者操作過的商品的中繼資料資訊和為使用者推薦的商品的中繼資料資訊展示出來，將這些資訊按照一定的範本組織成半結構化的資料，最終形成輸入給大模型的提示詞。大模型會從提示詞提供的資訊中找出使用者行為與推薦商品之間的內在關係，這個內在關係就可以作為推薦的原因。

鑑於大模型有非常好的邏輯推理、知識總結能力，讓大模型找出推薦的原因理論上是可行的，實際上也確實如此。下面舉出兩種實現方案：一種利用月之暗面的 API，另一種基於開放原始碼的 Qwen-14B-Chat 模型，下面是具體的實現程式[①]。

```
import os
import json
```

① 程式：src/e-commerce_case/recommend_explain/llm_icl_explain.py。

```
import time
import torch
from openai import OpenAI
from transformers import AutoTokenizer, AutoModelForCausalLM
from dotenv_vault import load_dotenv  # pip install --upgrade python-dotenv-vault
load_dotenv()  # https://vault.dotenv.org/ui/ui1
MOONSHOT_API_KEY = os.getenv("MOONSHOT_API_KEY")
OUTPUT_NUM = 50  # 只針對前面的 OUTPUT_NUM 個使用者舉出推薦原因，避免計算時間太長
instruction = ("You are a product recommendation expert, and when recommending "
               "products to users, you will provide reasons for the recommendation.")

def formatting_prompt(history, recommendation):
    prompt = f"""The user purchased the following beauty products(in JSON format):
{history}
Based on the user's historical purchases, our recommendation system recommends the 
following
product to the user(in JSON format):
{recommendation}
Please provide a recommendation explanation in one sentence, which is why the 
recommendation system
recommends this product to users. Your reasons must be clear, easy to understand, and 
able to be recognized by users.
The reason you provide should be between 5 and 20 words.
"""
    return prompt

def llm_api_rec_reason(dict_path: str = 'data/icl_dict.json', store_path: str = 
'data/llm_api_reasons.json'):
    f = open(dict_path, "rb")
    dic = json.load(f)
    item_dict = dic['item_dict']
    train_user_dict = dic['rec_dict']
    generated_reasons = []
    print("total user number = " + str(len(train_user_dict.keys())))
    i = 0
    for user in list(train_user_dict.keys())[0:OUTPUT_NUM]:
        H = []
        dic = train_user_dict[user]
        action_list = dic['action']
```

```
        for h in action_list:
            info = item_dict[h]
            if 'description' in info:
                del info['description']  # description 欄位太長，消耗的 token 太多，剔除
            H.append(info)
        history = json.dumps(H, indent=4, ensure_ascii=False)
        item = dic['recommendation']
        info = item_dict[item]
        info['item_id'] = item
        if 'description' in info:
            del info['description']  # description 欄位太長，消耗的 token 太多，剔除
        recommendation = json.dumps(info, indent=4, ensure_ascii=False)
        prom = formatting_prompt(history, recommendation)
        client = OpenAI(
            api_key=MOONSHOT_API_KEY,
            base_url="https://api.moonshot.cn/v1",
        )
        llm_response = client.chat.completions.create(
            model="moonshot-v1-32k",  # moonshot-v1-8k 、moonshot-v1-32k、moonshot-v1-
128k
            messages=[
                {
                    "role": "system",
                    "content": instruction,
                },
                {"role": "user", "content": prom},
            ],
            temperature=0.1,
            stream=False,
        )
        reason = llm_response.choices[0].message.content.strip()
        explain = {
            "user": user,
            "prompt": prom,
            "reason": reason
        }
        i += 1
        print("------------- " + str(i) + " ----------------")
        print(json.dumps(explain, indent=4, ensure_ascii=False))
```

```
        generated_reasons.append(explain)
        if i % 7 == 0:
            time.sleep(1)  # 避免 moonshot 認為呼叫太頻繁不合法
    res = json.dumps(generated_reasons, indent=4, ensure_ascii=False)
    with open(store_path, 'a') as file:  # 將生成的訓練資料儲存起來
        file.write(res)

def openllm_rec_reason(model_path: str =
'/Users/liuqiang/Desktop/code/llm/models/Qwen1.5-14B',
                       dict_path: str = 'data/icl_dict.json',
                       store_path: str = 'data/openllm_reasons.json'):
    f = open(dict_path, "rb")
    dic = json.load(f)
    item_dict = dic['item_dict']
    train_user_dict = dic['rec_dict']
    model = AutoModelForCausalLM.from_pretrained(
        model_path,
        device_map="auto",
        torch_dtype=torch.float16,
    )
    tokenizer = AutoTokenizer.from_pretrained(model_path, trust_remote_code=True)
    tokenizer.padding_side = 'right'
    generated_reasons = []
    print("total user number = " + str(len(train_user_dict.keys())))
    i = 0
    for user in list(train_user_dict.keys())[0:OUTPUT_NUM]:
        H = []
        dic = train_user_dict[user]
        action_list = dic['action']
        for h in action_list:
            info = item_dict[h]
            if 'description' in info:
                del info['description']  # description 欄位太長，消耗的 token 太多，剔除
            H.append(info)
        history = json.dumps(H, indent=4, ensure_ascii=False)
        item = dic['recommendation']
        info = item_dict[item]
        info['item_id'] = item
        if 'description' in info:
```

```
            del info['description']  # description 欄位太長，消耗的 token 太多，剔除
        recommendation = json.dumps(info, indent=4, ensure_ascii=False)
        input = formatting_prompt(history, recommendation)
        prompt = f"""### Instruction:
                    {instruction}
                    ### Input:
                    {input}
                    ### Response:
                    """
        input_ids = tokenizer(prompt, return_tensors="pt", truncation=True).input_ids
        outputs = model.generate(input_ids=input_ids.to('mps'),
                                 max_new_tokens=500, pad_token_id=tokenizer.eos_token_id)
        reason = tokenizer.batch_decode(outputs,
skip_special_tokens=True)[0][len(prompt):]
        explain = {
            "user": user,
            "prompt": prompt,
            "reason": reason
        }
        i += 1
        print("-------------- " + str(i) + " ----------------")
        print(json.dumps(explain, indent=4, ensure_ascii=False))
        generated_reasons.append(explain)
        if i % 7 == 0:
            time.sleep(1)  # 避免 moonshot 認為呼叫太頻繁不合法
    res = json.dumps(generated_reasons, indent=4, ensure_ascii=False)
    with open(store_path, 'a') as file:  # 將生成的訓練資料儲存起來
        file.write(res)

if __name__ == "__main__":
    llm_api_rec_reason()
    openllm_rec_reason()
```

執行上述程式，輸出兩個 JSON 格式的文件，分別是這兩種方法提供的推薦解釋，下面舉出月之暗面的 API 的推薦解釋樣本，另一種方法的結果類似，這裡不再單獨展示。

```
[
    {
```

```
        "user": "A105A034ZG9EHO",
        "reason": "The user already purchased a Bath & Body Works body wash,
indicating a preference for this brand's products."
    },
    {
        "user": "A10JB7YPWZGRF4",
        "reason": "The user has a preference for moisturizing and refreshing body wash
products, as shown by their previous purchases."
    },
    {
        "user": "A10M2MLE2R0L6K",
        "reason": "The user has a preference for Pre de Provence products,
particularly those with lavender scents."
    },
    ...
]
```

8.8.3 模型微調

本節採用連續提示詞微調大模型進行推薦解釋。

1. 連續提示詞微調基本原理

上下文學習以文字方式向大模型輸入提示詞，大模型基於提示詞舉出對應的回答。一般的提示詞是人類可以理解的文字。但實際上，提示詞不必嚴格限定為文字，也可以是隨機初始化的向量，還可以是由另一個模型生成的向量，這種類型的提示詞被稱為連續提示詞 [10]。ID 向量也可以直接用作生成推薦解釋的連續提示詞。

- 連續提示詞微調的原理。

針對推薦解釋場景，如何建構連續提示詞呢？可以將使用者 ID 和商品 ID 兩種類型的 ID 編碼轉為連續提示詞。我們可以構造一個提示詞範本，透過該範本將 ID 向量輸入預先訓練的模型中，用於解釋生成。

```
U I E
```

其中，U 和 I 分別是使用者和商品的連續嵌入表示，E 是推薦解釋文字。這個範本中加入了使用者和商品的嵌入，主要目的是將使用者和商品的資訊整合到模型中，讓最終獲得的解釋更加個性化。

在模型微調階段，可以基於微調樣本建構上述範本對應的輸入①，然後微調大模型。在推理階段，將 ***U***、***I*** 對應的向量輸入，就可以利用大模型生成對應的推薦解釋。具體的大模型架構如圖 8-16 所示②。

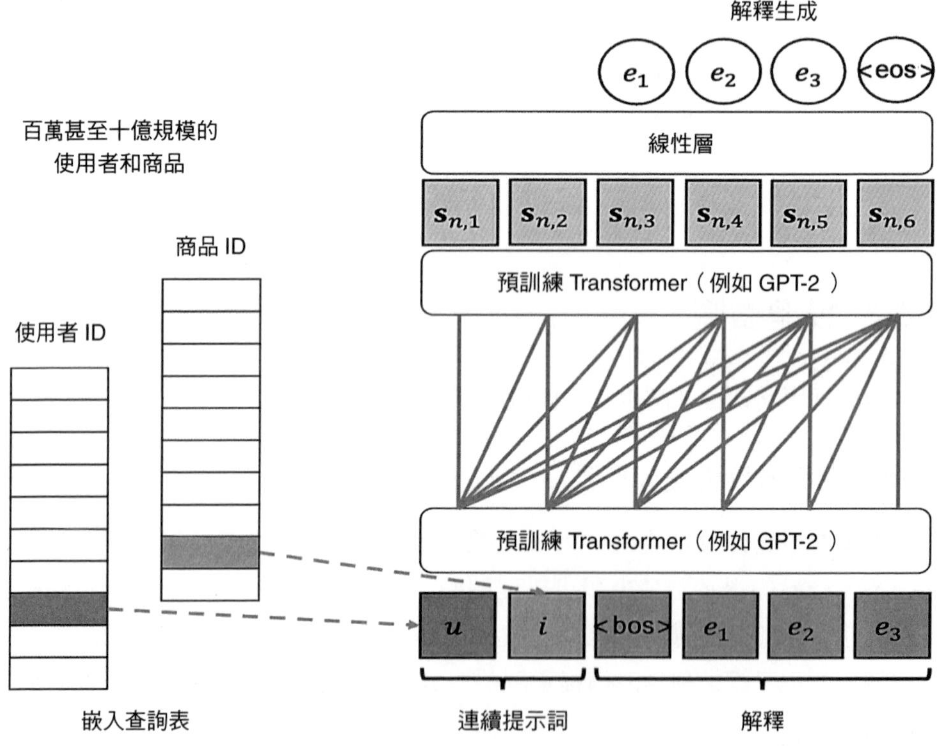

▲ 圖 8-16 將使用者 ID 和商品 ID 作為連續提示詞進行推薦解釋

圖 8-16 的右下方是大模型的輸入：$S=\left[u,i,e_1,e_2,\cdots,e_{|E_{u,i}|}\right]$。$u$，$i$ 是使用者 ID 和商品 ID 的嵌入，其中，$\boldsymbol{U}\in\mathbf{R}^{|U|\times d}$，$\boldsymbol{I}\in\mathbf{R}^{|I|\times d}$，嵌入的維度為 d。$|U|$ 和 $|I|$

① 將使用者 ID 和商品 ID 嵌入，解釋拼接成一段完整的文字。

② 這裡將 GPT-2 作為基底模型。

分別是使用者數和商品數，e_i 是文字 token，$e_i \in v$（v 是詞庫）。在微調時，u、i 隨機初始化，然後在演算法反向傳播迭代時持續更新直到收斂。

透過嵌入，輸入的 token 表示表示為 $\left[u,i,e_1,e_2,\cdots,e_{|E_{u,i}|}\right]$，加入位置向量，最初輸入大模型的就是 $S_0=\left[s_{0,1},s_{0,2},\cdots,s_{0,|S|}\right]$。透過大模型神經網路的處理在頂層輸出 $S_0=\left[s_{0,1},s_{0,2},\cdots,s_{0,|S|}\right]$。右上方增加了一個線性層，將 GPT-2 的輸出映射到一個機率向量，向量的維數就是詞庫的大小，向量的分量代表輸出對應詞的機率大小。採用負對數似然損失函式，目標函式可以表示為

$$\mathcal{L}_{\mathcal{C}}=\frac{1}{|\mathcal{T}|}\sum_{(u,i)\in\mathcal{T}}\frac{1}{|E_{u,i}|}\sum_{t=1}^{|E_{u,i}|}-\ln c_{t+2}^{e_t} \tag{8-1}$$

式（8-1）中，$\left(u,i,E_{u,i}\right)$ 是一筆訓練樣本，u、i 分別是使用者 ID、商品 ID，$E_{u,i}$ 是對應的文字解釋組成的 token 集合。這裡 $c_t^{e_t}$ 是解釋 token e_t 的機率分佈③，具體公式如下。

$$c_t=\text{softmax}\left(W^{\text{v}}s_{n,t}+b^{\text{v}}\right) \tag{8-2}$$

透過上面的損失函式微調模型，微調完成後，可以形式化地利用式（8-3）進行推理，下面的 |prompt| 在這個場景是 2，表示向右移動兩位。$\hat{\mathcal{E}}$是生成的所有可能解釋的組合，可以利用集體搜尋進行推理，這裡不再贅述。

$$E^*=\arg\max_{E\in\hat{\mathcal{E}}}\sum_{t}^{|E|}\ln c_{t+|\text{prompt}|}^{e_t} \tag{8-3}$$

- 對連續提示詞模型進行微調。

有了目標函式，如何對模型進行微調呢？ GPT-2 是訓練好的基底模型，它的參數 Θ_{LM} 處於最佳狀態，而 u、i 是引入的新參數（記為 Θ_P），是隨機初始化的，GPT-2 的參數和 u、i 參數處於不同的訓練階段。為了解決這個問題，可以

③ 式（8-1）中加 2 表示向右移 2 位，因為前面兩個 token 分別是 u 和 i，它們作為待最佳化的參數，不參與損失函式的計算過程。

採用管線的微調策略，先將模型參數 Θ_{LM} 固定，只微調新參數 Θ_P，當 Θ_P 微調到最佳狀態時，同時最佳化 Θ_{LM} 和 Θ_P，式（8-4）可以形象地說明這個過程。

$$\mathcal{J} = \min_{\Theta_{\mathrm{P}}} \mathcal{L}_{\mathcal{C}} \xrightarrow{\text{followed by}} \mathcal{J} = \min_{\Theta=\{\Theta_{\mathrm{LM}},\Theta_{\mathrm{P}}\}} \mathcal{L}_{\mathcal{C}} \tag{8-4}$$

關於演算法原理更詳細的介紹，詳見參考文獻 [11]，這也是本節的基本想法來源。

2. 提示詞微調進行推薦解釋程式實現

微調嚴格按照上述原理實現並參考了參考文獻 [2]，採用兩步微調的策略①。

```
import json
import os
import math
import torch
import argparse
from torch.optim import Adam
from transformers import GPT2Tokenizer
from module import ContinuousPromptLearning
from utils import (rouge_score, bleu_score, Batchify, now_time,
                   ids2tokens, unique_sentence_percent, feature_detect,
                   feature_matching_ratio, feature_coverage_ratio, feature_diversity)
parser = argparse.ArgumentParser(description='PErsonalized Prompt Learning for
Explainable Recommendation (PEPLER)')
parser.add_argument('--data_path', type=str, default=None,
                    help='path for loading the explain data')
parser.add_argument('--dict_path', type=str, default=None,
                    help='path for loading the user and item index dict data')
parser.add_argument('--lr', type=float, default=0.001,
                    help='learning rate')
parser.add_argument('--epochs', type=int, default=100,
                    help='upper epoch limit')
parser.add_argument('--batch_size', type=int, default=128,
                    help='batch size')
parser.add_argument('--log_interval', type=int, default=20,
```

① 程式：src/e-commerce_case/recommend_explain/prompt_learning.py。

```
                    help='report interval')
parser.add_argument('--checkpoint', type=str, default='./models/',
                    help='directory to save the final model')
parser.add_argument('--output_path', type=str, default='generated.txt',
                    help='output file for generated text')
parser.add_argument('--endure_times', type=int, default=5,
                    help='the maximum endure times of loss increasing on validation')
parser.add_argument('--words', type=int, default=20,
                    help='number of words to generate for each sample')
args = parser.parse_args()
if args.data_path is None:
    parser.error('--data_path should be provided for loading data')
print('-' * 40 + 'ARGUMENTS' + '-' * 40)
for arg in vars(args):
    print('{:40} {}'.format(arg, getattr(args, arg)))
print('-' * 40 + 'ARGUMENTS' + '-' * 40)
device = 'mps'
if not os.path.exists(args.checkpoint):
    os.makedirs(args.checkpoint)
model_path = os.path.join(args.checkpoint, 'model.pt')
prediction_path = args.output_path
###############################################################################
# 載入資料
###############################################################################
print(now_time() + 'Loading data')
bos = '<bos>'
eos = '<eos>'
pad = '<pad>'
tokenizer = GPT2Tokenizer.from_pretrained('/Users/liuqiang/Desktop/code/llm/models/
gpt2',
                                          bos_token=bos, eos_token=eos, pad_token=pad)
f = open(args.data_path, "rb")
corpus = json.load(f)
f = open(args.dict_path, "rb")
dic = json.load(f)
feature_set = set()
train = corpus['train']
val = corpus['val']
test = corpus['test']
```

```
feature_set.union([d['feature'] for d in train])
feature_set.union([d['feature'] for d in val])
feature_set.union([d['feature'] for d in test])
train_data = Batchify(train, dic, tokenizer, bos, eos, args.batch_size, shuffle=True)
val_data = Batchify(val, dic, tokenizer, bos, eos, args.batch_size)
test_data = Batchify(test, dic, tokenizer, bos, eos, args.batch_size)
print(now_time() + 'Loading data Finished')
########################################################################
# 建構模型
########################################################################
nuser = len(dic['user_dict'])
nitem = len(dic['item_dict'])
ntoken = len(tokenizer)
model =
ContinuousPromptLearning.from_pretrained('/Users/liuqiang/Desktop/code/llm/models/gpt
2', nuser, nitem)
model.resize_token_embeddings(ntoken)  # three tokens added, update embedding table
model.to(device)
optimizer = Adam(model.parameters(), lr=args.lr)

########################################################################
# 訓練程式
########################################################################

def train(data):
    # 開啟訓練模式，Dropout 生效
    model.train()
    text_loss = 0.0
    total_sample = 0
    while True:
        user, item, seq, mask = data.next_batch()  # data.step += 1
        user = user.to(device)  # (batch_size,)
        item = item.to(device)
        seq = seq.to(device)  # (batch_size, seq_len)
        mask = mask.to(device)
        # 在每個批次開始時，將隱藏狀態與之前的生成方式分離
        # 如果不這樣做，模型將嘗試反向傳播到資料集的起始位置
        optimizer.zero_grad()
        outputs = model(user, item, seq, mask)
```

```
        loss = outputs.loss
        loss.backward()
        optimizer.step()
        batch_size = user.size(0)
        text_loss += batch_size * loss.item()
        total_sample += batch_size
        if data.step % args.log_interval == 0 or data.step == data.total_step:
            cur_t_loss = text_loss / total_sample
            print(now_time() + 'text ppl {:4.4f} | {}/{}
batches'.format(math.exp(cur_t_loss),
                                                                                    data.step,
data.total_step))
            text_loss = 0.
            total_sample = 0
        if data.step == data.total_step:
            break

def evaluate(data):
    # 開啟賦值模式，將會關閉 Dropout
    model.eval()
    text_loss = 0.
    total_sample = 0
    with torch.no_grad():
        while True:
            user, item, seq, mask = data.next_batch()  # data.step += 1
            user = user.to(device)  # (batch_size,)
            item = item.to(device)
            seq = seq.to(device)  # (batch_size, seq_len)
            mask = mask.to(device)
            outputs = model(user, item, seq, mask)
            loss = outputs.loss
            batch_size = user.size(0)
            text_loss += batch_size * loss.item()
            total_sample += batch_size
            if data.step == data.total_step:
                break
    return text_loss / total_sample

def generate(data):
```

```
    # 開啟賦值模式，將會關閉 Dropout
    model.eval()
    idss_predict = []
    with torch.no_grad():
        while True:
            user, item, seq, _ = data.next_batch()  # data.step += 1
            user = user.to(device)  # (batch_size,)
            item = item.to(device)
            text = seq[:, :1].to(device)  # bos, (batch_size, 1)
            for idx in range(seq.size(1)):
                # 每步生成一個單字
                outputs = model(user, item, text, None)
                last_token = outputs.logits[:, -1, :]  # the last token, (batch_size,
ntoken)
                word_prob = torch.softmax(last_token, dim=-1)
                token = torch.argmax(word_prob, dim=1,
                                     keepdim=True)
# 選擇機率最大的 token
                text = torch.cat([text, token], 1)  # (batch_size, len++)
            ids = text[:, 1:].tolist()  # remove bos, (batch_size, seq_len)
            idss_predict.extend(ids)
            if data.step == data.total_step:
                break
    return idss_predict

print(now_time() + 'Tuning Prompt Only')
# 根據 epochs 迴圈
best_val_loss = float('inf')
endure_count = 0
for epoch in range(1, args.epochs + 1):
    print(now_time() + 'epoch {}'.format(epoch))
    train(train_data)
    val_loss = evaluate(val_data)
    print(now_time() + 'text ppl {:4.4f} | valid loss {:4.4f} on
validation'.format(math.exp(val_loss), val_loss))
    # 如果驗證損失是我們目前見過的最好的，則儲存模型
    if val_loss < best_val_loss:
        best_val_loss = val_loss
        with open(model_path, 'wb') as f:
```

```
            torch.save(model, f)
    else:
        endure_count += 1
        print(now_time() + 'Endured {} time(s)'.format(endure_count))
        if endure_count == args.endure_times:
            print(now_time() + 'Cannot endure it anymore | Exiting from early stop')
            break
# 載入儲存的最佳模型
with open(model_path, 'rb') as f:
    model = torch.load(f).to(device)
print(now_time() + 'Tuning both Prompt and LM')
for param in model.parameters():
    param.requires_grad = True
optimizer = Adam(model.parameters(), lr=args.lr)
# 基於 epochs 迴圈
best_val_loss = float('inf')
endure_count = 0
for epoch in range(1, args.epochs + 1):
    print(now_time() + 'epoch {}'.format(epoch))
    train(train_data)
    val_loss = evaluate(val_data)
    print(now_time() + 'text ppl {:4.4f} | valid loss {:4.4f} on
validation'.format(math.exp(val_loss), val_loss))
    # 如果驗證損失是我們目前見過的最好的，則儲存模型
    if val_loss < best_val_loss:
        best_val_loss = val_loss
        with open(model_path, 'wb') as f:
            torch.save(model, f)
    else:
        endure_count += 1
        print(now_time() + 'Endured {} time(s)'.format(endure_count))
        if endure_count == args.endure_times:
            print(now_time() + 'Cannot endure it anymore | Exiting from early stop')
            break
# 載入最佳模型
with open(model_path, 'rb') as f:
    model = torch.load(f).to(device)
print(now_time() + 'Generating text')
idss_predicted = generate(test_data)
```

```
# 執行測試資料
test_loss = evaluate(test_data)
print('=' * 89)
print(now_time() + 'text ppl {:4.4f} on test | End of
training'.format(math.exp(test_loss)))

tokens_test = [ids2tokens(ids[1:], tokenizer, eos) for ids in test_data.seq.tolist()]
tokens_predict = [ids2tokens(ids, tokenizer, eos) for ids in idss_predicted]

text_test = [' '.join(tokens) for tokens in tokens_test]
text_predict = [' '.join(tokens) for tokens in tokens_predict]

text_out = []
for (real, predicted) in zip(text_test, text_predict):
    sample_predict = {
        "real": real,
        "predicted": predicted
    }
    text_out.append(sample_predict)
with open(prediction_path, 'w', encoding='utf-8') as f:
    res = json.dumps(text_out, indent=4, ensure_ascii=False)
    f.write(res)
print(now_time() + 'Generated text saved to ({})'.format(prediction_path))
```

可以透過下面的指令稿執行上面的程式，具體的參數解釋在上面的程式中。

```
python prompt_learning.py \
--data_path ./data/explain.json \
--dict_path ./data/finetune_dict.json \
--epochs 5 \
--lr 1e-3 \
--batch_size 32 \
--log_interval 10 \
--output_path ./data/generated.json \
--words 20 \
--checkpoint ./models/ >> ./logs/finetune.log
```

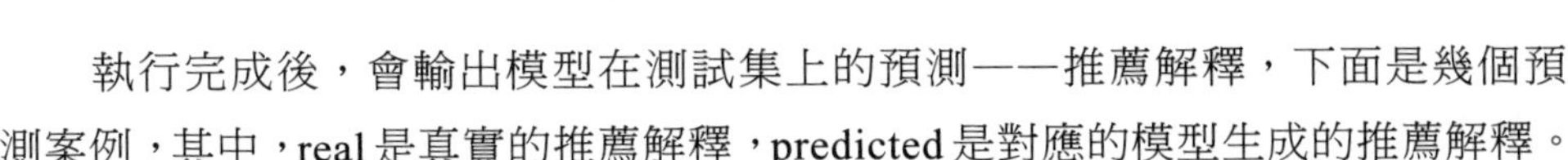

執行完成後，會輸出模型在測試集上的預測——推薦解釋，下面是幾個預測案例，其中，real 是真實的推薦解釋，predicted 是對應的模型生成的推薦解釋。

```
[
    {
        "real": "as an amazon prime member this slip qualified for free two day
shipping and arrived on time in great",
        "predicted": "the length is perfect"
    },
    {
        "real": "have good traction",
        "predicted": "the price is reasonable"
    },
    {
        "real": "very short in length",
        "predicted": "the price is great"
    },
    ...
]
```

8.8.4 模型效果評估

模型生成的是文字，可以採用機器翻譯和文字摘要領域兩個最常用的評估指標——BLEU 和 ROUGE 來評估生成的推薦解釋的文字品質。關於這兩個指標，你可以自行補充學習，另外，我們在 GitHub 的程式倉庫中也舉出了具體的計算程式，這裡不再贅述。

```
# 計算 BLEU-1、BLEU-4
tokens_test = [ids2tokens(ids[1:], tokenizer, eos) for ids in test_data.seq.tolist()]
tokens_predict = [ids2tokens(ids, tokenizer, eos) for ids in idss_predicted]
BLEU1 = bleu_score(tokens_test, tokens_predict, n_gram=1, smooth=False)
print(now_time() + 'BLEU-1 {:7.4f}'.format(BLEU1))
BLEU4 = bleu_score(tokens_test, tokens_predict, n_gram=4, smooth=False)
print(now_time() + 'BLEU-4 {:7.4f}'.format(BLEU4))

# 計算 ROUGE
text_test = [' '.join(tokens) for tokens in tokens_test]
text_predict = [' '.join(tokens) for tokens in tokens_predict]
```

```
ROUGE = rouge_score(text_test, text_predict)  # a dictionary
for (k, v) in ROUGE.items():
    print(now_time() + '{} {:7.4f}'.format(k, v))
```

執行上面的程式可以輸出對應的指標。

```
[2024-03-18 22:15:25.016199]: BLEU-1 14.8556
[2024-03-18 22:15:25.442129]: BLEU-4  0.7141
[2024-03-18 22:15:26.395553]: USR  0.1942 | USN    3481
[2024-03-18 22:15:40.051302]: DIV  0.0959
[2024-03-18 22:15:40.064965]: FCR  0.1902
[2024-03-18 22:15:40.070101]: FMR  0.1227
[2024-03-18 22:15:40.725882]: rouge_1/f_score 14.7837
[2024-03-18 22:15:40.725914]: rouge_1/r_score 15.6023
[2024-03-18 22:15:40.725919]: rouge_1/p_score 16.3132
[2024-03-18 22:15:40.725921]: rouge_2/f_score  1.5362
[2024-03-18 22:15:40.725924]: rouge_2/r_score  1.7167
[2024-03-18 22:15:40.725926]: rouge_2/p_score  1.6831
[2024-03-18 22:15:40.725928]: rouge_l/f_score 11.2061
[2024-03-18 22:15:40.725930]: rouge_l/r_score 13.7274
[2024-03-18 22:15:40.725932]: rouge_l/p_score 13.4428
```

8.9 利用大模型進行對話式推薦

本節介紹一個利用大模型革新傳統推薦系統的架構——對話式推薦，也可以稱為互動式推薦。8.2 節提到的淘寶問問就是阿里在這一領域的初步探索。

大家已經非常熟悉如何與大模型進行高品質的互動，對於推薦系統，也可以採用對話的方式來建構。本節提供一個通用的、實現較簡單的方案，讓你可以直觀感受對話式推薦。

首先，提供一個統一的架構說明對話式大模型推薦系統包含哪些核心模組，它們之間是如何協作的；其次，介紹作為案例展示的亞馬遜的 Beauty 資料集相關的資料；接下來，針對對話或大模型推薦系統的各個元件和模組，提供具體的程式實現；最後，基於 Gradio 架構實現一個簡單的對話式推薦的 UI 互動介面。

本節的框架和程式參考了參考文獻 [12]，對程式進行了較多裁剪和最佳化，方便你理解。

8.9.1 對話式大模型推薦系統的架構

對話式大模型推薦系統將大模型作為控制中樞（類似人的大腦），借助各種工具（類似人的眼耳手足）來實現個性化推薦。由於可以利用傳統推薦的召回、排序及搜尋等能力（作為工具），對話式大模型推薦系統大大拓展了傳統推薦系統的疆域，可以將搜尋、推薦、建議、知識問答等多種互動形式融合到一個統一的系統中，成為你的專業顧問。

8.2 節對互動式推薦範式的架構和模組進行了簡單說明，下面針對本節的實現方案進行詳細的介紹。圖 8-17 是對話式大模型推薦系統核心模組和互動流程。

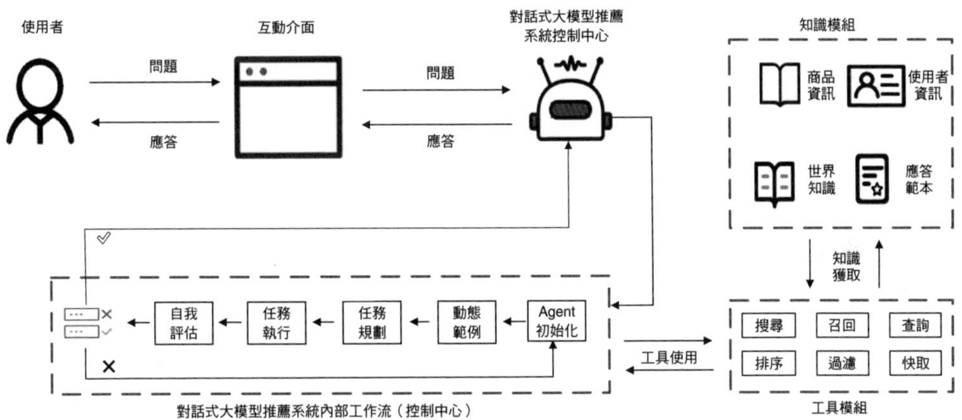

▲ 圖 8-17 對話式大模型推薦系統核心模組和互動流程

圖 8-17 中有 3 個比較重要的模組：一是知識模組，包含 Beauty 資料集、使用者等方面的知識，這些是大模型推理的原料；二是工具模組，大模型使用工具將相關知識注入它的推理流程，透過上下文學習回答使用者與 Beauty 推薦、搜尋相關的問題；三是工作流，這是大模型內部作業的流程，類似工廠廠房的管線，大模型透過管線的方式科學地解決使用者的問題。下面對工作流說明。

- 智慧體初始化。

我們將大模型看作一個智慧體（Agent），這一步將使用者的輸入、對話歷史、快取資訊等初始化，方便後續任務的執行。

- 動態範例。

由於對話式大模型推薦系統採用上下文學習，如果能夠為大模型提供幾個相關的範例（few-shot），那麼大模型就可以更進一步地完成任務。這一步基於範例範本、結合本次使用者的問題建構幾個與問題相關的範例並輸入大模型的上下文學習知識中。

- 任務規劃。

工具模組中提供了召回、排序、檢索、過濾等工具，這些工具可以協助使用者解決關於 Beauty 資料集的問題。任務規劃主要基於使用者當前的問題和歷史對話，將使用者的問題拆解為合適的步驟，不同步驟用不同的工具實現。這一步的作用類似蓋房的設計階段，也類似 IT 專案的詳細專案規劃書。

- 任務執行。

有了上面的任務規劃，大模型就知道如何做了。任務執行是具體的執行過程，有些步驟是對資料庫進行查詢，有些步驟是召回相關的 Beauty 商品，有些步驟是對商品進行過濾，等等。

- 自我評估。

任務執行完就可以獲得結果了，這個結果一定可靠嗎？在最終回饋給使用者之前，有沒有辦法評估結果的正確性呢？答案是肯定的。這一步是利用大模型強大的邏輯推理能力對獲得的結果進行評估，如果認可評估結果，就可以將結果直接回饋給使用者，否則大模型會提供最佳化的建議，然後重新執行。

8.9.2 資料準備

本節的資料基於 Beauty 資料集，這裡忽略了具體的原始資料處理過程，僅對結果資料說明，如圖 8-18 所示。

```
(train-llm) liuqiang@liuqiangdeMBP interactive_rec % tree -L 2 data
data
└── beauty_product
    ├── SASRec-SASRec.pth
    ├── columns.json
    ├── item_sim.npy
    ├── products.ftr
    └── settings.json

2 directories, 5 files
```

▲ 圖 8-18 對話式大模型推薦系統依賴的資料

資料一共包含 4 個檔案。columns.json 是對 Beauty 中繼資料的欄位描述，方便大模型更進一步地理解，也方便生成對應的 SQL 敘述執行查詢操作。

```
{
  "id": "beauty product id.",
  "title": "product title.",
  "category": "a string concatenated with several categories about beauty products, one product may have multiple tags, the seperator is a comma.",
  "price": "price of product",
  "description": "detailed description about the product.",
  "brand": "brand of beauty products",
  "visited_num": "how many times the products has been purchased, indicates how popular the product is. Product with larger visited number are more popular."
}
```

settings.json 是對相關資料的配置說明，包含多個類資料，例如中繼資料有哪些列、哪些欄位是類別欄位①等。

```
{
  "BEAUTY_INFO_FILE": "products.ftr",
  "TABLE_COL_DESC_FILE": "columns.json",
  "MODEL_CKPT_FILE": "SASRec-SASRec.pth",
  "ITEM_SIM_FILE": "item_sim.npy",
  "USE_COLS": [
    "id",
    "title",
```

① 類別欄位是用於搜尋過濾的。

```
      "category",
      "price",
      "description",
      "brand",
      "visited_num"
    ],
    "CATEGORICAL_COLS": [
      "category",
      "brand"
    ]
}
```

products.ftr 是 Beauty 的中繼資料，採用 Feather 格式儲存。item_sim.npy 是商品與商品的相似資料，方便使用者查詢相關商品。SASRec-SASRec.pth 是基於 SASRec 演算法進行排序的模型，是利用 Beauty 資料集提前訓練好的。

8.9.3 程式實現

有了前面的框架說明和資料準備，本節實現對話式大模型推薦系統的核心功能。

1. 控制中心工作流程

控制中心是圖 8-17 中左下側的部分，它是管線的架構，透過串聯 5 個元件實現使用者問題的拆解和逐步解決。本節利用月之暗面的 API 充當大模型的大腦。

具體來說，控制中心將當前的輸入（使用者的問題）、對話歷史（對話方塊中使用者和推薦系統的對話記錄）、範例說明（基於使用者的問題，系統提供的最相關的解決方案）、工具描述（在上下文學習中提供每個工具的描述，例如這個工具是做什麼的、能解決什麼問題、局限性是什麼）這 4 類資訊作為大模型上下文學習的「原材料」，大模型利用自身的邏輯推理能力自行選擇合適的工具解決具體的問題。程式詳見本書配套程式，路徑為 src/e- commerce_case/interactive_rec/agent_plan_first_openai.py。

程式中包含 3 個類別：ToolBox 是具體的工具使用類別，該類別提供 run 方法，基於提供的資料呼叫對應的工具來解決對應的問題並獲得最終的結果；DialogueMemory 是對使用者與大模型的對話歷史進行管理的類別，提供獲取歷史、增補歷史、清空歷史等功能；CRSAgentPlanFirstOpenAI 是最核心的類別，它就是控制中心，基於前面提到的 4 類資訊，呼叫各種工具和月之暗面的 API 解決使用者的問題，並最終傳回結果。

2. 控制中心核心模組

控制中心的模組比較多，這裡重點講解動態範例生成模組和自我評估模組，這對解決使用者問題並提升答案準確性非常關鍵，當然，增加了自我評估，模型會更慢、消耗更多 token，所以成本更高。

- 動態範例生成模組。

動態範例生成基於下面提供的 15 個解決問題的範例，從中選擇 *k* 個（一般為 2~3 個）與使用者的問題最相關的範例作為樣本，指導大模型更進一步地解決相似問題。

```
{"request": "Hello? How is it today", "plan": "Don't use tool"}
{"request": "Tell me about the first product.", "plan": "1. LookUpTool"}
{"request": "Recently I want to buy some TYPE products, do you have some
suggestions?", "plan": "1. HardFilterTool (TYPE); 2. RankingTool (by popularity); 3.
MapTool"}
{"request": "PRODUCT1 is suitable for me, I want to find some similar products
with it.", "plan": "1. SoftFilterTool (PRODUCT1); 2. RankingTool (by similarity);
3.MapTool"}
{"request": "Do you know any TYPE products? I want one like PRODUCT1.", "plan": "1.
HardFilterTool (TYPE); 2. SoftFilterTool (PRODUCT1); 3. RankingTool (by similarity);
4.MapTool"}
{"request": "My history is PRODUCT1, PRODUCT2, PRODUCT3, please suggest products for
me to buy next.", "plan": "1. RankingTool (by preference); 2. MapTool"}
{"request": "I bought PRODUCT1, PRODUCT2, PRODUCT3 in the past, please suggest
products for me to play next.", "plan": "1. RankingTool (by preference); 2. MapTool"}
{"request": "I tried PRODUCT1, PRODUCT2, PRODUCT3, PRODUCT4 in the past, which
one is the most suitable in above products?", "plan": "1. BufferStoreTool (store
possible products mentioned in conversation before as candidates);  2.RankingTool (by
```

```
preference); 3. MapTool"}
{"request": "I used to use PRODUCT1, PRODUCT2, PRODUCT3 in the past, now I want some
TYPE product with price lower than PRICE.", "plan": "1.HardFilterTool (TYPE, PRICE);
2.RankingTool (by preference); 3. MapTool"}
{"request": "Can you recommend me some products based on my previous history?",
"plan": "Don't use tool, ask for more information"}
{"request": "I like PRODUCT1, PRODUCT2, now I want some products similar with PRODUCT1
but cheaper.", "plan": "1.HardFilterTool (cheaper than PRODUCT1);  2.RankingTool (by
preference); 3. MapTool"}
{"request": "In above products, which is the cheapest?", "plan": "1.LookUpTool (price
of products mentioned before)"}
{"request": "Which of the above products is more similar to PRODUCT?", "plan":
"1.BufferStoreTool (store possible products mentioned in conversation before as
candidates); 2. SoftFilterTool (calculate similarity score); 3. RankingTool (by
similarity); 4. MapTool"}
{"request": "Is there any TYPE1 or TYPE2 products?", "plan": "1. HardFilterTool (TYPE1
or TYPE2); 2. RankingTool (by popularity); 3.MapTool"}
{"request": "Some of my favorite products include PRODUCT1, PRODUCT2. I'm looking for
TYPE1 products with TYPE2", "plan": "1.HardFilterTool (TYPE1, TYPE2); 2.RankingTool (by
preference); 3.MapTool"}
```

範例越多、可能覆蓋的使用者提出的問題就越多，從而能更進一步地指導大模型。另外，如何選擇與使用者的問題最相關的範例也非常重要，如果提供了「牛頭不對馬嘴」的範例，那麼對大模型解決問題不但沒有幫助，還可能起誤導作用。程式實現中使用了 LangChain 中的 SemanticSimilarityExampleSelector 查詢相關範例①。

```
import json
import os
from typing import *
import sys
from langchain.embeddings import HuggingFaceEmbeddings
from langchain.prompts import FewShotPromptTemplate, PromptTemplate
from langchain.prompts import example_selector
from langchain.vectorstores import Chroma
sys.path.append(os.path.dirname(os.path.dirname(os.path.dirname(__file__))))
```

① 程式：src/e-commerce_case/interactive_rec/module/dynamic_demo.py。

```
from utils.util import replace_substrings_regex
example_prompt = PromptTemplate(
    input_variables=["request", "plan"],
    template="User Request: {request} \nPlan: {plan}",
)

def read_jsonl(fpath: str) -> List[Dict]:
    res = []
    with open(fpath, 'r') as f:
        for line in f:
            data = json.loads(line)
            res.append(data)
    return res

class DemoSelector:
    def __init__(self, mode: str, demo_dir_or_file: str, k: int, domain: str) -> None:
        assert mode in {'fixed',
                        'dynamic'}, f"Optional demonstration selector mode: 'fixed',
'dynamic', while got {mode}"
        self.mode = mode
        self.k = k
        examples = self.load_examples(self, demo_dir_or_file)
        self.examples = self.fit_domain(self, examples, domain)
        input_variables = {
            "example_prompt": example_prompt,
            "example_separator": "\n-----\n",
            "prefix": "Here are some demonstrations of user requests and corresponding
tool using plans:",
            "suffix": "Refer to above demonstrations to use tools for current request:
{in_request}.",
            "input_variables": ["in_request"],
        }
        if self.mode == 'dynamic':
            selector = example_selector.SemanticSimilarityExampleSelector.from_
examples(
                # 這是可供選擇的範例列表
                self.examples,
                # 這是用於生成嵌入的嵌入類別，用於測量語義相似性
                HuggingFaceEmbeddings(),
```

```
                # 這是用於儲存嵌入並進行相似性搜尋的 VectorStore 類別
                Chroma,
                # 這是生成的例子數量
                k=self.k
            )
            input_variables["example_selector"] = selector
        else:
            input_variables["examples"] = self.examples[: self.k]
        self.prompt = FewShotPromptTemplate(**input_variables)
    @staticmethod
    def load_examples(self, dir: str) -> List[Dict]:
        examples = []
        if os.path.isdir(dir):
            for f in os.listdir(dir):
                if f.endswith("jsonl"):
                    fname = os.path.join(dir, f)
                    examples.extend(read_jsonl(fname))
        else:
            if dir.endswith('.jsonl'):
                examples.extend(read_jsonl(dir))
        assert len(
            examples) > 0, ("Failed to load examples. Note that only .jsonl file format
is supported for demonstration "
                                "loading.")
        return examples
    def __call__(self, request: str):
        return self.prompt.format(in_request=request)
    @staticmethod
    def fit_domain(self, examples: List[Dict], domain: str):
        # fit examples into domains: replace placeholder with domain-related words,
like replacing item with movie, game
        domain_map = {'item': domain, 'Item': domain.capitalize(), 'ITEM': domain.
upper()}
        res = []
        for case in examples:
            _case = {}
            _case['request'] = replace_substrings_regex(case['request'], domain_map)
            _case['plan'] = replace_substrings_regex(case['plan'], domain_map)
            res.append(_case)
```

```
        return res

if __name__ == "__main__":
    selector = DemoSelector("dynamic",
                            "./demonstration/seed/", k=3, domain="beauty_product")
    request = "I want some farming games."
    demo_prompt = selector(request)
    print(demo_prompt)
    print("passed.")
```

- 自我評估模組。

自我評估模組針對生成的結果進行自我反省、自我評估，透過自我評估確保結果的準確性。具體的做法是：將使用者問題、大模型舉出的答案（需要評估）、對話歷史、之前的工具使用記錄透過一個特定的 critic 範本輸入大模型，然後讓大模型基於範本和輸入的資料評估答案是否可靠，如果不可靠，則提供解決問題的建議，最終在控制中心中重新執行。自我評估模組的程式以下[①]。

```
import os
import re
from typing import *
from dotenv_vault import load_dotenv  # pip install --upgrade python-dotenv-vault
from prompt.critic import CRITIC_PROMPT_USER, CRITIC_PROMPT_SYS
from prompt.tool import OVERALL_TOOL_DESC, TOOL_NAMES
from utils.open_ai import OpenAICall
load_dotenv()  # https://vault.dotenv.org/ui/ui1
MOONSHOT_API_KEY = os.getenv("MOONSHOT_API_KEY")

def parse_output(s: str) -> Tuple[bool, str]:
    yes_pattern = r"Yes.*"
    no_pattern = r"No\. (.*)"
    yes_matches = re.findall(yes_pattern, s)
    if yes_matches:
        need_reflection = False
        info = ""
    else:
```

① 程式：src/e-commerce_case/interactive_rec/module/critic.py。

```
        no_matches = re.findall(no_pattern, s)
        if no_matches:
            need_reflection = True
            info = no_matches[0]
        else:
            need_reflection = False
            info = ""
    return need_reflection, info

class Critic:
    def __init__(
            self,
            model,
            engine,
            buffer,
            domain: str = "beauty_product",
            bot_type="chat_completion",
            temperature: float = 0.0,
            timeout: int = 120,
    ):
        self.model = model
        self.engine = engine
        self.buffer = buffer
        self.domain = domain
        self.timeout = timeout
        assert bot_type in {
            "chat",
            "completion",
            "chat_completion",
        }, f"'bot_type' should be 'chat', 'completion' or 'chat_completion', while got
{bot_type}"
        self.bot_type = bot_type
        self.bot = OpenAICall(
            model=self.engine,
            api_key=os.environ.get("OPENAI_API_KEY", MOONSHOT_API_KEY),
            api_type=os.environ.get("OPENAI_API_TYPE", "open_ai"),
            api_base=os.environ.get("OPENAI_API_BASE", "https://api.moonshot.cn/v1"),
            api_version=os.environ.get("OPENAI_API_VERSION", None),
            temperature=temperature,
```

```
            model_type=self.bot_type,
            timeout=self.timeout,
        )
    def __call__(self, request: str, answer: str, history: str, tracks: str):
        output = self._call(request, answer, history, tracks)
        need_reflection, info = parse_output(output)
        if need_reflection:
            info = f"Your previous response: {answer}. \nThe response is not 
reasonable. Here is the advice: {info} "
        return need_reflection, info
    def _call(self, request: str, answer: str, history: str, tracks: str):
        sys_msg = CRITIC_PROMPT_SYS.format(domain=self.domain)
        item_map = {
            "item": self.domain,
            "Item": self.domain.capitalize(),
            "ITEM": self.domain.upper(),
        }
        tool_names = {k: v.format(**item_map) for k, v in TOOL_NAMES.items()}
        usr_msg = CRITIC_PROMPT_USER.format(
            domain=self.domain,
            tool_description=OVERALL_TOOL_DESC.format(**tool_names, **item_map),
            chat_history=history,
            request=request,
            plan=tracks,
            answer=answer,
            **TOOL_NAMES,
        )
        reply = self.bot.call(
            sys_prompt=sys_msg,
            user_prompt=usr_msg,
            max_tokens=128
        )
        return reply
```

3. 工具使用

工具使用模組是大模型的「眼、耳、手、足」，有了這些工具的協助，大模型才能更進一步地解決使用者的問題。這些工具包括查詢工具、過濾工具、排序工具等，限於篇幅原因，下面只介紹查詢工具，其他工具詳見 GitHub 程式倉庫。

查詢工具的作用是將使用者輸入的問題中可能與 Beauty 產品類目相關的關鍵字轉為 SQL 查詢敘述，然後從 Beauty 中繼資料中找出與關鍵字相關的商品[①]，方便後續步驟從中選擇出最合適的推薦給使用者。舉例來說，輸入下面的問題。

```
please recommend some beauty product to me, thanks
```

查詢工具可能將這個問題轉為下面的 SQL 敘述。

```
SELECT * FROM beauty_product_information WHERE category LIKE "%skincare%" OR category
LIKE "%makeup%"'
```

下面是具體的程式實現，注意，最初的 SQL 查詢敘述是大模型提供的，查詢工具的作用是將之轉為適合相關資料的查詢[②]。

```
import json
import random
import re
from copy import deepcopy
from module.corpus import BaseGallery
from utils.sql import extract_columns_from_where
from utils.util import num_tokens_from_string, cut_list
from loguru import logger

class QueryTool:
    def __init__(self, name: str, desc: str, item_corpus: BaseGallery, buffer,
result_max_token: int = 512) -> None:
        self.item_corpus = item_corpus
        self.name = name
        self.desc = desc
        self.buffer = buffer
        self.result_max_token = result_max_token
```

① 是一種基於標籤的召回。

② 程式：src/e-commerce_case/interactive_rec/tools/query_tool.py。

```
        self._max_record_num = self.result_max_token // 5  # each record at least 5
tokens. If too many records,
        # sample randomly
    def run(self, inputs: str) -> str:
        logger.info(f"\nSQL from AGI: {inputs}")
        info = ""
        output = "can not search related information."
        try:
            inputs = self.rewrite_sql(inputs)
            logger.info(f"Rewrite SQL: {inputs}")
            info += (f"{self.name}: The input SQL is rewritten as {inputs} because "
                     f"some {self.item_corpus.categorical_col_values.keys()} are not
existing. \n")
        except:
            info += f"{self.name}: some thing went wrong in execution, the tool is
broken for current input. \n"
            return info
        try:
            res = self.item_corpus(inputs, return_id_only=False)
            try:
                res = res.to_dict('records')  # list of dict
            except:
                pass
            _any_cut = False
            if len(res) > self._max_record_num:
                _any_cut = True
                res = random.sample(res, k=self._max_record_num)
            if num_tokens_from_string(json.dumps(res)) > self.result_max_token:
                token_limit_per_record = self.result_max_token // (len(res) + 1)
                # shorten each record in the result list
                cut_res = [None] * len(res)
                for i, record in enumerate(res):
                    overflow_token = num_tokens_from_string(json.dumps(record)) -
token_limit_per_record
                    _cut_last = True
                    res_record = deepcopy(record)
                    # if not cut last time, the loop would end due to the cut
operation in the loop
                    # would not shorten the res anymore
```

```
                    while (overflow_token > 0) and _cut_last:
                        key2token = {}
                        all_token = 0
                        for k, v in res_record.items():
                            key2token[k] = num_tokens_from_string(str(v))
                            all_token += key2token[k]
                            res_record[k] = str(v)
                        cut_off_token = {k: (overflow_token * v // all_token) for k, v
in key2token.items()}
                        for k, v in res_record.items():
                            words = v.split(" ")
                            cut_word_cnt = min(len(words) * cut_off_token[k] //
key2token[k], len(words) - 3)
                            if (cut_word_cnt >= 1) and (len(words) > 10):
                                words = words[:-cut_word_cnt]
                                _suffix = "..."
                                _cut_last = True
                            elif (cut_word_cnt >= 1) and (len(words) > 5):
                                words = words[:-1]
                                _suffix = "..."
                                _cut_last = True
                            else:
                                _cut_last = False
                                _suffix = ""
                            res_record[k] = ' '.join(words) + _suffix
                        overflow_token = num_tokens_from_string(json.dumps(res_record))
- token_limit_per_record
                    cut_res[i] = res_record
                # END FOR LOOP
                if _any_cut:
                    info += f"{self.name}: The search result is too long, some are
omitted. \n"
                # double-check the token limit and shorten the result list
                if num_tokens_from_string(json.dumps(cut_res)) > self.result_max_
token:
                    cut_res = cut_list(cut_res, self.result_max_token)
                output = json.dumps(cut_res)
            else:
                output = json.dumps(res)
```

```
            info += f"{self.name} search result: {output}"
        except Exception as e:
            logger.info(e)
            info += f"{self.name}: some thing went wrong in execution, the tool is 
broken for current input. \n"
        self.buffer.track(self.name, inputs, "Some item information.")
        logger.info(info)
        return output
    def rewrite_sql(self, sql: str) -> str:
        """Rewrite SQL command using fuzzy search"""
        sql = re.sub(r'\bFROM\s+(\w+)\s+WHERE', f'FROM {self.item_corpus.name} WHERE', 
sql, flags=re.IGNORECASE)
        # groudning cols
        cols = extract_columns_from_where(sql)
        existing_cols = set(self.item_corpus.column_meaning.keys())
        col_replace_dict = {}
        for col in cols:
            if col not in existing_cols:
                mapped_col = self.item_corpus.fuzzy_match(col, 'sql_cols')
                col_replace_dict[col] = f"{mapped_col}"
        for k, v in col_replace_dict.items():
            sql = sql.replace(k, v)
        # grounding categorical values
        pattern = r"([a-zA-Z0-9_]+) (?:NOT )?LIKE '\%([^\%]+)\%'"
        res = re.findall(pattern, sql)
        replace_dict = {}
        for col, value in res:
            if col not in self.item_corpus.fuzzy_engine:
                continue
            replace_value = str(self.item_corpus.fuzzy_match(value, col))
            replace_value = replace_value.replace("'", "''")  # escaping string for 
sqlite
            replace_dict[f"%{value}%"] = f"%{replace_value}%"
        for k, v in replace_dict.items():
            sql = sql.replace(k, v)
        return sql
```

8.9.4 對話式推薦案例

下面提供一個大模型賦能的對話式推薦介面，方便你更進一步地體驗互動式推薦帶來的及時回饋能力，程式路徑為 src/e-commerce_case/interactive_rec/app.py。

直接執行程式，就可以啟動對話式推薦的服務了，即圖 8-19 中的 http://127.0.0.1:7860。

```
(train-llm) liuqiang@liuqiangdeMacBook-Pro interactive_rec % python app.py
[2024-03-23 13:19:36 INFO agent_plan_first_openai: 382] Mode changed to accuracy.
Running on local URL:  http://127.0.0.1:7860

To create a public link, set `share=True` in `launch()`.
```

▲ 圖 8-19 背景啟動對話式推薦服務

在瀏覽器中輸入 http://127.0.0.1:7860 就可以看到圖 8-20 所示的介面，這是一個對話式推薦介面，你可以輸入與 Beauty 相關的問題與大模型進行互動，大模型會為你提供相關的推薦。這個介面還具備多輪對話的能力，你可以自己部署嘗試一下①。

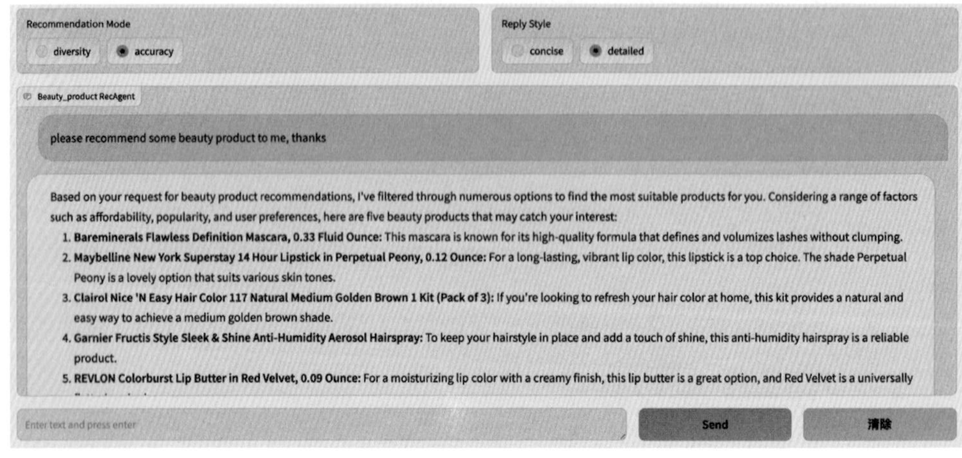

▲ 圖 8-20 對話式推薦介面

① 你也可以用中文與之對話，月之暗面的 Kimi 大模型是支援中英文雙語的。

8.10 總結

本章介紹了大模型應用於電子商務推薦系統的各種核心場景，讓讀者感受到了大模型給推薦系統帶來的新想法、新視野。本章的核心場景如下。

- 利用大模型生成使用者興趣畫像和個性化商品描述。
- 利用大模型進行召回、排序、做個性化推薦和商品連結推薦。
- 利用大模型進行推薦解釋與冷啟動。
- 利用大模型進行對話式推薦。

其中，生成使用者興趣畫像和個性化商品描述利用了大模型的生成能力，可以極佳地建構電子商務中最重要的兩類場景——猜你喜歡和連結推薦。大模型可以與傳統召回等方法極佳地配合，借助檢索增強生成的想法，優雅地賦能傳統推薦系統。

利用大模型解決傳統推薦系統的冷啟動問題的想法與傳統的解決冷啟動問題的想法不太一樣，有一些具有創新性的處理方法，例如生成冷啟動樣本、直接透過上下文學習解決冷啟動問題等。

推薦解釋在提供給使用者推薦時也提供推薦的原因，可以極大地提升推薦系統的說服力和使用者的滿意度，是最佳化推薦體驗的好想法。之前的推薦解釋更多基於簡單範本實現，有了大模型的賦能，可以生成更自然流暢、更有說服力的推薦解釋。本章舉出了利用大模型生成推薦解釋的兩種方法。其中上下文學習方法比較簡單，容易實現。如果呼叫商業的 API，那麼效果會很好。另外一個基於連續提示詞的方法比較複雜，提示詞學習是一種比較特殊的微調模型的方法，區別於 LoRA 方法，希望你可以掌握它的核心思想。

本章還實現了一個簡單的對話式大模型推薦系統，借助大模型的強大泛化能力、邏輯推理能力，我們可以用非常優雅、統一的方式將查詢、搜尋、推薦統一。這部分內容非常重要，你需要掌握相關的原理，特別是控制中心的工作流程、核心模組的作用和具體工具的使用方法。

目前，已經出現了非常多的大模型相關產品，推薦系統是網際網路行業最重要的資訊分發方式和營運手段，我相信，在不久的將來，大模型一定會革新傳統的推薦系統，讓推薦系統產生更大的業務價值，我們拭目以待！

9

專案實踐：大模型落地真實業務場景

作為最後一章，本章分析大模型落地企業級推薦系統相關的專案問題。先講解大模型如何進行高效預訓練和推理，再講解如何將大模型更進一步地落地到企業級推薦系統中。

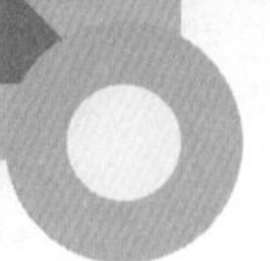

9.1 大模型推薦系統如何進行高效預訓練和推理

在具體業務場景中落地大模型不僅需要演算法和程式，還需要考慮預訓練、微調、部署和服務品質。本節講解如何在真實業務場景中高效率地使用大模型。

大模型從開始建構到服務於某個產品通常需要經歷預訓練、微調、服務部署 3 個階段，如圖 9-1 所示，與各種大模型推薦範式對應。

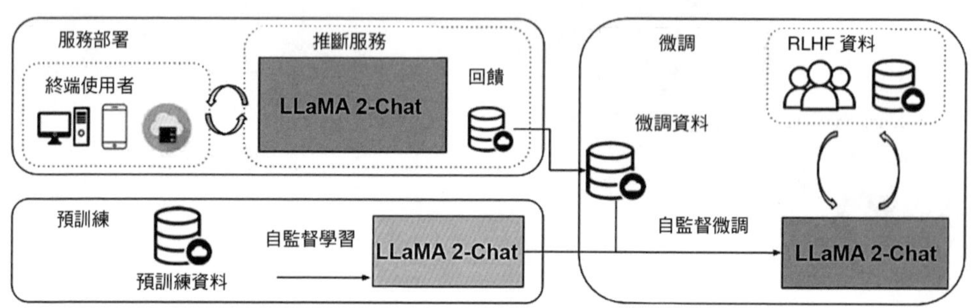

▲ 圖 9-1 大模型應用的 3 個階段

除了演算法和軟體，預訓練模型還需要特定的硬體支援，目前主流的預訓練大模型的硬體是輝達的 GPU 系列。下面聚焦大模型推薦系統應用，從模型的訓練（包括預訓練和微調）、推理、服務部署和硬體選擇 4 個維度詳細說明。

9.1.1 模型高效訓練

從零開始預訓練大模型，是非常具有挑戰性的事情。由於預訓練樣本較多（很多模型需要上兆個 token），參數較多，通常需要非常多的運算資源和合適的框架來協助。一般來說，做基礎大模型的公司才需要從零開始預訓練大模型，而做中間層或應用的公司可以對開放原始碼的大模型進行微調或直接部署。

1. 高效訓練的方法

針對高效訓練大模型，目前有非常多的開放原始碼框架可供選擇，這些框架都整合了一些加速訓練的技術手段，本節主要介紹透過平行計算和最佳化記憶體來提升訓練的吞吐量。

（1）平行計算。

平行計算是提升訓練效率最有效的方法，通常採用多 GPU 並行的方式。具體的實現方式有資料並行、模型並行、狀態並行等。我們在訓練推薦模型時，可以根據模型大小、資料大小選擇合適的並行方式，例如單一 GPU、多個 GPU 或多個伺服器。另外，可以調整 batch 的大小，batch 越大佔用的資源越多。為了加快訓練速度，還可以採取調整迭代次數（epochs）、提前終止等方法。

（2）最佳化記憶體。

大模型在訓練過程中將資料加入記憶體，可以透過最佳化記憶體（或顯示記憶體）使用的方法來提升訓練的效率，下面介紹 3 種最佳化方法。

- ZeRO。

ZeRO 的主要目的是最佳化大模型訓練中的記憶體效率。透過跨 GPU 分配模型的最佳化器狀態（Optimizer States）、跨 GPU 分配梯度和處理用於啟動的模型並行性，減少用於儲存這些資料的記憶體。

- 量化技術。

量化技術使用低位元的資料格式來表示權重或啟動函式值，透過這種方式縮減記憶體大小和計算時間，包含 8bit、6bit、4bit、2bit 等不同精度的量化方法。位元數越少，模型佔用空間越小，精度越低。我們可以根據電腦的配置和精度要求，選擇下載不同精度的模型。

- FlashAttention。

FlashAttention 是為了解決 Transformer 在處理大量序列時所面臨的內在挑戰而設計的演算法。這種演算法是 I/O 敏感的，它最佳化了 GPU 記憶體層級之間的相互作用，利用延展技術（Tiling）減少 GPU 的高頻寬記憶體（HBM）與晶片上的靜態隨機存取記憶體（SRAM）之間的記憶體讀取 / 寫入操作，以提高注意力機制的效率。

2. 高效微調的方法

對大模型進行微調通常不是微調整個模型，而是微調模型部分參數，這種方法被稱為參數有效的微調。一般有兩個常用的方法，下面分別介紹。

- LoRA 與 QLoRA。

前面章節中的大部分模型採用 LoRA 進行微調。QLoRA 是 LoRA 的量化版本，它將預訓練的模型轉為特定的 4 位元資料型態，從而大幅減少記憶體使用並提高計算效率，同時在量化過程中保持資料完整性。

- 提示詞調優。

提示詞調優是一種新穎的技術，專門用於使凍結的語言模型適應特定的下游任務。具體而言，提示詞調優強調透過反向傳播學習「軟提示」，允許使用標籤範例對其進行微調。我們在推薦解釋部分（見 8.8.3 節）的程式實現中採用了提示詞調優技術。

高效訓練的方法非常多，這裡不做過多介紹。關於高效預訓練、微調，以及在不同模型規模上的詳細對比分析，你可以從參考文獻 [1] 中獲得更多資訊。

3. 預訓練與微調的框架

目前針對大模型進行預訓練、微調的框架非常多，主流的框架包括 ColossalAI、DeepSpeed、Megatron-LM、MLX 等，這些在前面都有簡單介紹，這裡不再贅述。你可以參考它們的 GitHub 程式倉庫和官網進行深入了解，並根據自己的資源情況和偏好選擇使用。

9.1.2 模型高效推理

模型推理是一個需求更大的使用場景，目前有很多創業公司聚焦於高效的大模型推理領域，例如 Lepton AI、矽基流動等。

1. 高效推理的方法

目前主流的大模型採用與 GPT 類似的 decode-only 的架構，如圖 9-2 所示，提升解碼的效率是高效推理的前提。

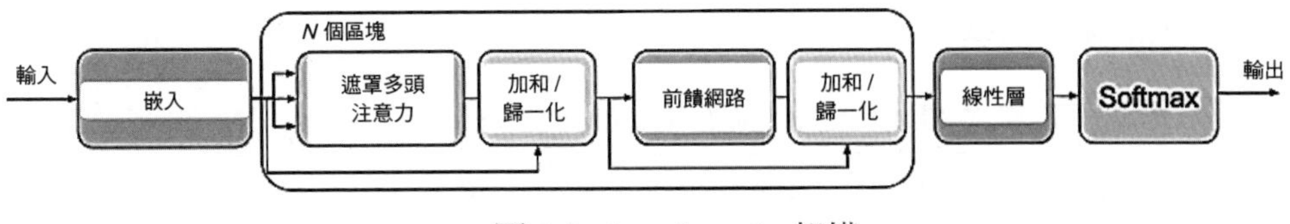

▲ 圖 9-2 decode-only 架構

高效推理有非常多的技術方案，可以從演算法層面、模型層面、系統層面等多種維度進行。以下是最常用的方法，更多細節可以閱讀參考文獻 [2]。

- 解碼策略最佳化。

GPT 系列模型基於前面的 token 預測下一個 token，這是一個自回歸的過程，這個過程是順序的，所以影響最終的解碼速度。可行的解決方案有非自回歸的方式或推測解碼（Speculative Decoding）策略。

非自回歸是放寬預測下一個 token 的條件，假設預測下一個 token 與前面的 token 條件無關，這樣就可以透過一定程度的並行化進行解碼。推測解碼是利用一個草稿（Draft）模型（更小的模型）來生成下一個 token，再快速評估這個 token 是否正確。

- 架構最佳化。

架構最佳化是對 GPT 的架構動手術，透過調整架構部分「元件」達到加速推理的效果。舉例來說，MoE 架構就是透過多個專家建構的統一的大模型，在預測時，只有部分專家被啟動，從而提升推理的速度。

- 模型壓縮。

模型壓縮最常用的手段是知識蒸餾，透過老師（Teacher）模型監督學生（Student）模型進行訓練，將知識從較大的老師模型「傳授」給學生模型。還有一個比較常用的方法是剪枝，例如剔除模型的部分層來減少參數，進而提升預測速度。

- 系統最佳化。

前面提到的量化方法、並行化方法、記憶體最佳化方法都是這類方法。還有一些偏底層的方法超出了簡單應用的範圍，這裡不詳細說明。

2. 高效推理的框架

前面提到的很多訓練框架本身就具備推理能力，例如 DeepSpeed、Colossal-AI 等，還有一些框架聚焦於推理方向，比較常見的有 vllm、TG（Text-Generation-Inference）、LightLLM 等。

針對 MacBook，如果在開發環境中部署服務進行測試等，則可以選擇 MLX、llama.cpp 等推理框架。

9.1.3 模型服務部署

對於大模型推薦系統的應用，在預訓練或微調好模型後，就需要進行高效的服務部署，以便提供給使用者更好的服務。

上面提到的很多框架可以直接將大模型服務部署成 Web 服務，如果你想自己調整業務邏輯，則可以選擇一些合適的 Python Web 框架，例如 FastAPI、Tornado、Flask 等。

如果利用大模型的上下文學習能力進行推薦，那麼可以選擇第三方的 API 服務或將大模型部署成類似於 ChatGPT 的 API 服務，可以使用的框架有 Ollama、FastChat、SGlang、Lepton AI 等。

傳統推薦系統的一些最佳化服務體驗的技術方案也可以應用到大模型推薦系統中，例如預計算、快取、部署多個等效的 Web 服務，再透過 Nginx 代理進行服務的水平擴充等。

9.1.4 硬體選擇建議

對於大模型，訓練和推理的硬體也是非常重要的。當預訓練的大模型參數較多（例如超過 30B）時，需要性能比較好的輝達的 GPU，例如 A800、A100、

H100 等。對於參數更小的模型，可以採用輝達的消費級顯示卡，例如 RTX 4090、RTX4080 等。國產的替代方案還不夠成熟，華為昇騰 910B 是可行的選擇。

對於大模型的微調和推理，相比於 A800、A100、H100 這些比較貴並且難買到的硬體，C/P 值更高的方式是使用輝達的 RTX 4090、RTX4080 等消費級硬體。

MacBook 使用者可以選用 M 系列的晶片對大模型進行預訓練和微調。如果處理比較大的模型（例如 30B、70B 等），M2 Ultra 192G、M3 Max 128G 等是可行的選擇。

由於推薦系統的資料量相對較小，即使是預訓練和微調，目前也不需要特別大的模型，通常使用 2B、7B、14B 的模型就可以達到較好的效果，因此對硬體的要求沒有常規的大模型高。

9.2 大模型落地企業級推薦系統的思考

經過近十年的發展，推薦系統已經相當成熟，它是幫助公司營利的核心技術，在企業中具有極高的「地位」，因此，革新它是非常困難的。新的技術要取代和顛覆原有的成熟技術系統是需要時間的。現在的大模型在推薦系統的應用屬於早期嘗試階段，還需要 1 ～ 2 年時間才能進入變革的深水區。大模型在推薦場景的落地需要克服哪些困難？又需要面對哪些挑戰呢？筆者試圖基於自己過去十幾年的推薦系統最前線經驗，結合對大模型技術的深度學習，談一些自己的思考，希望能為你提供一些新的角度，讓你更進一步地評估什麼時候及如何在推薦系統中引入大模型相關技術。

9.2.1 如何將推薦演算法嵌入大模型框架

推薦系統與大模型的建模空間有較大差別，傳統的推薦系統基於 ID 特徵建模，ID 之間的內在協作關係對推薦效果非常重要。另外，推薦系統是一個非常個性化的系統，如何將這種個性化的特性整合到大模型系統中呢？

傳統的大模型基於文字語料預訓練，token 詞庫的規模在幾萬到十幾萬等級。推薦系統的使用者 ID、商品 ID 的規模是億等級的，如果將使用者 ID、商品 ID 作為 token，那麼模型複雜度多了幾個數量級，預訓練和推理成本將極大提高。這個問題有很好的解決方案嗎？

前面的章節主要介紹了兩種方法：一種是不直接使用 ID 特徵，而是使用商品的標題等文字特徵，使用者的行為也透過購買過的商品的文字特徵來表示，這種方法將推薦問題映射到傳統的大模型擅長的文字建模領域；另一種是加入少量新的 token，例如將 item_id 拆分為「item」加後面的數字 ID，這樣不至於增加太多的 token。

上面的方法當然可以處理推薦問題，但是可能不是最好的。有沒有一種方法可以更加統一、自然地將推薦演算法映射到一個更加適合大模型處理的系統下呢？這是推薦行業後續努力的方向，參考文獻 [3] 就是這方面的探索。

正是因為有一套比較完整的、適合推薦系統的系統，深度學習才成為在推薦系統領域佔統治地位的核心演算法，為推薦系統帶來了極大的商業價值。大模型時代也遵循一樣的規律，如果有一個適合推薦系統的大統一的大模型框架，那麼推薦系統更有可為。

9.2.2 大模型特性給落地推薦系統帶來的挑戰

9.1 節提到大模型目前在訓練和推理中面對的挑戰，這些挑戰同樣存在於基於大模型的推薦系統中。其中，需要巨量的資料、訓練成本高、推理速度慢是比較重要的問題。

推薦系統有巨量的使用者、巨量的互動行為，資料量並不少，但是推薦的資料非常稀疏，這也是大模型落地推薦系統的很大的挑戰。

當軟硬體更加成熟後，訓練成本會遵循莫爾定律，每 1 ～ 2 年至少降低 50%。輝達發佈的 Blackwell 架構的 GB200 系列晶片的 GPU 算力就有極大的提升，功耗也降低了不少。

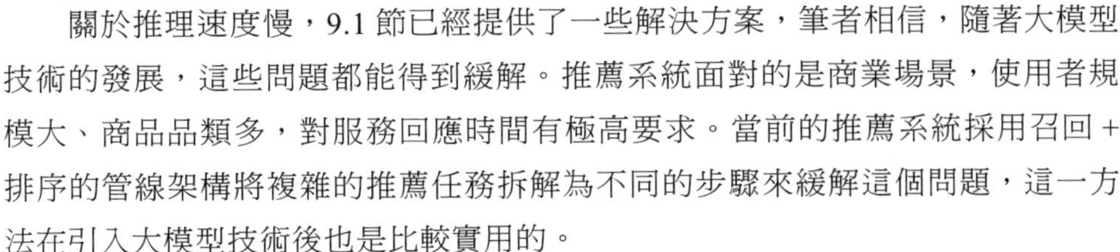

關於推理速度慢，9.1 節已經提供了一些解決方案，筆者相信，隨著大模型技術的發展，這些問題都能得到緩解。推薦系統面對的是商業場景，使用者規模大、商品品類多，對服務回應時間有極高要求。當前的推薦系統採用召回 + 排序的管線架構將複雜的推薦任務拆解為不同的步驟來緩解這個問題，這一方法在引入大模型技術後也是比較實用的。

9.2.3 大模型相關的技術人才匱乏

大模型的技術難度較大，有陡峭的學習曲線。筆者從 2023 年年初開始學習大模型，經過大半年才算正式進入這個領域。筆者有 15 年 AI 相關的實踐經驗，也需要經過較長時間的學習才能入門大模型，對於經驗不足的讀者，入門的學習成本可能更大。

推薦系統被大模型革新，人才是非常重要的。大模型相關人才比較匱乏，目前這些人才都在大模型創業公司和網際網路大廠。也許再經過 1 ～ 2 年，懂大模型的技術人才會有爆發式增長，那時才是真正的大模型落地推薦系統的黃金時間。

9.2.4 大模型推薦系統與傳統推薦系統的關係

在演算法層面，很難有一種演算法可以取代所有演算法，推薦演算法也一樣。傳統的推薦演算法（例如協作過濾、基於標籤的推薦等）有其適用場景。8.9 節將傳統的推薦演算法用於召回，大模型有著控制中心的作用。

在架構層面，經過十幾年的驗證，傳統推薦系統經典的召回 + 排序架構是一種非常高效的架構，大機率會持續發揮價值。

在產品層面，傳統推薦系統中最重要的是資訊流推薦、連結推薦，只要對話模式不變（例如電腦、手機），其產品形態就會在商業系統中持續產生價值。唯一可能產生的新的產品形態是對話式推薦，這是大模型對推薦系統產品最大的革新。

總之，大模型對傳統推薦系統造成的是賦能的作用，而非取代。有了大模型，推薦系統在應用場景、商業價值上會取得更大的成功。

9.2.5 大模型推薦系統的投資回報率分析

任何一門新技術只有能帶來極大的商業價值，才會被廣泛採納，大模型在推薦系統中的應用也不例外。大模型的學習成本、實踐成本、訓練成本、部署成本都比較高，這是將大模型應用於推薦業務的最大的障礙。

大模型為推薦系統帶來的價值主要有 3 個方面：一是宣傳公司技術實力；二是為公司帶來更多收入；三是創新產品形態，可以利用大模型的互動能力，在產品的各種場景中引入對話式大模型推薦系統。

企業的技術負責人（或推薦系統負責人）在引入大模型推薦系統時必須對投入和產出有比較好的了解，並且有比較合理的預期。

9.2.6 大模型落地推薦場景的建議

筆者相信，在 1 ～ 2 年內，大模型會革新現在的推薦系統，因此建議推薦行業的從業者及早關注並且實踐大模型在推薦系統中的應用，只有躬身入局，才能更進一步地了解大模型能為推薦系統帶來的價值。

學習大模型推薦系統的資料不少，除了大量的學術論文，本書也可以幫助你入門大模型推薦系統。

對於企業中的推薦演算法相關從業者，可以找一些推薦場景探索大模型的應用，筆者建議的探索場景包括首頁的「猜你喜歡」和商品連結推薦，這些是公司最核心的推薦場景。為了避免給業務收入帶來影響，可以採用 A/B 測試的方式，先給大模型推薦系統比較小的流量，透過資料逐步驗證大模型推薦系統帶來的價值，並動態調整分配給大模型推薦系統的流量。

對於對話式推薦場景，一定要大膽嘗試，這是大模型能夠帶來的最大的創新，在具體實現上，可以透過在首頁搜尋關鍵字進入對話式推薦頁面。具體的演算法原理在 8.9 節中有比較全面的講解，這裡不再贅述。

9.3 總結

本章簡單總結了大模型訓練、推理、部署的技術和框架。你可以選擇適合自己的框架深入了解和實踐。這個方向發展較快，可以多關注、了解行業動態。

本章整理了大模型落地企業級推薦系統可能面對的困難和挑戰，以及大模型可能為推薦系統帶來的巨大變革。

任何一種新的技術被大規模應用都需要一定時間，只有儘早入局才能獲得最大的紅利，希望你屬於最早「吃到螃蟹」的那一批人。讓我們一起努力，加速大模型在商業推薦系統中的應用！

後記

我們的學習之旅就馬上結束了。筆者從 2023 年年初開始學習、跟進大模型相關的進展，本書是筆者在大模型方向上探索的經驗總結。

ChatGPT 發佈以來，大模型相關的技術呈爆發之勢，大量科技公司重金投入大模型的研究、開發、實踐。蘋果公司放棄了 10 年的造車計畫，準備投入大模型的創新和落地，並宣佈了與百度合作的計畫 [1]，在 2024 年發佈的硬體產品中增加生成式 AI 能力。蘋果公司通常會在行業比較成熟、有非常大的商業前景時做出決策，可見其相信大模型會帶來產業顛覆和巨大的商業價值。

大模型越來越有革新所有行業的勢頭：Sora 的發佈帶來了視訊生產的革命，OpenAI 希望能與好萊塢合作 [2]，透過生成式 AI 革新影視製作 [3]；OpenAI 投資的 Figure 01 機器人 [4] 可以理解人的意圖，將蘋果遞給表示饑餓的測試人員；初創公司 CognitionLabs 推出的全球第一個 AI 程式設計師 Devin[5] 掌握了全端技能，不僅可以寫程式、debug、訓練模型，還可以去美國最大的求職網站 Upwork 上搶單；專注於研發無程式遊戲引擎的初創公司 BuildBox AI 發佈了新一代 AI 遊戲引擎——Buildbox 4 Alpha，輸入提示即可為遊戲增加資產和動畫，甚至提供幾個字就能生成整個場景 [6]；2024 年 3 月底大紅的 Suno，可以透過一段提示詞生成高品質的音樂，讓人人都成為作曲家 [7]。

新行業、新產品和新應用場景層出不窮，新的變化每天都在我們身邊出現，我們不能無視這些變化，否當我們的職務被 AI 取代時，我們被還蒙在鼓裡。

AI 取代的永遠是那些不會使用 AI 技術的人，最好的應對方法就是儘快掌握 AI 技術，而掌握 AI 技術的第一步是開始學習。推薦系統的從業人員也不例外，學習大模型在推薦場景中的應用是必須經歷的過程。

推薦系統是過去 10 年網際網路最重要的技術手段之一，是網際網路公司最核心的商業化工具。近年來，出現了抖音等以推薦系統為核心引擎的、具備極大變現效率的產品。通常情況下，越是具備商業價值的場景，人類越會利用新技術去變革它。推薦系統的商業價值高，是以大模型為驅動引擎的新技術革新的方向，這正是學術界和產業界正在努力推進的！

阿里巴巴於 2023 年上半年在淘寶開啟了淘寶問問的內測，百度將大模型應用到了最核心的廣告行銷場景，Meta 已經開始嘗試利用大模型打造兆級參數的新一代推薦系統……這些大公司已經走在了前面，推薦系統的從業者也不能落後，需要馬上出發。

本書是大模型推薦系統學習的導航，可以幫你順應 AI 變革大勢，進行嘗試和探索。希望透過對本書的學習，你可以儘早入門並實踐大模型推薦系統。

MEMO

MEMO

MEMO